Advanced Concepts and Technologies for Electric Vehicles

This book explains the basic and advanced technology behind the Power Electronics Converters for EV charging, and their significant developments, and introduces the Grid Impact issues that underpin the grid integration of electric vehicles. *Advanced Concepts and Technologies for Electric Vehicles* reviews state-of-the-art and new configurations and concepts of more electric vehicles and EV charging, mitigating the impact of EV charging on the power grid, and technical considerations of EV charging infrastructures.

The book considers the environmental benefits and advantages of electric vehicles and their component devices. It includes case studies of different power electronic converters used for charging EVs. It offers a review of PFC-based AC chargers, WBG-based chargers, and Wireless chargers. The authors also explore multistage charging systems and their possible implementations. The book also examines the challenges and opportunities posed by the progressive integration of electric drive vehicles on the power grid and reported solutions for their mitigation.

The book is intended for professionals, researchers, and engineers in the electric vehicle industry as well as advanced students in electrical engineering who benefit from this comprehensive coverage of electric vehicle technology. Readers can get an in-depth insight into the technology deployment in EV transportation and utilize that knowledge to develop novel ideas in the EV area.

Advanced Concepts and Technologies for Electric Vehicles

Edited by
Akshay Kumar Rathore
Arun Kumar Verma

CRC Press
Taylor & Francis Group
Boca Raton London New York

CRC Press is an imprint of the
Taylor & Francis Group, an **informa** business

First edition published 2024
by CRC Press
6000 Broken Sound Parkway NW, Suite 300, Boca Raton, FL 33487-2742

and by CRC Press
4 Park Square, Milton Park, Abingdon, Oxon, OX14 4RN

CRC Press is an imprint of Taylor & Francis Group, LLC

ISBN: 978-1-032-36073-7 (hbk)
ISBN: 978-1-032-36074-4 (pbk)
ISBN: 978-1-003-33013-4 (ebk)

DOI: 10.1201/9781003330134

Typeset in Times
by MPS Limited, Dehradun

Contents

Editor Biographies

Akshay Kumar Rathore (IEEE Fellow) is a professor (Engineering) at the Singapore Institute of Technology, Singapore. Prior to this appointment, he served as an associate professor at Concordia University, Montreal, Canada and as an assistant professor at the National University of Singapore, Singapore. He received his Master of Technology degree in electrical machines and drives from the Indian Institute of Technology (BHU) Varanasi, India, in 2003, and his Ph.D. in power electronics from the University of Victoria, British Columbia, Canada in 2008. Dr. Rathore is a recipient of the 2013 IEEE IAS Andrew W. Smith Outstanding Young Member Achievement Award, 2014 Isao Takahashi Power Electronics Award, 2017 IEEE IES David Irwin Early Career Award, 2020 IEEE Bimal Bose Award for Industrial Electronics Applications in Energy Systems, and 2021 Nagamori Award.

Arun Kumar Verma is a senior member of IEEE, received his master's degree in Energy Studies and his Ph.D. in Power Electronics from the Indian Institute of Technology Delhi, New Delhi, India, in 2010 and 2015, respectively. He was a visiting graduate researcher with the Smart Grid Energy Research Centre(SMERC), University of California, Los Angeles, CA, USA, during 2014–2015. He was a postdoctoral research fellow with the Energy Research Institute, Nanyang Technological University, Singapore, during 2015–2016. He is currently an Associate Professor in the Department of Electrical Engineering at the Indian Institute of Technology, Jammu, India. Prior to this appointment, he served as an Assistant Professor with the Department of Electrical Engineering, Malaviya National Institute of Technology Jaipur, Jaipur, India. He won the prestigious BASE fellowship for advanced solar energy research in 2014. He has been conferred with the prestigious POSOCO Power System Research Award 2016.

Contributors

Abhinandan Dixit
GaN Systems
Ottawa, ON, Canada

Monika Dabkara
Malaviya National Institute
  of Technology
Jaipur, Raj, India

N. K. Kandasamy
Lite-On Pte Ltd
Singapore

Sivaneasan Bala Krishnan
Singapore Institute of Technology
Singapore

Dhivya Sampath Kumar
Singapore Institute of Technology
Singapore

Kirti Mathuria
Malaviya National Institute
  of Technology
Jaipur, Raj, India

Sandeep N
Malaviya National Institute
  of Technology
Jaipur, Raj, India

Saravana Prakash P
Malaviya National Institute
  of Technology
Jaipur, Raj, India

Karan Pande
Power Integrations
Montreal, QC
Canada

Nil Patel
Gina Cody School of Engineering and
  Computer Science
Concordia University
Canada

Akshay Kumar Rathore
Engineering Department
Singapore Institute of Technology
Singapore

C. D. Rodríguez-Gallegos
Solar Energy Research
  Institute of Singapore (SERIS)
Singapore

Santhosh S
Malaviya National Institute
  of Technology
Jaipur, Raj, India

Suvendu Samanta
Indian Institute of Technology
Kanpur, UP, India

Anurag Sharma
Newcastle University
Singapore
Singapore

Asmita Singh
Malaviya National Institute
   of Technology
Jaipur, Raj, India

Kuan Tak Tan
Singapore Institute of Technology
Singapore

D. R. Thinesh
Singapore Institute of Technology
Singapore

Harpal Tiwari
Malaviya National Institute
   of Technology
Jaipur, Raj, India

Arun Kumar Verma
Indian Institute of Technology
Jammu, J&K
India

Elsa Feng Xue
Singapore Institute of Technology
Singapore

Pan Xuewei
Harbin Institute of Technology
Shenzhen, Guangdong, China

1 Introduction to Electric Vehicles

Kirti Mathuria, Monika Dabkara, Santhosh S Sandeep N, Arun Kumar Verma, and Harpal Tiwari

CONTENTS

1.1 A BRIEF HISTORY OF ELECTRIC VEHICLES

The issues of climate change or global warming have been rigorously discussed by many governments since the early 21st century. A great number of relevant reports have revealed that the negative impact of climate change is dominantly driven by human activities. With the globally increasing civilization and industrialization, a large number of fossil fuel burnings in industries have led to the acute problem of air pollution. Simultaneously, the exhaust emissions from automotive vehicles cannot be

DOI: 10.1201/9781003330134-1

ignored. Vehicle emissions, which mainly include CO_2, CO, NO_x, and particulate matter (PM10 and PM2.5), have been considered the major contributors to the effect of greenhouse gases, also leading to the increase in different forms of cancers and other serious diseases.

The ever-rapidly growing transportation sector consumes about 49% of oil resources. Following the current trends of oil consumption and crude oil sources, the world's oil resources are predicted to be depleted by 2038. Therefore, replacing non-renewable energy resources with renewable energy sources and use of suitable energy-saving technologies seems to be mandatory. Electric Vehicles (EVs) as a potential solution for alleviating traffic-related environmental problems have been investigated and studied extensively.

Because they offer a cleaner and more effective substitute for gasoline-powered cars, electric vehicles are significant. They are known to reduce greenhouse gas emissions and enhance air quality, which reduces global warming and climate change. By producing less waste, electric vehicles also aid in lowering electric pollution [1]. Due to this, sales of electric cars are estimated to account for nearly one-third of all vehicle sales by 2030. Since the 2000s, electric vehicles have rapidly increased in popularity. In addition, electric vehicles are becoming more widely available and affordable, making them a powerful force in the transportation sector [2].

1.1.1 History of Electric Vehicle

A Complete Guide:

1828 – Invention of the Electric Motor: The electric vehicle was first invented in 1828 by Hungarian inventor Ányos Jedlik. He developed the motor using electric current from a battery to power an electric train.

1832 – First Small-Scale Electric Vehicle Produced: In 1832, William Morrison of Scotland produced the first electric vehicle. It was a small-scale electric car that could only travel for a distance of about 12 miles on one charge.

1881 – Electric Trams Introduced: In 1881, electric trams were introduced in the city of Berlin. These electric trams ran on overhead wires and could travel up to 16 miles per hour.

1889 – Electric Vehicles Hit US Market: In 1889, electric vehicles were brought to the United States market by William Morrison. Morrison first encountered electric cars in Europe and noticed their popularity, decided to introduce the technology to US consumers. This was a defining moment in the electric vehicle boom that was to come in the early 1900s.

1900 – Electric Vehicle Increase in Popularity: The electric vehicle boom started in the early 1900s and lasted until around 1920. This was when electric vehicles became more popular than gasoline-powered cars. The main reason for this was that electric vehicles were more affordable and required less maintenance than gasoline cars. In addition, electric vehicles were quieter and cleaner, so electric cars became the preferred mode of transport for women.

1901 – First Hybrid Car Invented: In 1901, a Canadian electric car manufacturer named Henry Seth Taylor produced the first hybrid electric vehicle. This electric car was powered by both gas and batteries.

1908 – Ford Model T Introduced: In 1908, Henry Ford introduced the Model T electric car. The electric vehicle was a great success as it offered an alternative to gasoline-powered cars at a lower price point than people could afford.

1909 – Electric Vehicles Took Up to One-Third of US Market Share: By 1909, electric vehicles took up to one-third of the market share in the USA due to their growing popularity.

1920 – Electric Vehicle Boom Ended: The electric vehicle boom ended in 1920 when the cost of gasoline dropped, and electric vehicles became more expensive to produce than gasoline cars. In addition, the rise of the automobile industry led to a decrease in demand for electric vehicles. Many electric vehicle companies stopped producing electric cars and switched over to gasoline-powered cars. This led many electric car manufacturers out of business as they could not compete with the automobile industry.

1947 – First Mass Production of Electric Car Introduced: In 1947, electric vehicles were reintroduced to the market by Henry Ford's son Edsel and Ford Motor Company. The electric car was named "Edsel" after its creator and, sitting at an affordable price point at $650 per vehicle.

1971: NASA's Electric Lunar Rover Landed on the Moon: In 1971, electric vehicles came back when NASA's electric Lunar Rover was sent to land on the Moon. This electric rover ran on solar energy and traveled up to a distance of 400 miles. This was a key point in laying the foundation for the comeback of the electric vehicle, as NASA's Lunar Rover provided some essential market exposure.

1973: New Generation of Electric Vehicles Ushered In: In 1973, electric vehicles were reintroduced to the market by General Motors with their electric car model, the EV-01. The EV-01 was a small two-seater electric car that ran on lead-acid batteries.

1975: Sebring-Vanguard became sixth-largest automaker off the back of the CitiCar: In 1975, electric car manufacturer Sebring-Vanguard became the sixth largest automaker in the USA due to their electric car model, the CitiCar. The CitiCar was a small two-seater electric car that ran on lead-acid batteries and traveled up to a distance of 40 miles on a single charge. This electric car was popular with commuters who didn't need to travel far distances regularly and wanted an electric vehicle that wouldn't cost them much money in maintenance costs or gasoline charges. However, the small lift in growth for electric vehicles died off in 1979.

1979 – Electric Vehicle Interest Died Off: As electric cars became more popular, the market demand for electric vehicles grew. However, in 1979 this growth died off as electric car manufacturers could not keep up with the increasing demands of the consumers. In addition, gasoline prices decreased, which made electric cars less affordable compared to gasoline-powered cars.

1996 – EV1 Produced: In 1996, General Motors produced its electric car model, the EV-01. The EV-01 was a small two-seater electric car that ran on lead-acid batteries and traveled up to a distance of 40 miles on a single charge. However, the EV-01 was only leased to consumers and not sold, which limited its market exposure.

1997 – First Mass-Produced Hybrid Car: In 1997 electric car manufacturer Honda released their electric car model, the Insight. The Insight was a two-door hatchback electric vehicle that ran on an electric battery and gasoline fuel source. The hybrid electric vehicle became popular with eco-conscious consumers who wanted to reduce their carbon footprint while still using gasoline cars as they were more affordable in comparison to electric cars.

1998 – Toyota Prius Introduced: In 1998 electric car manufacturer Toyota released their electric vehicle, the Prius. The Prius was a four-door electric hybrid car that ran on an electric battery and gasoline fuel source. This electric vehicle became popular with eco-conscious consumers who wanted to reduce their carbon footprint while still using gasoline cars as they were more affordable than electric cars. However, the EV-01 was only leased to consumers and not sold, which limited its market exposure.

2008 – Tesla Roadster Produced: In 2008 electric car manufacturer Tesla produced their electric sports car, the Roadster.

2011 – Nissan LEAF Released: In 2011, Nissan released their electric car model, the LEAF. The LEAF was a five-door electric hatchback that ran on an electric battery and could travel up to 100 miles on a single charge. This electric car became popular with consumers who wanted an electric car that could travel long distances without recharging.

2013 – Costs to Produce Electric Batteries Dropped: Due to improved electric battery technology and a mining boom in precious metals used in electric cars, manufacturers could produce electric vehicles at a lower cost than ever before. This made electric cars more affordable for consumers and increased their popularity.

2016 – Norway Backed Electric Vehicle's Growth: In 2016 Norway led the world to become the first country to have 5% of all registered cars as plug-in electric models. The government supported this growth, which offered a number of incentives for citizens to purchase electric vehicles, such as tax exemptions and charging infrastructure investments.

2018 – Plug-In Electric Vehicles Skyrocketed: In 2018, electric car sales reached a new record high. The plug-in electric car segment represented just about 1 out of every 250 motor vehicles on the world's roads at the end of 2018. This was largely due to the decreasing cost of electric batteries, the increasing availability of charging infrastructure, and the growing popularity of electric vehicles among consumers.

2020 – Tesla Model Y Released: In 2020 electric car manufacturer Tesla released their electric SUV model, the Model Y. The Tesla Model Y was an electric battery-powered SUV that could travel up to 316 miles on a single charge. This electric vehicle became popular with consumers who wanted an electric vehicle that had the performance of a sports car but also offered the practicality of an SUV.

2021 – The Future of Electric Vehicles: The industry continues to grow due to the decreasing price of electric batteries and government efforts to increase the accessibility of electric vehicles. Therefore, making electric vehicles more affordable than ever before for consumers. As electric vehicles become cheaper and more accessible, electric vehicle sales will likely continue to grow, and electric vehicles will become the dominant form of transportation globally [2].

1.1.2 Recent Development in Electric Vehicles

As the world is alarmed by pollution and environment-related issues, EVs with zero emissions should be considered as a boon. The upcoming decades will be called as the decades of Electric Mobility due to recent development in semiconductor devices, battery technologies, advanced communication techniques, extended driving range, charging stations, etc. As EVs come under various configurations like PEV (Plug-in Electric Vehicle), HEV (Hybrid Electric Vehicle), FCEV (Fuel Cell Electric Vehicle), etc. which provide a wide choice criterion for people according to their requirements. Governments are also taking all initial and beneficial steps to make Electric Vehicle popular among the people. Since from early 1800s Electric vehicles paved a long way to mark their presence on the roads similar to Internal combustion (IC) vehicles. Due to advanced battery technologies, the price of the batteries has been reduced by 50% in the last four years with a simultaneous improvement in battery performance like energy, power life cycle, etc. On the other hand, EV manufacturers are putting all head-to-toe efforts for a durable, sustainable electrical drive with minimum cost thus overall making an EV affordable for everyone [3].

The major automobile industries Tesla, BMW, Nissan, Honda, etc. are playing a key role to revolutionize the EV market globally. For making the EV popular the charging infrastructure is an integral part that needs to be considered therefore, various smart charging ways permit the driver to charge the EV battery by controlling the rate and time of charging according to energy demands of the hours like vehicle to grid (V2G) charging which provides energy to the grid during high demand hours from the parked EV, thus making the EV drivers prosumer who can consume and provide the electricity as well. Wireless charging uses the principle of electromagnetic induction, where the electricity is transferred through magnetic coils which are installed as receiving coils in the vehicle and transmitting coils on the road; allows to charge the vehicle in both dynamic and static ways and also helps in improving the charging speed, allows the driver to charge the vehicle on the streets and signals and the wireless charging allows to charge the multiple vehicles at a time. The first wireless charging station will be established in Norway by 2023 for electric taxis with a capacity of 75 KW. The global wireless electric vehicle charging market was valued at $6,857.80 thousand in 2020 and is projected to reach $207,415.10 thousand by 2030 [4]. The Ev ecosystem is given in Figure 1.1.

Ultrafast charging takes only a few minutes to charge the vehicle which is equal to regular driving breaks and some electric vehicles like Kia EV6 and Hyundai IONIQ 5 provide 60 miles of driving for just 5 minutes of charging by a 250 KW charger. Battery swapping technology enables the consumer to swap the discharged battery with charged one in just a few minutes like fueling the ICE vehicle and renewable charging allows to charge the vehicle through abundant renewable resources, and in grid-to-vehicle charging (G2V), the grid is allowed to charge the vehicle. According to the recent pace of EVs, it is predicted to have 145 million EVs on the world's road by 2030 as the sales of EVs will be about 7 million per year by 2030.

In the past year, Tesla Model 3 was one of the highest-selling EVs with worldwide sales of 500,000 with a driving range of 358 miles with ultrafast charging, and another most popular EV is the Tesla Model Y which has a driving

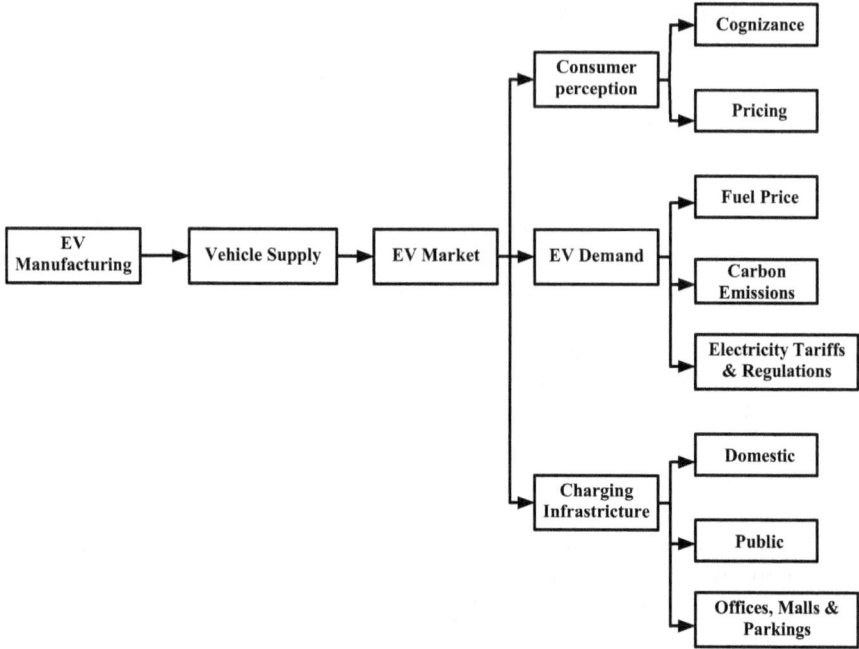

FIGURE 1.1 Scope of the EV ecosystem strategy.

range of 244 miles and is a battery-electric compact cross over, Volkswagen ID.4 is also a battery-electric compact crossover with a driving range of 260 miles, BYD Qin plus is the only PHEV which becomes widely popular as it comes with blade battery technology with a driving range of 620 miles. Nissan Leaf Plus is also the most popular EV due to its less cost with a driving range of 215 miles. At present consumers have a wide range of choices for the EV as recently there are 23 plug-in and 36 hybrid models available in the EV market with variable sizes and pricing. As the pricing of gasoline vehicles is continuously rising while EV prices are declining day by day, making an EV as the first choice for buying the vehicle. Today the world is moving toward clean and efficient energy as it is required to save the environment and thus EVs will play a major role.

1.1.3 OPPORTUNITIES, SCOPE, AND CHALLENGES IN ELECTRIC VEHICLES

The transportation sector will get electrified in the upcoming decades as per the recent scenario of EVs in the world. While shifting to the EVs market a well-established charging infrastructure should be maintained in such a way that fuel stations should be superseded by EV charging stations on the roads. The challenge of designing and operating the financially viable and sustainable charging infra for EVs with different levels of charging like home charging, commercial charging, and ultrafast charging is still formidable for EV manufacturers. Over the past few years battery technology has improved dramatically as a result of a significant reduction in the cost of the battery but the size of the battery pack has increased and also the downslope for the cost of the battery is slower than the cost per KWh, thus the compact battery pack is still an optimistic sector for growth. The battery swapping technology is also getting widespread attention due to its ease and minimum time requirement process but the major challenge for these markets is again a battery! As all the EV manufacturers have their own battery mechanisms making battery swapping difficult for different manufacturers as there are no common criteria for EV batteries and these techniques still need to be improved with a sound solution. Battery lifecycle and cost also drag the EV market but to reduce this drawback the recycling of EV batteries in a greener way can be the biggest opportunity for lifting up the EV market in terms of batteries [5].

Unlike, ICE vehicles maintenance and services are not available for EVs, the challenge is giving training to consumers or service engineers for operating and maintaining the EV, thus an opportunity arises for the training and service centers of EV to aware the public about EVs. Dealing with surplus energy requirements for EVs is both an opportunity and a challenge for electric utilities. Electrical utilities have secured the surplus energy demand during peak hours as the home charging for EVs during the night can overload the transformers and distribution utilities. Smart or net-metering has to be deployed for managing the peak loads and providing additional discounts for the off-peak charging hours can balance the energy demands [6,7].

Grid energy storage modules should be introduced for charging the EVs directly from the grid and again when energy is needed during peak hours the energy can be regained from the EV which we can call as Vehicle to Grid charging (V2G). An orientation of new electricity tariffs is needed for making the EV successful as ICE and making the EV market an opportunistic at every level. The requirement of additional power can be fulfilled by renewable resources and among them solar integrated EV charging will be the biggest achievement as developing the solar-powered charging infrastructure for EVs will help in balancing the energy demands from the utilities and consuming the abundant solar power for charging the EV will be a greener way to save the environment. The opportunities and challenges in EV are given in Figure 1.2.

1.2 ENVIRONMENTAL IMPACT OF ELECTRIC VEHICLES

1.2.1 Reduction of Pollutant Gases in the Environment

According to a report by Electric Power and Research Institute (EPRI) the emission of greenhouse gases (GHG) and other air pollutants can be significantly reduced by using electric transportation compared to gasoline vehicles. Passenger vehicles contribute about 60% of the total carbon emission from the transportation sector and thus by electrifying those vehicles from renewable resources by 2050, we can reduce environmental changes. Addressing about 23% of global energy-related GHG emissions, the carbon emission from the transportation sector, which mostly contributes to climate change, can be reduced. Not surprisingly, an EV can reduce pollution while using brakes as EV uses a regenerative braking system which restores the energy back to the battery and on the other hand ICE vehicle mostly uses a disk brake which creates pollution. The tires of ICE vehicles also emit particle mass (PM) which

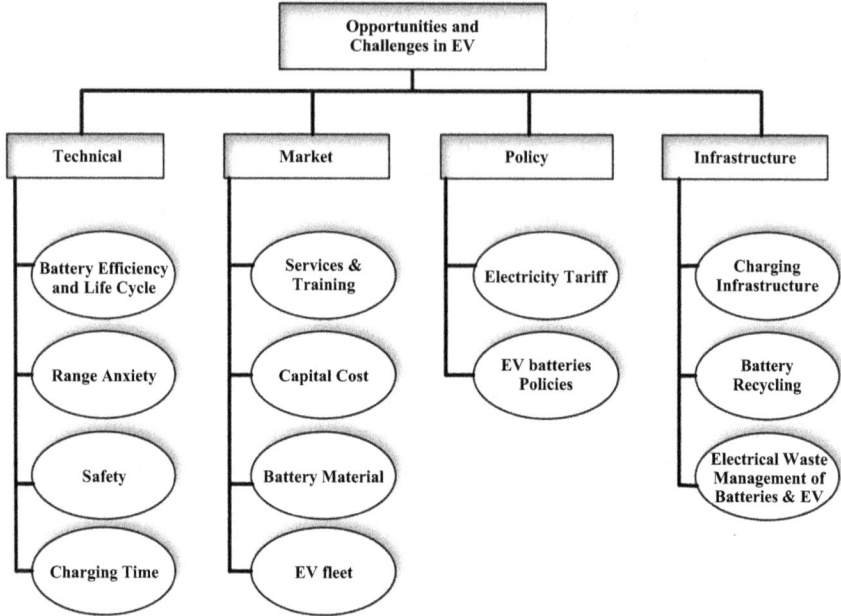

FIGURE 1.2 Flowchart for the opportunities and challenges in EV.

will react with the other pollutants in the air and form harmful gases, on the other hand, the tires of EV are designed and manufactured with special features to tolerate the weight of the batteries and thus they will emit less particle mass.

An ICE vehicle emits all tailpipe pollution e.g., NO_2, CO, HC, etc. which are responsible for many respiratory and cardiovascular diseases.

The ICE vehicle also emits C_6H_6 (benzene) and aromatic hydrocarbons responsible for human cancer, therefore the ultimate solution for eliminating these pollutants is to switch to EV from ICE and take steps forward for a cleaner and healthier environment. Although EVs are charged from the fuel-based electric grid, the GHG emission will only be 1/3rd while producing this electricity as compared to using ICE vehicles. The GHG emissions will be significantly reduced if we electrify the future vehicles on the road ranging from light-duty vehicles to medium-duty commercial vehicles and to heavy-duty vehicles. By penetration of EV into the market, carbon emissions will reduce by 8.4 million tons in 2030, and by 2050 it will be reduced by 49.5 million tons. For realizing a greener environment there must be a transition of fossil energy to electrical energy which can be achieved in maximum EVs run on the road.

1.2.2 GLOBAL AND LOCAL ENERGY SAVINGS

According to a report by IEA, the energy consumption of the transportation sector is increasing by 1.4% annually. Globally Electric Vehicle will account for 4% of global electricity demand by 2030. As the world is moving toward the target of a greener environment and sustainable development there will be a total of 25 million

EVs on the roads which will share the electricity for charging and as a result the demand will increase from 4% to 10% globally, thus managing the charging of the off-peak hours charging for EVs will be given preference with the modified tariffs. On the other hand, while looking at the brighter side of sustainable development, there will be 16,000 GWh that can be stored in those 25 million EV batteries by 2030 and can be utilized by the grid at suitable times. EVs will be entirely beneficial only when charged with a renewable source of electricity and to see these fruitful environmental changes there are decades to go; but steps have to be taken right now for the future generation to live in a healthy, clean, and greener environment.

Although in V2G system for EVs, there are some technical constraints related to batteries and consumer participation which will limit the energy withdrawal to 5% only, and thus 600 GWh of energy can be consumed globally from the estimated EVs target by 2030. According to a study an owner is expected to save about $800 to $1000 per year with EV which also helps to build the economy. A gasoline engine is responsible for about 64% to 75% of energy losses while EV will have energy loss of about 15%–20% only. The regenerative braking system used for EVs makes it about 77% to 100% efficient when calculated for vehicle-to-fuel efficiency. About 1.5 million gram of CO_2 can be saved by just one electric vehicle on the road which make a huge difference in energy consumption, supply, and demand and thus making a balance energy bar graph with a cleaner and greener environment [8].

1.2.3 DEPENDENCY ON FUEL

The highly efficient electric drive components of an EV can make a sharp decline in fuel cost. Unlike conventional vehicles, the fuel economy of every EV is measured in **Miles per gallon of gasoline-equivalent** (MPGe) and **Kilowatt hour** (KWh) per 100 miles metrics as all EVs depend completely or partially on electric power. Expanding the EV fleet over time, the well-to-wheel (W2W) GHG emission will get reduced and the dependency on petroleum by the transportation sector reduces which makes other sectors to touch the crown of success. As fuel economy gets improved by saving the fuel taxes, the revenue can be used to build up the other infrastructure for EVs like charging stations and discounted electricity tariffs. It is estimated that by 2040 the EV penetration to the market will go up to 60% against the conventional vehicle which will decline the dependency of transportation on fuel to a remarkable extent. If the miles covered by EV are for the short-drive only and longer drives commute through only conventional vehicles then the consumption of oil will remain the same and it makes no difference in introducing EVs [9].

1.3 HYBRID ELECTRIC VEHICLE (HEV)

A hybrid electric vehicle (HEV) is a type of hybrid vehicle that combines a conventional internal combustion engine (ICE) system with an electric propulsion system (hybrid vehicle drivetrain). The presence of the electric powertrain is intended to achieve either better fuel economy than a conventional vehicle or better performance. Modern HEVs make use of efficiency-improving technologies such as regenerative brakes which convert the vehicle's kinetic energy to electric energy,

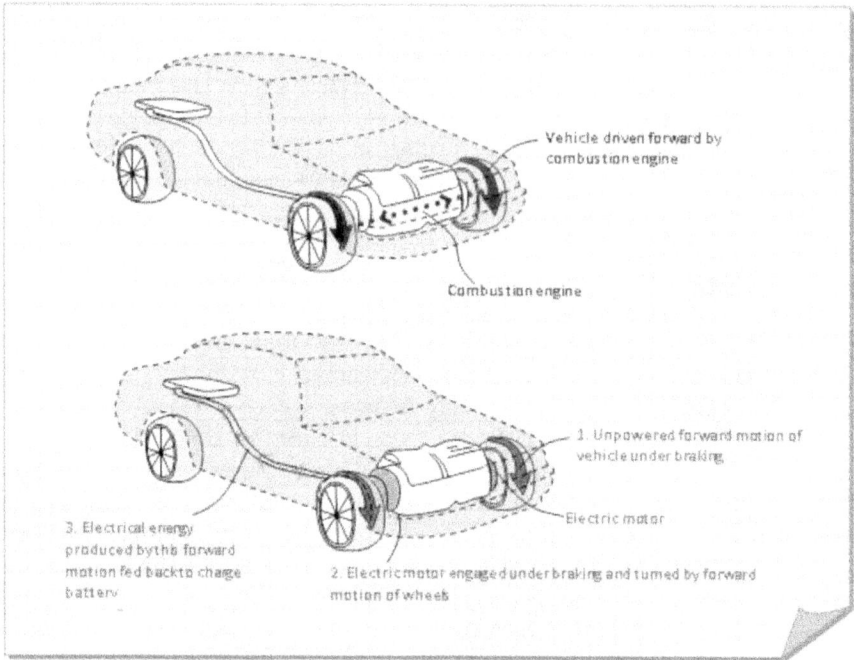

FIGURE 1.3　Schematic of a hybrid electric vehicle.

which is stored in a battery or supercapacitor.The Schematic of a hybrid electric vehicle is given in Figure 1.3.

Some varieties of HEV use an internal combustion engine to turn an electrical generator, which either recharges the vehicle's batteries or directly powers its electric drive motors; this combination is known as a motor–generator. Many HEVs reduce idle emissions by shutting down the engine at idle and restarting it when needed; this is known as a start-stop system.

1.3.1　DRIVETRAIN

A drivetrain is a collection of components that deliver power from a vehicle's engine or motor to the vehicle's wheels. In hybrid-electric cars, the drivetrain's design determines how the electric motor works in conjunction with the conventional engine. The drivetrain affects the vehicle's mechanical efficiency, fuel consumption, and purchasing price.

1.3.1.1　Series Drivetrains

Series drivetrains are the simplest hybrid configuration. In a series hybrid, the electric motor is the only means of providing power to the wheels. The motor receives electric power from either the battery pack or from a generator run by a gasoline engine. A computer determines how much of the power comes from the battery or the engine/generator. Both the engine/generator and the use of regenerative braking recharges the battery pack. The engine is typically smaller in a series drivetrain because it only has to meet certain power demands; the battery pack is generally more powerful than the one in parallel hybrids to provide the remaining power needs. This larger battery and motor, along with the generator, add to the vehicle's cost, making series hybrids more expensive than parallel hybrids.

1.3.1.2   Parallel Drivetrains

In vehicles with parallel hybrid drivetrains, the engine and electric motor work in tandem to generate the power that drives the wheels. Parallel hybrids tend to use a smaller battery pack than series drivetrains, relying on regenerative braking to keep it recharged. When power demands are low, parallel hybrids also utilize the motor as a generator for supplemental recharging, much like an alternator in conventional cars. Since the engine is connected directly to the wheels in parallel drivetrains, the inefficiency of converting mechanical power to electricity and vice versa is eliminated, increasing the efficiency of these hybrids on the highway. This reduces but does not eliminate, the efficiency benefits of having an electric motor and battery in stop-and-go traffic.

1.3.1.3   Series/Parallel Drivetrains

Series/parallel drivetrains merge the advantages and complications of parallel and series drivetrains. By combining the two designs, the engine can both drive the wheels directly (as in the parallel drivetrain) and be effectively disconnected, with only the electric motor providing power (as in the series drivetrain). The Toyota Prius helped make series/parallel drivetrains a popular design. With gas-only and electric-only options, the engine operates at near-optimum efficiency more often. At lower speeds, it operates more as a series vehicle, while at high speeds, where the series drivetrain is less efficient, the engine takes over and energy loss is minimized. This system incurs higher costs than a pure parallel hybrid since it requires a generator, a larger battery pack, and more computing power to control the dual system. Yet its efficiencies mean that the series/parallel drivetrain can perform better – and use less fuel – than either the series or parallel systems alone.

1.3.2   Plug-in Hybrid Electric Vehicle (PHEV)

A plug-in hybrid electric vehicle (PHEV), also known as a plug-in hybrid, is a hybrid electric vehicle with rechargeable batteries that can be restored to full charge by connecting a plug to an external electric power source. A PHEV shares the characteristics of both a conventional hybrid electric vehicle, having an electric motor and an internal combustion engine, and of an all-electric vehicle, also having a plug to connect to the electrical grid. PHEVs have a much larger all-electric range as compared to conventional gasoline-electric hybrids and also eliminate the "range anxiety" associated with all-electric vehicles because the combustion engine works as a backup when the batteries are depleted.

The introduction of HEVs into the transportation sector can be viewed as a good start, but the range (the distance that can be traveled with one charging cycle) is not adequate. So PHEVs have started penetrating the market, in which the batteries can be charged at any point where a charging outlet is available. For HEVs, the impact on the grid is not a matter of concern, since HEVs are charged from their internal combustion engine by regenerative braking, whenever the driver applies a brake. As a result, batteries in HEVs maintain a certain amount of charge (70%–80%). In the case of PHEVs, the car batteries are used steadily while driving in order to maximize fuel efficiency and the battery charge decreases over time. The vehicle thus needs to be connected to the power grid to charge its batteries when the vehicle is not in use. During its charging time, the plug-in vehicle more than doubles the average household load. Hence, for PHEVs, a major concern is the impact on the grid, since they can be plugged in for charging at any point in the distribution network regardless of time. PHEVs will be posed as a new load on the primary and secondary distribution network, where many of these circuits are already being operated at their maximum capacity. With the increase in the number of PHEVs, the additional load has the potential to disrupt the grid stability and significantly affect the power system dynamics as a whole [10].

1.3.3 Renewable Energy Resources

Buildings and cars have a significant impact on the increasing energy demand, global warming, and air quality. Hence, research is quickly gearing toward interconnected renewable energy-based systems for transportation and residential/commercial buildings. PHEVs benefit from an existing infrastructure that can directly use renewable energy. Even though the present grid system is capable of supplying power during peak hours, energy can be stored in the storage systems to provide it for peak hours. This would increase the reliability of the system. This stored energy can be obtained from renewable energy sources. The storage medium can be the ESU (Energy Storage Unit) of the PHEV itself. At a household level, this system can be designed to provide power to the household utilities as well as the PHEV. If the capacity of the ESU is more than required of the average of a vehicle i.e., 33 miles/day, this extra energy can be used to provide to the grid. Figure 1.4 shows that a Photovoltaic Array can provide power to the ESU as well as the grid. Vehicle battery recharging takes place during off-peak hours, taking advantage of cheaper rates and the available energy stored in the residential batteries from renewable sources during daylight. The residential batteries are used as a buffer for surplus and needs. If neither the renewable source nor the batteries are able to supply the requested energy, some energy is purchased from the grid. The exchange of energy between the users with self-generating capability is based on Net Metering System. If the electricity provider

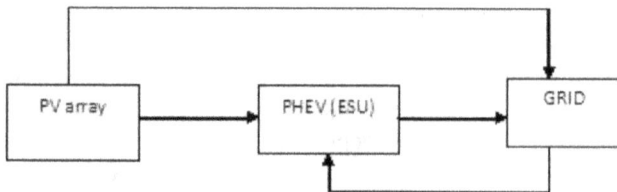

FIGURE 1.4 A typical PV array that can provide power to both PHEV as well as the grid.

provides more kWh to the customer than what the customer-generating facility feeds back to the electricity provider's system, the customer's energy charges shall be calculated using the customer's net energy usage for a billing period. On the other hand, if the customer's generating facility feeds more kWh back to the electricity provider's system, then the customer's net billing is negative during the billing period. The negative net billing shall be allowed to accumulate as a credit to offset billing in the next billing cycle [10].

1.4 BATTERY ELECTRIC VEHICLE

Electric vehicles with only batteries to provide power to the drive train are known as Battery Electric Vehicles (BEVs). BEV has entered a high-speed developing period due to new regulations and homologations which aim to limit emissions and lead automakers to focus on CO_2-free transportation. The range of such vehicles depends directly on the battery capacity. Typically, they can reach 100 km to 250 km on one charge depending on driving conditions, style, vehicle configuration, climate, and road conditions [11]. Architecture of battery electric vehicle is given in Figure 1.5.

Because the vehicle is powered only by batteries or other electrical energy sources, zero-emission can be achieved. However, the high initial cost of BEVs, as well as their short driving range and long refueling time has limited their use. Still, new BEV architectures have been proposed that use several energy sources (e.g., batteries, supercapacitors, and even reduced-power fuel cells) connected to the same dc bus, which should eventually reduce the refueling time, expand the driving range, and drive down the price.

Electronic components have a significant diffusion inside not only electric vehicles but also combustion engine vehicles. In the past, there were only low-voltage electronic components that were supplied by 12 V batteries. But today BEV has an electric propulsion system that is supplied by High Voltage (HV) battery unit that provides around 355 V to HV components. The voltage in the high voltage link can be reached up to 900 V in some electric vehicles. There are various propulsion architectures for BEV such as single-centered e-machine, in-wheel hub motors, etc.

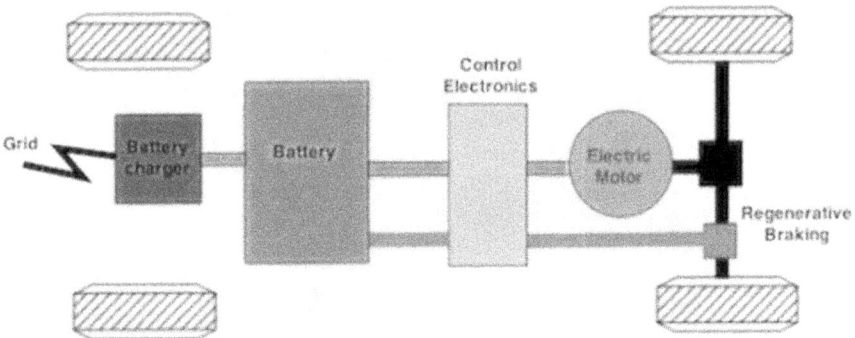

FIGURE 1.5 Architecture of battery electric vehicle.

1.5 FUEL CELL ELECTRIC VEHICLE

The development of substitute green power sources is essential to replace the existing need for fossil fuels in order to minimize its harmful effects on the environment and health. Among the substitutes, fuel cell technology is an encouraging replacement source of energy for sustainable development. Conventional fuel sources are responsible for rapid climate change as well as global warming. On the contrary, fuel cell does not emit any harmful gases or substances such as carbon monoxide, carbon dioxide, etc. Fuel cell converts chemical energy into electrical energy, unlike conventional engines which convert fossil fuels into heat and light energy.

Fuel cell vehicles (FCVs) use fuel cells to generate electricity from hydrogen and air. The electricity is either used to drive the vehicle or stored in an energy-storage device, such as a battery pack or supercapacitors. FCVs emit only water vapor and have the potential to be highly efficient. However, they do have some major issues: 1) the high price and the life cycle of the fuel cells; 2) onboard hydrogen storage, which needs improved energy density; and 3) a hydrogen distribution and refueling infrastructure that needs to be constructed [2]. FCVs could be a long-term solution. Although prototypes have already been proposed by manufacturers, the potential of FCVs, including hydrogen production and distribution facilities, has yet to be proven.

Among the several kinds of fuel cells available, the proton exchange membrane fuel cell (PEMFC) is used widely because of its characteristics such as high efficiency, fast startup, and ability to perform at low temperatures. Having a lifespan of 5000 h and a high-power density (0.3–$0.8 W/cm^2$), PEMFC is perfectly suited for electric vehicles. However, the PEMFC has the limitations of having sensitivity to carbon monoxide (CO) and the use of costly platinum catalysts. A hydrogen storage tank in fuel cell electric vehicle is given in Figure 1.6.

1.6 EXTENDED RANGE ELECTRIC VEHICLE

The extended-range electric vehicle (E-REV) is essentially an all-electric vehicle, with an electric motor providing all motive power and a tiny internal combustion engine (ICE) providing additional electric power. It might also be considered a

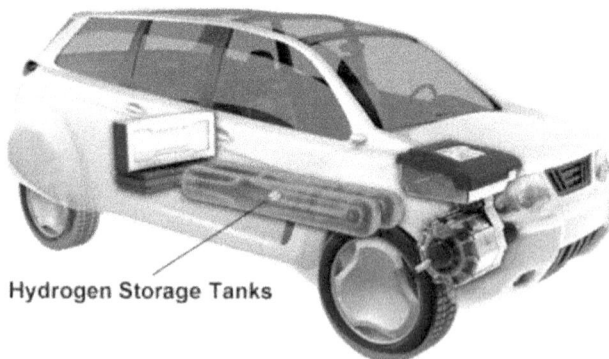

Hydrogen Storage Tanks

FIGURE 1.6 Hydrogen storage tanks in fuel cell electric vehicle.

series hybrid with a considerably bigger battery. When the battery reaches a certain degree of discharge, the ICE activates a generator, which delivers energy to the motor and recharges the battery. The natural range restriction of a BEV can be solved using this setup.

E-REVs, unlike parallel hybrids and other series hybrids with their smaller batteries and restricted electric range, can operate in the full-electric mode for short distances, clean and energy-proficient as BEVs. E-REVs use the ICE to keep the battery charged over longer distances, yet they use far less fuel than standard ICEVs for two reasons:

- An E-engine REV is substantially lesser than that of a standard ICEV; it just needs to fulfill typical power demands as the battery pack provides peak demand. An ICEV's engine, on the other hand, must be able to handle peak-power surges, such as accelerations.
- An E-engine REV runs at a steady, high-efficiency rotation speed, whereas an ICEV's engine frequently runs at low or high rotation speeds, with low efficiency in both cases.

Figure 1.7 schematically illustrates the different forms of E-REV operation. The battery state-of-charge (SoC) is close to 100% when the car starts its journey. The

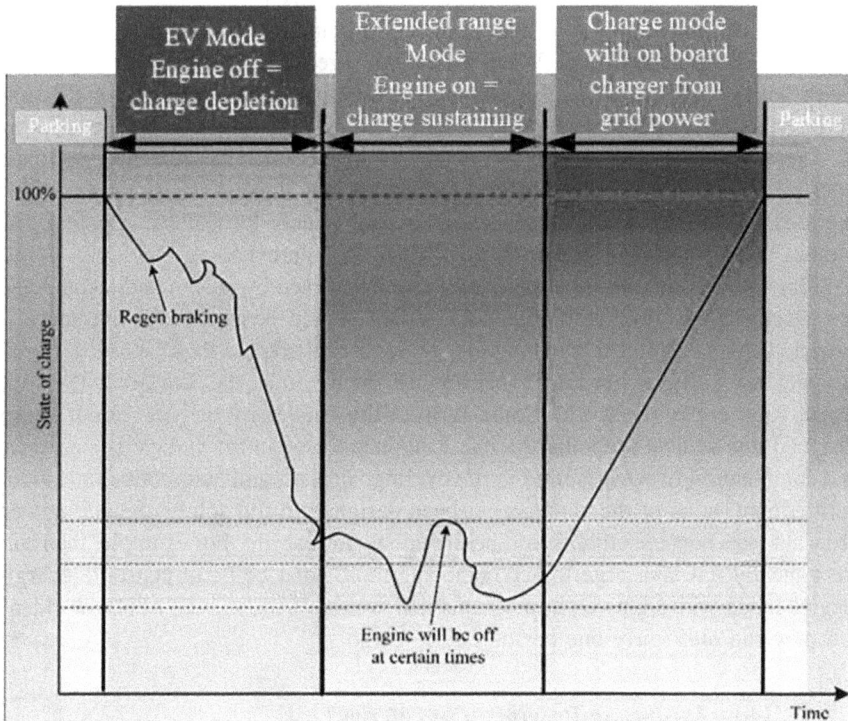

FIGURE 1.7 Battery state-of-charge (SoC) in different extended-range electric vehicle (E-REV) operation modes.

electric motor provides all of the vehicle's power, drawing energy solely from the battery, and there are no local exhaust emissions. Each regenerative braking operation partially recharges the battery. The vehicle shifts to the extended-range mode when the battery is drained to a pre-determined SoC – shown in Figure 1.7 as three levels of increasing severity, green, orange, and red. The ICE is turned on as needed while the car is in this mode to keep the battery within the SoC range indicated by the green and red dashed lines.

With electricity taken from the grid, the battery SoC is recovered to 100% after the journey. A tiny gas turbine could be used to replace the piston engine as a range extender in the future. The Jaguar C-X75 hybrid concept car is an E-REV with two tiny gas turbines (each weighing 35 kg) that charge the battery (15-kWh lithium-ion). The 1350-kg vehicle can be driven up to 205 mph (330 km h1) with a total torque of 1600 Nm thanks to four 145-kW electric motors, one at each wheel. The C-X75 has a 70-mile (113-kilometer) electric-only range and a 60-liter gasoline tank.

1.7   SOLAR-POWERED CARS

As awareness of clean and sustainable energy supplies has developed, solar-powered cars have become more prominent. Many automakers are developing solar cars, and the technology can change the automotive industry's future. Solar cars are electric vehicles that transform sunshine into electricity using photovoltaic panels. These cars can store some solar energy in batteries and function properly at night or when there is no natural light. When used on a large scale, solar-powered cars help lessen both ecological and noise pollution. Numerous solar-powered car concepts are being created right now. Several small and large car manufacturers are working on solar hybrid cars. According to some estimates, solar car sales may surpass $689 billion by 2027. To capitalize on the idea, automakers are already creating intermediary energy sources such as solar roof panels for battery charging and internal systems. A solar powered car is given in Figure 1.8.

Solar automobiles, on the other hand, are constrained by design and technology. For a vehicle like this, numerous solar panels would be necessary, but space is restricted. As a result, the vast majority of solar cars created thus far are designed for solar car racing rather than everyday use. At the time, the Sun Swift IV is the fastest solar car in the world. Students from the University of New South Wales designed this vehicle for Solar Racing. Equipment used in the vehicle is connected to a combination of what is used in the cycling, aircraft, and automotive industries. Furthermore, none of the cars has yet been designed to run solely on solar power. They also use concepts that would be unrealistic in real life. For example, the roofs are typically flat and large to accommodate more solar cells; to conserve energy, they're made with lightweight materials that wouldn't resist even a little accident, and they can only carry one person but no cargo.

1.7.1   How Do Solar-Powered Cars Work?

Photons, or light particles, excite electrons in a solar panel, resulting in an electric current. A network of linked photovoltaic cells is what solar panels are. Each solar

FIGURE 1.8 Solar-powered car.

cell is made up of two semiconductor layers of silicon (mostly). To give each "layer" a negative or positive electrical charge, silicon is combined with additional elements, most commonly phosphorus and boron. At the intersection of the two layers, an electric field is created. When a photon of light knocks an electron loose, it is pushed out from the silicon junction by the electric field. On the cell's sides, metal conducting plates gather and transfer electrons to wires. The electrons then can flow freely, just like any other energy source. The following are some of the advantages of utilizing solar cars:

- It conserves fuel and hence, is cheaper to run
- It is environmentally friendly and sustainable
- Air and Noise pollution is not an issue for solar-powered electric vehicles
- Noise and air pollution are also lessened by the use of solar electric vehicle

1.7.2 CURRENT DEVELOPMENTS IN SOLAR-POWERED CARS

One of the earliest solar-powered automobiles was built by General Motors and shown at a conference in Chicago in 1955. The "Sun mobile," a solar-powered 15-inch (38-cm) concept "vehicle," was made up of a tiny Pooley electric motor and 12 selenium photovoltaic cells. The back wheel shaft was rotated by the Pooley electric motor with a pulley. The first solar car was constructed, but it was too small to be driven. Another amazing solar-powered car is the Sono Motors Sion. According to the company, this is the first commercially available hybrid solar-electric car. It has a range of up to 160 miles (255 kilometers) and charges itself using solar energy. In 2019, Toyota developed a solar-powered Prius prototype that generated 180 watts of electricity per hour and had a range of 3.8 miles (6.1 kilometers) after a day of charging. Later versions, on the other hand, claimed to give 860 W of power and a single-charge range of 27.6 miles (44.5 kilometers). Because the solar panels could only power the third battery, the automobile had to be charged at a charging station.

1.8 FUTURE OF ELECTRIC VEHICLES

By the end of 2020, there will be 10 million electric vehicles on the road, following a decade of rapid growth. Despite the epidemic, worldwide automobile sales fell by 16% in 2020, despite a 41% increase in electric car registrations. Europe eclipsed China as the world's largest electric vehicle (EV) market for the very first time, selling about 3 million units (equivalent to 4.6% market share). Electric bus and truck registrations also increased in key countries, hitting 600,000 and 31,000 units globally, respectively. In 2020, consumer expenditure on electric vehicles will reach USD 120 billion. Governments worldwide spent USD 14 billion to promote electric vehicle sales in 2019, which is up 25% from the year before, owing primarily to higher incentives in Europe. However, the share of government aid in total expenditure on electric cars has decreased over the previous five years, showing that EVs are becoming much more desirable to buyers. According to a recent projection, the EV market would be worth at least $475 billion by 2025. Automobile manufacturers have stated more ambitious goals for electrification. Eighteen of the world's top 20 automakers, which will account for more than 90% of new car registrations in 2020, have announced intentions to extend their model line-ups and rapidly increase the production of light-duty electric vehicles. Electric heavy-duty vehicle models are becoming more widely available, with four major vehicle companies predicting an all-electric future [10].

Three main factors underpin the viability of electric vehicle sales in the foreseeable years:

Strong regulatory frameworks: several countries have enacted important legislation, such as CO_2 emission limits and zero-emission vehicle (ZEV) mandates. By the end of 2020, more than 20 countries will have imposed limitations on traditional automotive sales or will have mandated that all new sales be ZEVs. Additional incentives to safeguard EV purchases from the economic recession: certain European countries extended purchase incentives, while China delayed the end of its subsidy program; the number of electric car models has increased, and battery prices have reduced [11,12]. Projected growth for EV sales in India is given in Figure 1.9.

Existing policies throughout the world suggest considerable growth over the next decade: in 2030, the worldwide EV stock will reach 145 million, contributing to 7% of the road vehicle fleet, according to the Stated Policies Scenario. The Electric Vehicles Initiative (EVI) was created in 2010 by the Clean Energy Ministerial. Recognizing the advantages of electric cars, the EVI is dedicated to promoting their adoption worldwide. Its goal is to help governments better grasp the policy challenges underlying electric mobility, as well as serve as a knowledge-sharing platform.

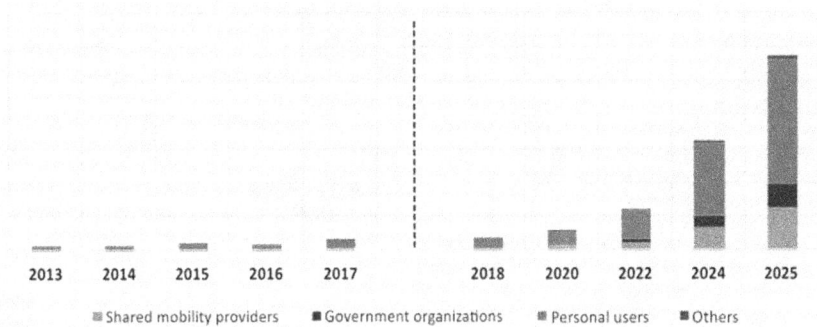

FIGURE 1.9   Projected growth for EV sales in India by 2025.

- Support and monitor the rollout of electric car chargers.
- Encourage governments and businesses to commit to using electric cars in their fleets.
- Expand the breadth of policy study and exchange of knowledge.
- Assist governments with capacity building and training.

In India under the FAME II scheme,133 million USD is allotted to promote and incorporate infrastructure for charging stations in India. The Ministry of Heavy Industries issued an expression of interest in October 2020, inviting investors to participate in the project by installing at least one charging station every 25 kilometers along significant routes and every 100 kilometers to accommodate HDVs. The rapid development of EV mobility and industrial adoption of battery manufacturing will drastically change the scenario of EVs in the world.

REFERENCES

[1] M. Guarnieri, "When Cars Went Electric, Part One [Historical]," IEEE Industrial Electronics Magazine, vol. 5, no. 1, pp. 61–62, March 2011, doi: 10.1109/MIE.2011. 940248

[2] M. Guarnieri, "When Cars Went Electric, Part 2 [Historical]," IEEE Industrial Electronics Magazine, vol. 5, no. 2, pp. 46–53, June 2011, doi: 10.1109/MIE.2011. 941122

[3] M. R. Khalid, M. S. Alam, A. Sarwar, and M. S. J. Asghar, "A comprehensive review on electric vehicles charging infrastructures and their impacts on power-quality of the utility grid," *eTransportation*, vol. 1, Aug. 2019, Art. no. 100006.

[4] https://www.theverge.com/2019/3/21/18276541/norway-oslo-wireless-charging-electric-taxis-car-zero-emissions-induction

[5] C. C. Chan, A. Bouscayrol and K. Chen, "Electric, Hybrid, and Fuel-Cell Vehicles: Architectures and Modeling," *IEEE Trans. Veh. Technol.*, vol. 59, no. 2, pp. 589–598, Feb. 2010.

[6] M. Yilmaz and P. T. Krein, "Review of battery charger topologies, charging power levels, and infrastructure for plug-in electric and hybrid vehicles," *IEEE Trans. Power Electron.*, vol. 28, no. 5, pp. 2151–2169, May 2013.

[7] M. R. Khalid, I. A. Khan, S. Hameed, M. S. J. Asghar and J. -S. Ro, "A Comprehensive Review on Structural Topologies, Power Levels, Energy Storage Systems, and Standards for Electric Vehicle Charging Stations and Their Impacts on Grid," *IEEE Access*, vol. 9, pp. 128069–128094, 2021.

[8] M. H. Mobarak, R. N. Kleiman and J. Bauman, "Solar-Charged Electric Vehicles: A Comprehensive Analysis of Grid, Driver, and Environmental Benefits," *IEEE Trans. Transp. Electrification*, vol. 7, no. 2, pp. 579–603, June 2021.

[9] C. Ding, H. Li, W. Zheng, and Y. Wang, "Reconfigurable photovoltaic systems for electric vehicles," *IEEE Des. Test*, pp. 37–43, Nov./Dec. 2018.

[10] P. Mulhall, S. Lukic, S. Wirasingha, Y. Lee, and A. Emadi, "Solar assisted electric auto rickshaw three-wheeler," *IEEE Trans. Veh. Technol.*, vol. 59, no. 5, pp. 2298–2307, June 2010.

[11] S. S. Williamson, A. K. Rathore, and F. Musavi, "Industrial Electronics for Electric Transportation: Current State-of-the-Art and Future Challenges," *IEEE Trans. Ind. Electron.*, vol. 62, no. 5, pp. 3021–3032, May 2015.

[12] Husain *et al.*, "Electric Drive Technology Trends, Challenges, and Opportunities for Future Electric Vehicles," *Proc. IEEE*, vol. 109, no. 6, pp. 1039–1059, June 2021.

2 Introduction to EV Charging Systems

*Kirti Mathuria, Asmita Singh, Sandeep N,
Arun Kumar Verma, Saravana Prakash P, and
Harpal Tiwari*

CONTENTS

DOI: 10.1201/9781003330134-2

2.1 GENERAL INFORMATION ABOUT ELECTRIC VEHICLE CHARGING SYSTEM

A significant obstacle to the sustained advancement of human civilization is climate change. Governments throughout the world are striving to lower carbon emissions, of which road transportation frequently makes up a substantial percentage. As a result, many nations have implemented electric vehicle (EV) emission targets. The environmental benefits of EVs have garnered considerable attention in recent years.

Rapid improvements in battery production technology have increased energy density, increased charge cycles, improved safety, and lowered cost, all of which are contributing to the continued rise in EV adoption. Sales of EVs are booming worldwide due to government initiatives encouraging cleaner modes of mobility and less energy waste. Given the present and planned goals, it is expected that there will be 500 million EVs on the roads by 2030.

Infrastructure for the transport or charging of EVs is essential as the technology for these vehicles develops rapidly. Expanding the usage of electric vehicles necessitates not only technological improvements to the vehicles themselves but also the development of a reliable network of charging stations. Potentially a massive market on a global scale, this might prompt significant investments in supporting infrastructure. Governments worldwide have developed ambitious strategies to become world leaders in EV design, research, and production and to attract EV-related foreign direct investment.

Therefore, the charging infrastructure and EV technologies will act as the primary facilitator for this change in mobility away from ICE-based vehicles. Home, office, retail, recreational, and highway areas will all need EV charging stations. The distribution network must deliver the EV charging power at the lowest possible cost, with the least amount of support, and with the highest level of dependability.

A high percentage of EVs on the road can raise peak grid demand and potentially overburden distribution network infrastructure. Second, fossil fuels are used to power most of the current energy grid. A significant portion of the emission level is transferred between vehicles and power plants during charging. EVs are thus not environmentally friendly as one might think. Thus, it will be essential for future EVs to be charged by utilizing renewable energy sources like the sun or wind [1].

With their capacity to serve as quick-response storage for the grid and a regulated load at the same time, EVs have the potential to be a game-changer.

High EV costs, limited operating range, and rapid battery deterioration are the primary obstacles to the spread of EVs. To overcome these difficulties, efficient charging technologies are required, and the enhancement of the EV battery's life cycle and an increase in EV performance are required. The charging strategy and communication technologies are utilized to determine the efficiency of the charging technology. The power levels, the charger type, and the power quality (PQ) utilized for EV charging are all factors that might impact charging efficiency.

The existence of charging stations specifically designed for electric vehicles will be essential to the successful execution of a seamless transition to e-mobility. In this context, five technologies are essential to the infrastructure for charging electric vehicles: Smart charging, vehicle-to-grid (V2G) aspect, photovoltaic (PV) charging of EVs, on-road charging of EVs, and contactless charging [2].

2.1.1 THE STATUS OF CONDUCTIVE CHARGING FOR EVS

Cables using either alternating current (AC) or direct current (DC) may now be used to recharge EVs. There is no universally accepted standard for charging power, as there are many available options. The AC charger is an isolated AC/DC converter and is the easiest way to charge an electric vehicle. Three different AC charging systems are used worldwide: the Type 1 SAE J1772-2009 system utilized in the United States, the Type 2 Mennekes charger system implemented in Europe, and the charger by Tesla.

The European Type 2 connector uses a three-phase 400 V grid connection, allowing higher charging capabilities up to 43 kW compared to the North American Type 1 connector (up to 63 A). The control pilot (CP) and proximity pilot (PP) are used for interacting with the EV. Electric vehicle manufacturers have mostly shifted to using Type 2 EV chargers instead of the EV plug alliance's Type 3 EV chargers. The onboard AC charger in commercial EVs generally has a maximum charging output of 20 kilowatts (kW) due to space and weight constraints. The drivetrain propulsion power electronics are the lone exception, typically providing significantly greater power (80–500 kW). Onboard-integrated chargers are what these devices are termed. The integrated chargers used for electric vehicle charging combine the inverter for the drivetrain with the windings of the propulsion motor. A 43 kW integrated "Chameleon" charger is seen in the Renault Zoe. It is recommended to use an off-board charger while charging a high-power electric vehicle (one with a capacity of more than 50 kW).

DC charging was developed to alleviate the size and weight restrictions of an onboard charger and allow for quicker EV charging (up to 350 kW). DC chargers come in three varieties at the moment: Type 4 CHADeMO, CCS/COMBO (Combined Charging System), and Tesla's own domestic and international offerings. Table 2.1 provides more evidence of this [3].

2.1.2 SMART CHARGING OF EVS

EVs contribute to the grid because of their flexibility in charging power, ability to ramp up or down in power swiftly, and capacity to charge and discharge. But at the moment, this potential is untapped. The method of charging an EV at the moment is unregulated; once attached to a charger, fixed power charging takes place till the charging is finished. With smart charging, the power and the direction of EV charging can be continually adjusted (dynamic charging).

For both EV owners and the companies who offer the infrastructure for EV charging, smart charging of EVs can have the following advantages Figure 2.1:

TABLE 2.1

Power Adapters, Voltage, and Current in Europe and the United States

Type of Power Adapters	Total Connectors (Communication)	Level of Charging	Voltage, Current, Power Ratings
Type1 SAE JI772 USA	Power-3 [L1, N, E] Control-2 [CP, PP] (PWM over CP)	Level 1 (AC) Level 2 (AC)	1-phase 120V ≤ 16A, upto 1.9 KW 1-phase 240V ≤ 80A, upto 19.2 KW
Type 2 Mennekes Europe	Power-4 [L1, L2, L3, N, E] Control-2 [CP, PP] (PWM over CP)	Level 1 (AC) Level 2 (AC)	1-phase 230V, ≤ 32A, 7.4 KW 3-phase 400V, ≤ 63A, 43 KW
Type 4 CHADeMO	Power-3 [DC+, DC-, E] Control-7 (CAN communication)	Level 3 (DC)	200-500V ≤ 400A, 200 KW
SAE CCS Combo	Power-3 [DC+, DC-, E] Control-2 [CP, PP] (PLC over CP, PE)	Level 3 (AC)	200-1000V DC ≤ 350A, 350 KW
Tesla US	Power-3 [DC+, DC-, E or L1, N, E] Control-2 [CP, PP]	Level 3 (DC)	Model S, 400V ≤ 300A, 120 KW

1. Lower EV charging costs based on energy pricing
2. Establish new revenue sources, such as vehicle-to-grid
3. Use wind energy more at night and solar energy more during the day to charge EVs. Renewable energy can be stored in EV batteries, negating the need for additional storage.
4. Reduce losses in the distribution system.
5. Using demand-side management, EV charging can lower the grid's peak demand. This postpones or eliminates the requirement for distribution network infrastructure updates.

The conventional method for smart charging is taking one or a handful of these applications into account during optimization to get either direct or indirect financial advantages. Although this approach is more straightforward, it under-utilizes smart charging's potential and renders the benefits economically unappealing. In the future, it will be crucial to combine several intelligent charging applications into a unified framework. The cumulative effects will then become economically appealing for widespread application.

2.1.3 Solar Charging of EVs

The infrastructure for charging electric vehicles must receive all or most of its power from renewable resources to guarantee that using electric vehicles results in net zero CO_2 emissions. Here, the decreasing cost of PV over time and its simplicity

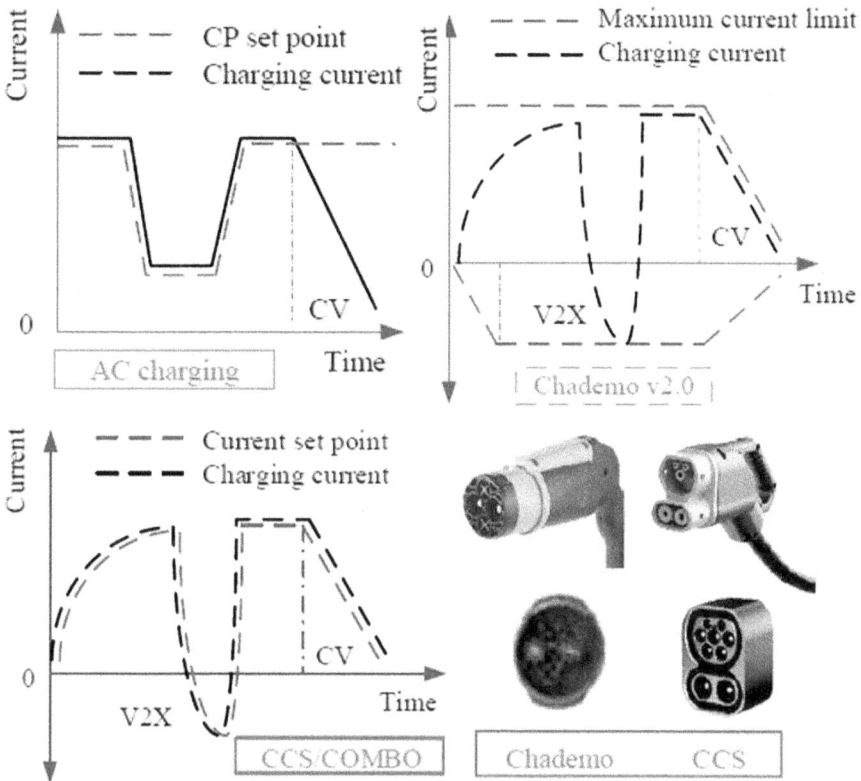

FIGURE 2.1 AC Dynamic charging_CHADeMO v 1.2 and CCS.

of integration into the distribution network become crucial. Office buildings and industrial zones make excellent locations for solar EV charging because they can have PV panels on their roofs and in their parking lots [4].

Beyond the overall reduction in CO_2 emissions, there are additional benefits of charging EVs from PV:

1. EV and PV can be deployed close to one another.
2. PV energy can be stored in an EV battery.
3. Lowered grid energy demand because the EV charging power comes locally from PV
4. Lower EV charging costs and a smaller impact from changes to PV feed-in tariffs

2.1.3.1 AC Charging from PV

A conventional PV inverter and an EV charger that is connected to the AC grid can be utilized to charge EVs from PV, as shown in Figure 2.2(a). However, compared to DC, this AC connection is less effective because

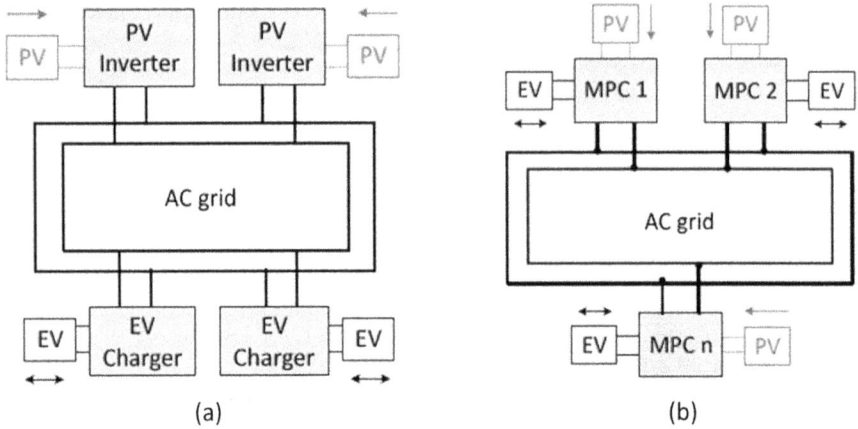

FIGURE 2.2 (a) Using a dedicated PV inverter and EV charger to charge an EV from solar panels. (b) A PV and AC grid-integrated multiport power converter (MPC) for recharging the EV.

1. Because PV and EV are DC devices, transferring power over AC involves different conversion stages and losses.
2. To power the PV and EV, two inverters will be required.
3. The converters must communicate to enable the EV to be charged using PV power.

2.1.3.2   EV-PV Power Converter Built-In with DC Charging

Connecting the EV, PV, and grid in a single integrated multi-port converter, as shown in Figure 2.2(b), is a better way to deal with the abovementioned problems. It comprises a DC-DC converter for photovoltaics, an isolated DC-DC converter for electric vehicles, and an AC grid connecting the DC-AC inverter. A DC link is used to connect these sub-converters. Because of safety concerns, the rules for charging EVs say that there is a need for these converters. The PV converter follows the solar array's maximum power point, and the charging current controls the EV charger.

2.1.4   CONTACTLESS CHARGING OF EVs

Contactless charging of EVs using inductive power transfer (IPT) is becoming increasingly common as a critical component of EV autonomy. This method transmits energy through air gaps between charge pads that are only weakly linked via electromagnetic coupling. Figure 2.3 depicts a schematic of the system.

These are the critical elements of this technology:

- Base power electronics: The rectified AC from the 3-phase or 1-phase grid, and DC is provided to an inverter that creates the AC input for IPT.
- Charging pads: Researchers are looking into several magnetic architectures that make magnetic coupling and power transfer as efficient as possible.

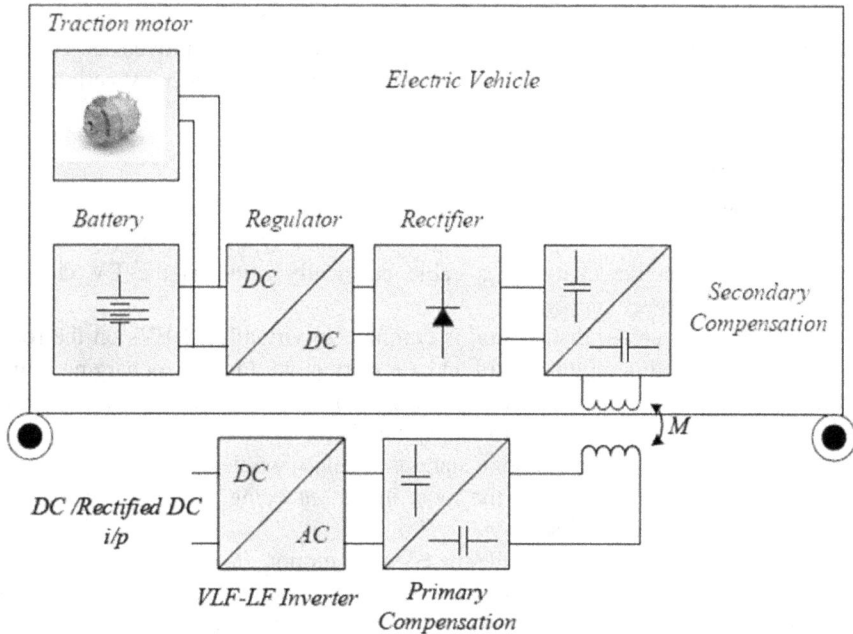

FIGURE 2.3 EV IPT-based system illustrating the many stages of power conversion.jpg.

- Compensation circuit: The use of resonant capacitors in the compensation circuitry is being looked into as a way to compensate for the primary and secondary inductive leakages. Researchers are looking at various combinations, like series-series (SS), series-parallel (SP), parallel series (PS), parallel-parallel (PP), etc.
- Vehicle electronics: Energy is first regulated by the inbuilt rectifier and dc voltage regulator before being stored in the battery.

IPT systems are useful for charging comfortably without plugging in a cable and removing the risk of electrocution, especially in bad weather. Additionally, it is naturally safe, reliable, and needs little maintenance. Additionally, independent inductive charging can be used to supplement advancements in autonomous vehicles [5].

2.1.5 TYPES OF CHARGING

2.1.5.1 Charging by On-Board and Off-Board

The charger's location determines whether an EV charger is considered off-board or on-board. In most cases, there is onboard AC level 1 and level 2 charging. The charging process moves more slowly due to the charger's mounting on the vehicle. DC-level charging is made possible by an off-board charger, which features quick power transfer and power conversion that happens outside of the EV. The integrated onboard charger is the term given to a charging system that makes use of already installed propulsion systems without the need for any extra charging infrastructure. The motor winding inductance is employed as a filter, and the inverter used for propulsion powers the built-in charger by acting as a bidirectional converter [6].

2.1.5.2   Wireless and Wired Charging

Depending on the power transfer method, wires (conductive charging) and wireless technologies (inductive charging) are the two main types of charging electric vehicles. In conductive charging, a cable physically connects the EV charger connector to the power supply.

This method of power distribution is common to virtually all EVs on the road today since it is both straightforward and very effective. Inductive charging eliminates the need for a direct wire connection between the EV and the charger. Electromagnetic induction is used to supply electricity to the battery. Inductive coils can be installed in EVs and charging stations to allow wireless charging. Wireless charging has several advantages, the most important being its longevity, convenience, speed, and electrical safety.

When the battery is charged while the EV is in motion, its size is smaller, and it takes less time to charge fully. Instead of using a corded or wireless charging method, one may change to a fresh battery. The batteries of electric vehicles can be swapped out with fully or partially charged ones at designated charging stations. Eliminating peak loads, charging batteries at a central hub during off-peak hours, charging batteries with renewable energy, boosting customer comfort by reducing charging station trips, prolonging battery life, and enhancing dependability are among the key advantages.

2.1.5.3   Uni and Bidirectional Charging

The direction of power flow determines whether a charging process is considered unidirectional or bidirectional. The principal controller of the vehicle determines the charging rate during unidirectional charging, in which electricity flows only in one direction, from the grid to the battery. In bidirectional charging, an aggregator sits between the EV charging network and the utility and controls the power flow in both directions Vehicle to grid (V2G) and Grid to Vehicle (G2V)(G-V and V-G). The exchange in energy rate for V-G is governed by the aggregator, which keeps a close eye on power usage,Time of Discharge (ToD) ToD price, availability of EV, arrival/departure duration, SoC, and generating capacity. Power from batteries is sent back to the grid when demand exceeds supply, and electric vehicles may be charged during times of surplus [7] Figures 2.4 and 2.5.

2.2   CLASSIFICATION OF EV CHARGING AND REQUIREMENTS

2.2.1   CLASSIFICATION OF EVs

EVs can be divided into three main types based on how their electrical and fuel energy sources are grouped:

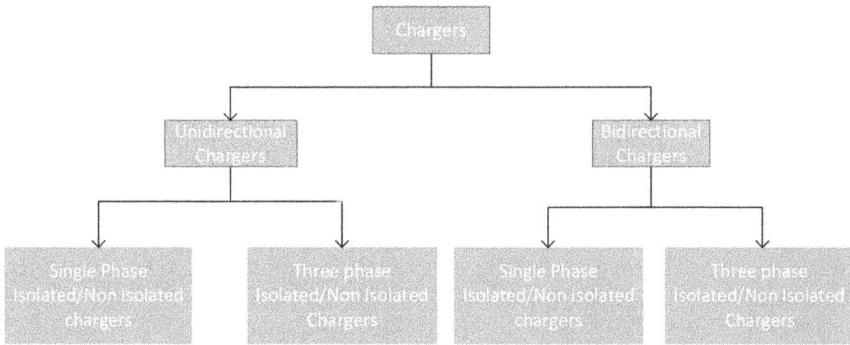

FIGURE 2.4 Classification of EV chargers.jpg.

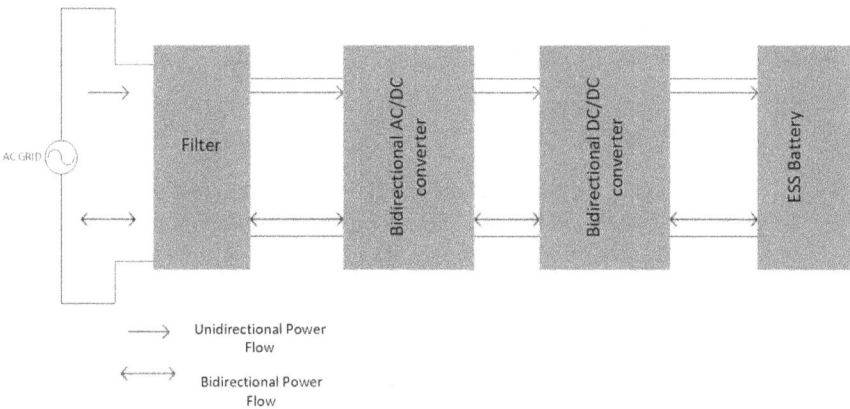

FIGURE 2.5 Types of power flow in EV.jpg.

2.2.1.1 Battery Electric Vehicle (BEV)

BEV needs a motor and an Energy Storage System (ESS) to work. It doesn't need a traditional Internal Combustion Engine (ICE) to support it. When the charge is gone, BEVs are plugged into a power source to charge their batteries. BEVs can also charge their batteries through a process called "regenerative braking," which uses the vehicle's electric motor to help slow it down and to get back the energy that the brakes usually turn into heat [8] Figure 2.6.

The main advantages of BEVs are

- Zero exhaust fumes
- No need for Gas/oil refueling
- Charging at home
- Rapid start-up and easy-going acceleration
- Relatively low operating expenses

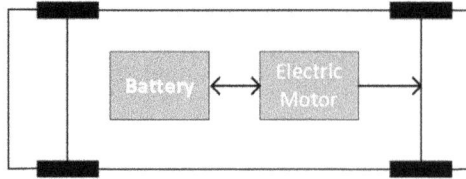

FIGURE 2.6 Architecture of BEV.jpg.

Some disadvantages of BEVs are

- Less range than comparable gas-powered vehicles;
- Higher initial cost; nevertheless, the payback period from fuel savings is just roughly two to three years.

2.2.1.2 Plug-in Hybrid Vehicle (PHEV)

The plug-in hybrid vehicle (PHEV) combines the ICE with an electric motor and ESS. Because it has an ICE, the PHEV is an excellent choice for long trips. Two distinct modes of operation—charge depletion (CD) and charge sustaining (CS)—make up a PHEV's overall operation. When the PHEV's battery state of charge (SoC) drops below a certain level, the ICE is disabled, and the EV is driven only by its battery. When the PHEV's SoC drops below the minimum required, it changes to CS mode, where the IC engine takes over propulsion duties while the battery's charge is maintained at a level just over the minimum required. A third mode, (CB) was developed to improve fuel economy. Optimally and dynamically using the electric motor and IC engine throughout a driving cycle in CB mode, both may operate for longer at their most efficient settings, resulting in total efficiency reductions [9] Figure 2.7.

The advantages of PHEVs are

- Superior Mileage Capability
- Reduced environmental damage from emissions
- Lower fuel usage compared to vehicles powered by internal combustion engines

The disadvantages of PHEVs are

- Not effective reduction of pollution
- Expensive to operate in comparison to BEVs

2.2.1.3 Hybrid Electric Vehicle

This vehicle has two ways to move: an ICE equipped with a fuel tank and a motor with an ESS. HEV is driven by both the ICE and the electric motor simultaneously. But HEVs don't have a way to charge from the power grid. Instead, they get all the energy they need to move from the fuel and the regenerative braking system Figure 2.8.

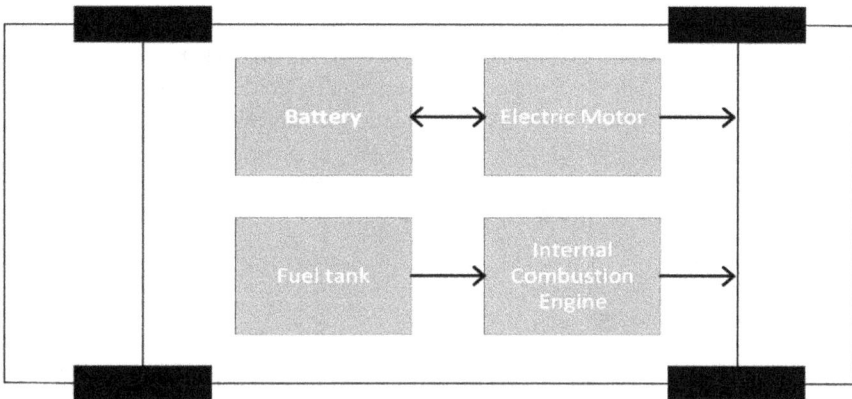

FIGURE 2.7   Architecture of PHEV.jpg.

The advantages of HEVs are-

- Longer driving range than BEV's
- Lesser fuel consumption compared to ICE-based vehicles.
- Lesser emission than ICE-based engines.

Some disadvantages of HEVs are

- Zero tailpipe emission is not attained
- The modes of operation are complicated
- High-priced in comparison to BEV's
- Inexpensive compared to ICE-vehicles

2.2.2   CLASSIFICATION OF EV CHARGING

EV chargers need to be made with several different things in mind. With the help of charging infrastructure and different charging levels, how charging affects the

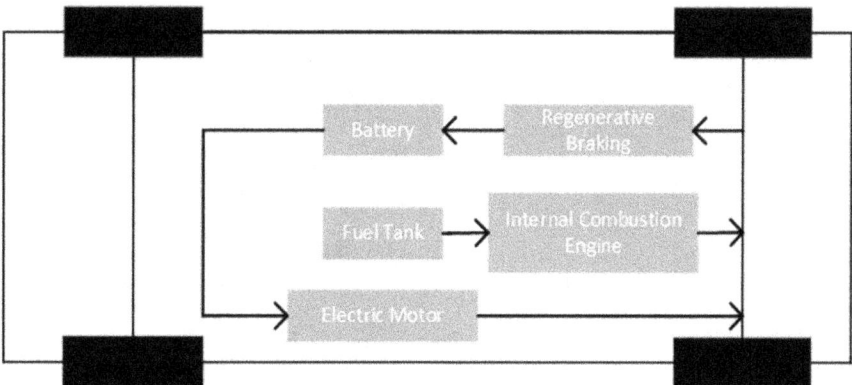

FIGURE 2.8   Architecture of HEV.jpg.

power grid, the price, the equipment, the site, the total charging time, and the amount of power. There are many things to look at that need to be taken into account when installing charging infrastructure and electric vehicle supply equipment (EVSE). These include: Charging station uniformity; charging time and length; charging policy distribution; and demand mechanisms Table 2.2.

- Charging station uniformity
- Charging time and duration
- Distribution and Demand Policies
- Regulatory workings

The availability of charging infrastructure helps minimize the price and need for on-board energy storage devices. A few of EVSEs most crucial parts are:

- Vehicle charging connectors and codes
- Charging stations in public and private spaces
- Assortment of plugs for making connections
- Power sources and safety gear

There are three power levels for EV charging: Level 1 charging, Level 2 charging, and Level 3 charging.

2.2.2.1 Level 1 Charging

This is a very gradual charging method. Typically, a NEMA 5-15R 120V/15A single-phase grounded outlet is used for Level 1 charging in the United States. The universal J1772 connector is used for this EV charging port. Onboard chargers often supply these and may produce as much as 1.9 kW of power from a 120-volt, single-phase ac source. Charging currents between 15 A and 20 A are acceptable. Since standard plugs may be found in most places. Level 1 charging can take 18–20 hours to fully charge an EV, depending on the ESS type and storage capacity. In most cases, this is most useful for whole-night charging in garages. In-vehicle chargers often support this charging standard.

2.2.2.2 Level 2 Charging

There is a standard way of payment in both public and private establishments. Private charging stations use a 220 V supply, whereas public stations use 208 V. An on-board 3.3-kilowatt-hour recharge allows for around 15 miles per hour of driving time. The exact amount of time spent charging a 6.6 KW onboard charger, however, will allow for around 30 miles of driving. They employ specialized equipment to speed up the charging process and need a professional electrical installation with a separate power supply. The initial setting up expenses are between $1000 and $3000. These chargers are convenient since they charge quickly and utilize a standard plug for connecting vehicles. One may anticipate being able to charge their vehicle through either alternating current (ac) or direct current (dc) with the standard's SAE J1772 AC charger connection on top and a DC pin connector at the bottom.

TABLE 2.2

Charging Station Classification Based on Charging Power Level

Levels	Type	Power Supply	Power rating	Protection	Time for Charging
AC Level 1	On-board (Home)	120/230V(AC);12A to 16A (Single phase)	From 1.4 KW to 1.9 KW	Breaker in cable	11–36 hours for EV (16–50 KWh)
AC Level 2	On-board (Home or workplace)	208/240V(AC);15A to 80A (single/split phase)	From 3.1 KW to 19.2 KW	Pilot function and breaker in cable	2–7 hours for EVs (16–30 KWh)
DC Level 3	Off-board (Public places similar to gas stations)	300V-600V(DC) (Max. 400A) (Three phase)	From 50 KW to 350 KW	Communication and monitoring between the charging station and the electric vehicle	Less than half hour (20–50 KWh)
DC ultra-fast	Off-Board (Public places similar to gas stations)	800V (DC) and higher; 400A and higher (Polyphase)	400 KW and more	Liquid cooling, monitoring, and communication between the charging station and the EV	Few minutes for EVs (20–50 KWh)

2.2.2.3 Level 3 Charging

In the future, this might increase concerns about the range of extended-range ESS for EVs. This suggests slower commercial rapid charging for electric vehicles (in an hour). They are often built beside the highway in a line with the gas stations [10]. Off-board chargers typically require a three-phase input voltage of 480 V or more. There is the option of using a direct dc connection to the vehicle. Level-1 and Level-2 EVSE are associated with SAE J1772 standards. However, level-3 EVSE is not (situated outside the EV). Increased losses in distribution transformers, frequency and voltage variations, harmonic distortion, peak demand, and thermal loading of the distribution and transmission system, particularly transformers, are only some drawbacks of level 2 and level 3 chargers. About 80–100 miles of the range may be restored in about 20–30 minutes with this charging method. The total price tag for this pricing structure ranges from $30,000 to $160,000.

2.3 SAFETY AND DESIGN STANDARDS

2.3.1 ELECTRICAL VEHICLE CHARGING STANDARDS

There are few regulations available globally that manage EV charging foundations. The SAE (Society for Automobile Engineers) and IEEE (Institute of Electrical and Electronic Engineers) standards are applied by the USA and based on markers. In contrast, IEC (International Electromechanical Commission) is limitlessly applied in Europe. Japan and China have their EV charging norms named CHAdeMO and Guobiao (GB/T) standards, respectively [11].

For charging levels 2 and 3, the National Electrical Code recommends ensuring the cables and connections are de-energized before plugging them into the EV. This comes at an extra cost to EVSE. Standards and charging protocols for EVs are currently being developed by several international organizations, including:

A. Society for Automobile Engineers
   SAE J1772: This conveys design and safety rules for charging control and connectors, and this is applicable for ac as well as dc charging stations
   SAE J2293: Guidelines are set for electric vehicles and the off-board EVSE that connects to a power grid to provide electricity for the vehicle's onboard electronics.
   SAE J1773: Inductive Coupled Charging
   SAE 2836: Communication purposes
   SAE 2894: Concerns about Power Quality
B. Institute of Electrical and Electronic Engineers (IEEE)
   IEEE 2030.1.1: Quick DC charging for EVs
   IEEE P2690: Vehicle Authorization, Charging Network Management
   IEEE P1809: Electric Transportation Guide
   IEEE 1547: Power grid integration/distribution system tie-ins
   IEEE 1901: Overnight charging data rates
   IEEE P2030: Smart grid interoperability

C. International Electro-mechanical Commission (IEC)
   IEC-1000-3-6: Concerns about the quality of power
   IEC TC 69: About charging-station infrastructure and mandatory safety
      features
   IEC TC 64: Electrical systems installation, safeguards against electric
      shock.
   IEC TC 21: Regarding BMS
D. International Organization for Standardization
   ISO 6469-1: 2009: Part of a portable, rechargeable power supply.
   ISO/CD 6469-3: 3: Safety Operations
E. Japan Electric Vehicle Association
   JEVS G101-109: Fast Charging
   JEVS D701: Batteries
   JEVS C601: EV Plugs for charging
F. National Fire Protection Association
   NFPA 70: Safety Measurement
   NEC 625/626: EV charging systems
   NFPA 70E: About Safety
   NFPA 70B: Electrical Equipment Maintenance
G. Underwriters Laboratories (UL) INC
   UL 2231: Safety purpose
   UL 2594/2251,2201: EVSE

2.3.2 Isolation Requirements and Safety Precautions for EV Chargers

All the primary components of EVs, including dc-dc converters, electric motor driving inverters, high voltage batteries, and the charging module, which are connected to the electrical grid, require isolation. As a result, the power transformer is the most critical component of the infrastructure linking EVSE to the larger grid. For both on-board and off-board chargers to work appropriately, the EV's chassis must be earthed. Despite their lack of isolation, the many advantages of dc-dc converters cannot be overstated. However, the dc-dc converter stage of low-frequency techniques lacks galvanic separation. That's why a line frequency transformer, which provides galvanic isolation between the grid and the batteries, is so crucial. To reduce the amount of magnetic material needed and the overall size of the device, the high operating frequency is the greatest possible motivation. Because of factors such as the need for converter cooling systems, inductors, capacitors, and an isolation transformer, battery chargers are generally kept in off-board locations.

To accommodate more comprehensive ranges of frequencies, the dc-dc converter step uses high-frequency transformers to provide galvanic isolation. The size, losses, and cost of transformers may all be significantly reduced with careful design.

The main merits of giving isolation by high-frequency transformers are:

- Better regulation of voltage
- Reduced size

- Safeguarding load equipment
- Utility across a range of applications

The difficulties observed while giving isolation by high-frequency transformers are:

- High snubber losses.
- Impact on soft switching operation for partial load conditions.

The current electrical voltage level must be dropped down to a level acceptable for Level 2 charging equipment, i.e., 208 V to 240 V, to meet the applicable power requirements of such devices. Installation of isolation transformers can do this for Level 2 charging processes and step-up operation for Level 3 charging schemes if this is impossible. A suitable quality isolation transformer will set you back anywhere around $8000.

For safety reasons, galvanic isolation is preferable to the more expensive isolated onboard charges, which can be used in charging circuits. Main concerns, including increased weights and space chargers and greater price, can be overcome if existing traction hardware is used, i.e., inverter for the charging circuit and traction motor. The primary concern is reducing the risk of electric shock to the vehicle's owner during the charging process for electric vehicles, regardless of whether the circuit is grounded Table 2.3 Figure 2.9.

TABLE 2.3
Isolation and Safety Standards for EVs

Technical Code for Various Standards	Details
SAE J2929	This concerns propulsion battery safety.
SAE J2910	This tests hybrid automobiles and bus electric safety.
SAE J2344	EV Safety rules
SAE J 2464	Defines recharging energy storage system safety regulations.
ISO 6469-1:2009(IEC)	RESS, EVs, within and outside a person's protection
ISO 6469-2:2009(IEC)	EV safety, prevent internal breakdowns
ISO 6469-2:2001(IEC)	Electric hazard protection
IEC TC 69/64	Safety of EV infrastructure, electrical installation, shock prevention
NFPA 70/70E	Workplace, charging system, branch protection safety
UL 2202	Charger protection
UL 2231	To protect power supply circuits.
UL 225a	Protects couplers, plugs, and receivables.
DIN V VDE V0510-11	Safety guidelines for secondary battery installation

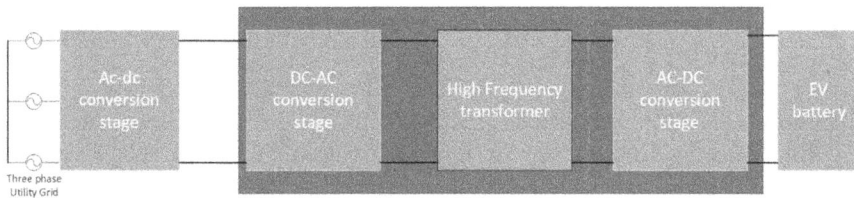

FIGURE 2.9 Off-board EV charger.jpg.

2.3.3 EV CHARGING PROTECTIONS

The home EVSE must also have these safety certifications and provide maximum charging protection for the device at all times, regardless of who is using it.

2.3.3.1 In-Cable Control Box (ICCB)

The ICCB for a home EVSE is between the power outlet and the vehicle connection. By transmitting data between the charging port and the EV, this gadget assures safe charging. The ICCB provides many layers of security for both the EV and the charger, including safeguards against over-voltage, over-current, under-current, and under-voltage. To guarantee the ICCB communicates the correct voltage and current range when charging the EV, protections play a critical role.

2.3.3.2 Controls on the Ground

For both the driver's and the EV's protection, the charging station must include a ground(earth) connection. To prevent the charger's circuitry from getting energized in the absence of grounding or to interrupt the circuit in the event that grounding is lost at any point in the charging process, the charger must be fitted with a device that monitors ground continuity inside the charging system.

2.3.3.3 Temperature Control

If the EVSE is to be used in a location with fire risk, it must have over-temperature protection for the ICCB and the AC power cord. This safeguard ensures that ICCB will shut down immediately to prevent the transmission of more charges in the event of overheating.

2.3.4 ELECTRICAL VEHICLE CHARGING SAFETY TIPS

- When charging an EV, be sure to stick to the guidelines provided by the manufacturer. If you need any further clarification, be sure to contact a nearby authorized retailer.
- Investing in a charger that has been validated as safe by an independent testing facility.
- Level 1 EV charging must be done at a wall outlet rated for at least as much current as the charger used. You must not use the usage of an extension cable or a multi-outlet adaptor.
- A residual current device should be connected to the charging system. If an issue is detected, power can be cut off to prevent a fire.

- When not in use, keep all parts of the charging device out of the reach of youngsters.
- Take care of your charging station's parts by following the guidelines provided by the manufacturer. Extreme wear and tear might be an indicator of a dangerous electric current. Using a damaged EV charger is dangerous and should be avoided.
- Water damage to your electric vehicle charging station may be avoided by simply covering the plug. Make sure it's okay to charge your EV in the rain by consulting the manual.

2.3.5 IMPACT OF EV CHARGERS ON UTILITY

Utility strain rises with the widespread installation of EV chargers. This is a bigger problem for utilities that haven't been adequately maintained.

2.3.5.1 Impact of RES

Due to the intermittent nature of RES, one of the challenges is incorporating them. Intermittency problems and difficulty in connecting electric vehicle chargers have arisen as a result of the development of power converters and high-density ESS. During the process of charging and discharging electric vehicles, the goal is to maintain tight control over the supply and demand dynamics of the utility.

2.3.5.2 Impact on Grid Stability

The load that electric vehicles put on the grid might potentially cause stability issues for the power system. Even without an electric vehicle load, many distribution systems operate dangerously close to the edge of stability; hence, conducting a stability study is essential before attaching electric vehicle chargers as load [12].

2.3.5.3 Impact on Grid Current Harmonics

The power electronics non-linearity present in electric vehicle chargers is the cause of the injection of current harmonics. The high magnitude of total harmonic distortion (THD) for the input current is dominated by odd harmonics. In most cases, EV chargers contain input filters that are inserted one step before the front-end rectifier. The purpose of these filters is to filter out the input line current and decrease its harmonic content. Electric vehicle chargers utilize high-frequency or moderated PWM approaches to reduce the current harmonics. In addition, matrix converters have a role in multi-phase EV chargers. These high-frequency converters make the circuit more complicated yet result in a reduction in THD. Even the power factor is impacted when there is a high harmonics content in the input line current. This, in turn, leads to an increase in the RMS value of the line current, which in turn degrades the different assets of the grid.

2.3.5.4 Impacts on Grid Loss

Losses in the grid caused by EV chargers make the I^2R losses high, where I is the RMS value of the current drawn and R is the equivalent resistance of the grid. The shortening of the life span of grid parts is also due to the high losses in the grid. To have low losses, the EV charger should take in-line current with fewer harmonics and a high-power factor.

2.4 COMMUNICATION AND REGULATORY FRAMEWORK

As the number of EVs that can communicate increases, the exchange of energy and its information, the robustness of EV aggregators, and other EVs will increase, being portable and secure. For user safety and the lifetime of the battery and charging connections, the EV, the EVSE, and the onboard charger must communicate with one another. Continual transmission of a sufficient battery voltage and current from EVs to EVSE is required. The electrical connection between the chassis and the battery is monitored by the EVSE, which performs routine ground fault checks. It communicates with the battery management system (BMS) to ensure proper charging and discharging, which is essential for maintaining the battery at the right SoC. EVSE communicates with the centralized management system (CMS) to adjust the rate of charging in response to fluctuations in system demand and tariff pricing (CMS).

When it detects that the plug is fully inserted, the vehicle will send a signal to the EVSE for protection. By delivering a signal to the on-board charger, which starts the power flow and begins charging the battery, arcing may be prevented. EV Smart Grid integration, data interchange, aggregator control, and energy flow can be improved by better management of the EV network. The EV network's heavy reliance brings up many issues with communication and technology.

Because of the wide variety of standards that are now in use, it is more challenging for electric vehicles and connected infrastructure to interact with one another. As a result, EV supply equipment has to implement standardized connections, charging stands, plugs, power outlets, and protective devices to adhere to the standards of various regulatory boards [13]. Electric vehicles are equipped with two connectors: one is used to plug-in into the EV outlet, and the other is used to plug into the charging station. The power rating of the charging socket and the kind of electric vehicle determine the type of connector used.

Type-1 and Type-2 connectors are used for slow and fast charging, whereas CHAdeMO and Combined Charging System (CCS) connections are used for DC rapid charging. The new CHADeMO charging standard is widely adopted since it delivers the maximum power and can fully charge electric vehicles in about half an hour. CCS includes both AC and DC connectors, thus the name "combination connection." High-gauge supply wires, a protective earthing system, a mechanism of bidirectional signaling, and safety connections are all features universal to all charging standards. Proximity Pilot and Charge Pilot wire pairs are used in IEC and SAE connector specifications, respectively. The CHAdeMO uses a complex set of signals and a Controller Area Network to enable communication between an electric vehicle and an EVSE (CAN bus).

Numerous international organizations, including IEEE, IEA, IEC, SAE, and ISO, enlisted various charging standards to standardize the charging procedure. Listed below are a few charging standards:

- IEC TC 69: Infrastructure for charging and standards for safety
- IEEE 2030.1.1: Rapid DC charging
- J1722: Types of Charging Connectors
- ISO/IEC 1511S: EVSE and controller communication for V-G.
- J1773: Wireless charging
- NEC625/626: System for EV charging

REFERENCES

[1] M. Y. Metwly, M. S. Abdel-Majeed, A. S. Abdel-Khalik, R. A. Hamdy, M. S. Hamad and S. Ahmed, "A Review of Integrated On-Board EV Battery Chargers: Advanced Topologies, Recent Developments and Optimal Selection of FSCW Slot/Pole Combination," in IEEE Access, vol. 8, pp. 85216–85242, 2020.

[2] M. Yilmaz and P. T. Krein, "Review of battery charger topologies, charging power levels, and infrastructure for plug-in electric and hybrid vehicles," *IEEE Trans. Power Electron.*, vol. 28, no. 5, pp. 2151–2169, May 2013.

[3] T. Chen et al., "A Review on Electric Vehicle Charging Infrastructure Development in the UK," in Journal of Modern Power Systems and Clean Energy, vol. 8, no. 2, pp. 193–205, March 2020.

[4] S. Pareek, A. Sujil, S. Ratra and R. Kumar, "Electric Vehicle Charging Station Challenges and Opportunities: A Future Perspective," 2020 International Conference on Emerging Trends in Communication, Control and Computing (ICONC3), Lakshmangarh, India, 2020, pp. 1–6.

[5] G. R. C Mouli, P. Venugopal and P. Bauer, " Future of Electric Vehicle Charging," in *19th International Symposium on Power Electronics*, 2017, pp. 1–7

[6] G. R. C. Mouli, J. Kaptein, P. Bauer, and M. Zeman, "Implementation of dynamic charging and V2G using Chademo and CCS/Combo DC charging standard," in *2016 IEEE Transportation Electrification Conference and Expo (ITEC)*, 2016, pp. 1–6.

[7] H. Lund and W. Kempton, "Integration of renewable energy into the transport and electricity sectors through V2G," *Energy Policy*, vol. 36, no. 9, pp. 3578–3587, 2008.

[8] C. E. Thomas, "Fuel cell and battery electric vehicles compared," *Int J. Hydrogen Energy*, vol. 34, no. 15, pp. 6005–6020, 2009.

[9] M. Yilmaz and P. T. Krein, "Review of battery charger topologies, charging power levels, and infrastructure for plug-in electric and hybrid vehicles," *IEEE Trans. Power Electron.*, vol. 28, no. 5, pp. 2151–2169, May 2013.

[10] N. Naik, M. G. M and C. Vyjayanthi, "Research on Electric Vehicle Charging System: Key Technologies, Communication Techniques, Control Strategies and Standards," 2021 IEEE 2nd International Conference on Electrical Power and Energy Systems (ICEPES), Bhopal, India, 2021, pp. 1–6.

[11] M. C. Falvo, D. Sbordone, M. Devetsikiotis, and I. S. Bayram, "EV charging stations and modes: International standards," in *Proc. Int. Symp. Power Electron., Elect. Drives, Autom. Motion (SPEEDAM)*, Jun. 2014, pp. 1134–1139.

[12] K. M. Tan, K. Vigna Ramachandaramurthy, and J. Y. Yong, "Integration of electric vehicles in smart grid: A review on vehicle to grid technologies and optimization techniques," *Renew. Sustain. Energy Rev.*, vol. 53, pp. 720–732, Jan. 2016.

[13] S. Habib, M. M. Khan, F. Abbas, L. Sang, M. U. Shahid and H. Tang, "A Comprehensive Study of Implemented International Standards, Technical Challenges, Impacts and Prospects for Electric Vehicles," in IEEE Access, vol. 6, pp., Mar, 2018.

3 Basic Power Factor Correction (PFC) Converters for Plug-in EV Charging

Abhinandan Dixit, Karan Pande, and Akshay Kumar Rathore

CONTENTS

3.1 INTRODUCTION

The automotive industry globally is witnessing a major transformation due to research and development on transportation electrification. Growing concerns for the environment and energy security clubbed with rapid advancements in technologies for powertrain electrification are transforming the automotive business. One of the key facets of such a change is the rapid development in the field of electric mobility which might transform the automotive industry like never before. Electric vehicles (EVs) for road transport enhance the energy efficiency, require no direct fuel combustion, and rely on electricity – the most diversified energy carrier, thereby contributing to a wide

DOI: 10.1201/9781003330134-3

41

ELECTRIC VEHICLES VS
INTERNAL COMBUSTION ENGINE

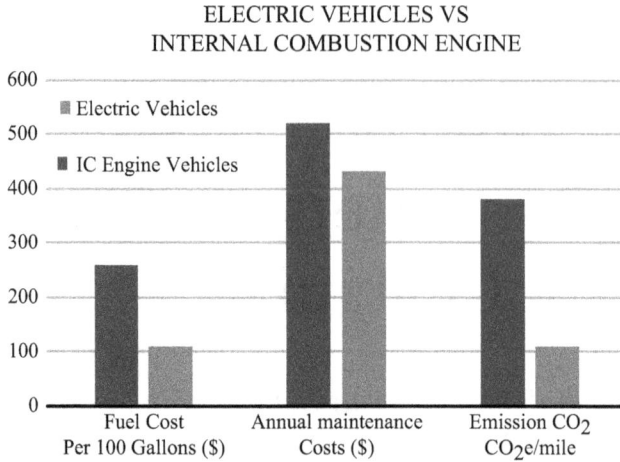

FIGURE 3.1   Comparison of EV and IC.

range of transport policy goals. Figure 3.1 shows a comparison between the classical IC engine vehicle and EVs [1]. It is observed that EVs are a realistic solution to offer reduced maintenance cost, better fuel economy, and limits CO_2 emissions as compared to the classical IC engines. Global EV sales figures have been growing rapidly and according to an analyst, the global fleet of EV sales could rise to 120 million in 2030 [2]. At the end of 2019, the global fleet of plug-ins was 7.5 million counting light vehicles [3]. Medium and heavy commercial vehicles add another 700,000 units to the global stock of plug-in vehicles [4]. EVs comprise a broad spectrum of vehicles right from two-wheelers, three-wheelers (rickshaws), golf carts, intra-logistics equipment, passenger cars, trucks, and electric buses.

Figure 3.2 illustrates the classification of EVs depending upon their battery capacity whereas Table 3.1 shows EV charging type. Three-wheelers are an intrinsic part of local transportation in Southeast Asian countries. E-rickshaw has gained popularity in the Asian market post-2010 because of their symbolic resemblance with traditional auto-rickshaw, they are also exhausting as they don't require all-day peddling unlike cycle rickshaws resulting in more rides in a day proving more profitability. E-rickshaw is hauled by an electric motor ranging from 850 W to 1400 W, which is supplied from a lead-acid battery pack of 100–120 Ah [4]. An article published by Bloomberg claims that the South Asian countries combined have 1.5 million electric three-wheeled rickshaws, which are more than the total number of electric passenger cars sold in China since 2011 [5]. India and China are the two biggest manufacturers of E-Rickshaw [5]. According to an analyst it is estimated that about 60 million Indians hop on an e-rickshaw every day [6].

3.1.1   Plug-In Chargers

As shown in Figure 3.3, plug-in chargers for the EVs are classified into two types namely on-board chargers and off-board chargers. An offboard charger is generally

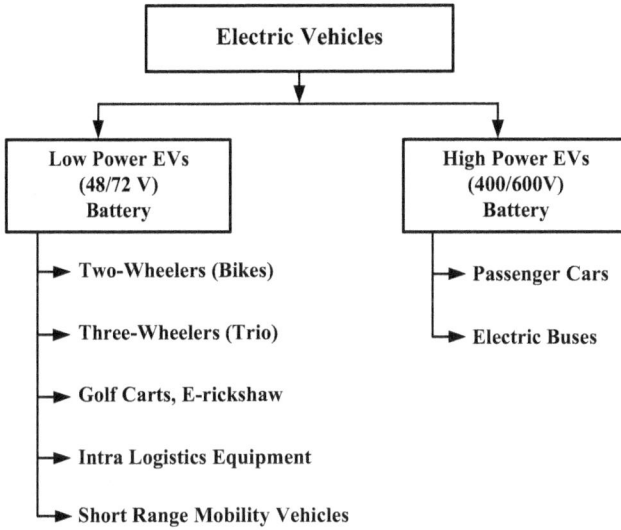

FIGURE 3.2 Classification of EVs.

TABLE 3.1
Types of EVs and Charging Time

EV	Battery Type	Charger Power	Charging Time
Two-Wheeler/ E-rikshaw/ Intra Logistics Equipment	SLI – 48 V	0.5–1 kW	6–8 Hrs
Trio/Golf Carts	Li-Ion / AGM 48 V	1–3 kW	3–4 Hrs
Short Range Mobility Vehicle	Li-Ion 120 V	3.3–7 kW	4 Hrs
Passenger cars/ Buses	Li-Ion 400–600 V	62–500 kW	30 mins

designed to transfer higher kilowatts of power and requires a more sophisticated battery management system (BMS) on plug-in hybrid electric vehicles (PHEV) and EVs. A high-power off-board charger (>100 kW) can charge the EV battery within 30 minutes. In addition, it removes significant weight from the PHEVs and EVs, which can increase the vehicle's overall efficiency. On the contrary, an onboard charger is generally designed for lower kilowatts of power transfer usually 3.3 kW, and is further classified into isolated and non-isolated chargers. The non-isolated topologies do not implement galvanic isolation as the output voltage is low. Therefore, the non-isolated topologies are more compact in weight and have high efficiency [7]. On the other hand, the isolated topologies end up bulky in overall weight since it uses galvanic isolation (usually achieved using an isolation trans-former) and possesses a complex control structure which reduces the overall reli-ability of On-Board Charger [8]. The isolated topologies are again further classified

```
┌─────────────────────────┐
│    Plug-in Chargers     │
└─────────────────────────┘
```

On-Board
(Slow Charging)

Off-Board
(Fast Charging)

Non-Isolated
Chargers

Isolated
Chargers

- 600 VDC/ 450 VAC, 400 Amp/ 200 Amp, 3-phase
- Generally higher KW transfer (62.5 -240 kW)
- Charge under 30 Mins
- Include more sophisticated BMS systems
- Manage battery heating
- Communications to home/grid energy management systems
- Removes weight from vehicle
- The higher the energy transfer rate

- 120 VAC/ 240 VAC, 20 Amp/ 80 Amp, 1-phase
- Generally lower KW transfer (1.0 -7.7 kW)
- Charge time – 4 to 12 hrs.
- BMS is managed by on board rectifier
- Less concern about battery heating
- Adds weight to vehicle

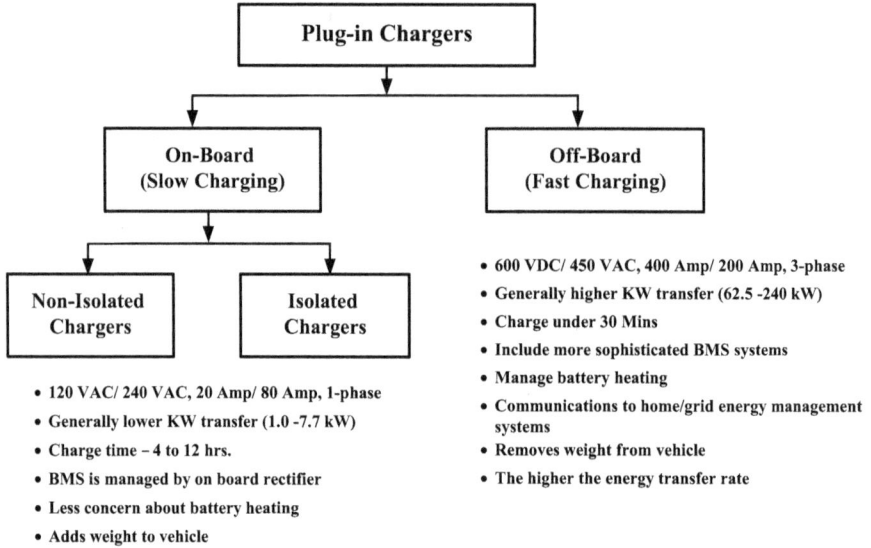

FIGURE 3.3 Classification of plug-in chargers.

into two types namely single-stage and two-stage topologies. Based on the on-board charger power level, they charge the battery in 4 to 7 hours [9]. Due to low current charging, it supports the specified lifetime of the battery and demands reduced maintenance. The most common E-Rickshaw battery is the lead-acid, SLI type as it is cheap to manufacture. Also, it can provide high currents (100–400 A) for turning the starter motor for short periods.

3.1.2 CONVENTIONAL OBC ARCHITECTURE

Figure 3.4(a) and (b) show the block diagram of the conventional isolated single-stage converter topologies. These isolated topologies consist of a diode bridge rectifier followed by an isolated flyback converter [10] or a half-bridge isolated DC/DC resonant converter [11]. The isolated flyback converter is a very simple and effective solution for battery charging applications. It has a low component count making it an inexpensive solution to provide a DC voltage from an AC source. This converter also does not require a complex control. The isolated flyback converter topologies suffer from the effect of leakage inductance resulting in high stress on switch voltage [11]. It is observed that when the switch is turn-off the leakage inductance of the transformer is discharged and a huge voltage spike across the switch appears. To clamp this voltage spike across the switch external RCD snubbers are connected [12]. In addition, the hard-switching operation of the switch also leads to high power losses in the converter. Moreover, connecting the diode at the secondary for the low-voltage battery charging application induces high conduction and turn-on losses for the diode leading to reduced efficiency. Though the topology is simple and easy to implement, it lacks the basic requirement of

(a)

(b)

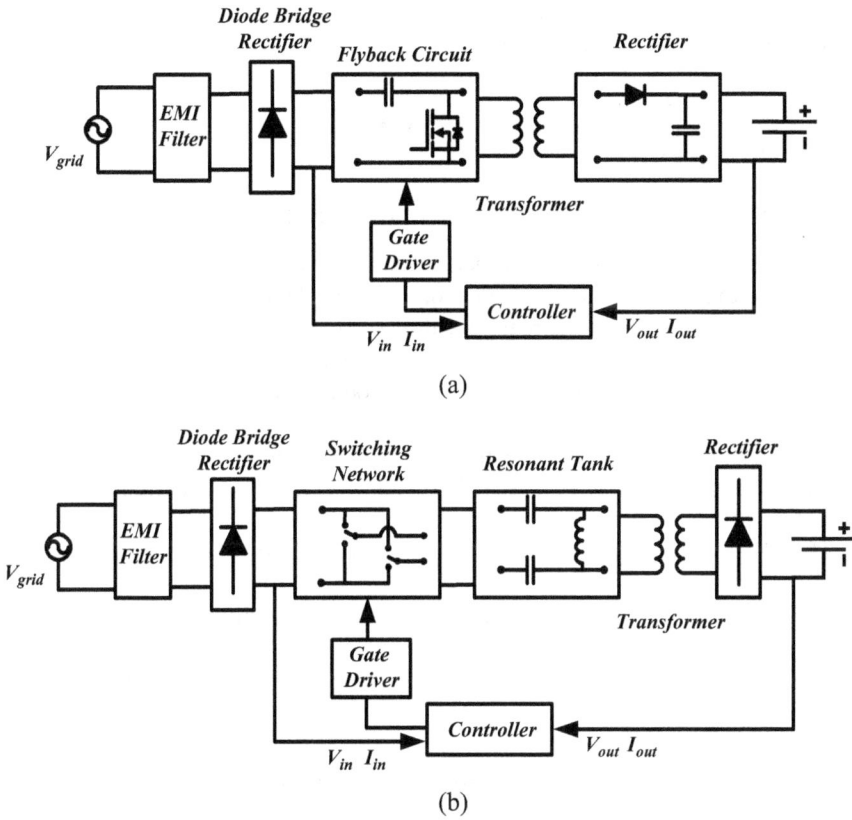

FIGURE 3.4 Single Stage (a) Isolated flyback converter; (b) half-bridge isolated DC/DC resonate converter.

achieving unity power factor (UPF) for various line voltages. Such topology presents poor power factor and THD as no active current wave shaping unit is present. An improved version of a low-voltage battery charger available in the market is shown in Figure 3.4(b). It is seen that though losses due to leakage inductance do not occur because of soft switching operation due to resonant tank, poor efficiency due to the output diode bridge rectifier losses is one of the main issues. Also, such battery charger configuration lacks PFC unit, which induces odd harmonics in the grid leading to poor power quality. The presence of an LC filter at the transformer secondary induces duty cycle loss on the transformer primary voltage. As the output of the inductor does not allow a sudden change in current, secondary diodes are shorted which leads to duty cycle loss and rectifier snubber losses [7]. It is observed that such a PFC unit becomes essential for chargers to improve power quality at the input and enhance converter performance.

The above-mentioned charger configurations come in the category of float chargers [11]. Such chargers are easy to implement in terms of control with one control loop to control the charging battery. Float chargers require current limiting resistors at the output and use a single sensor, making them a less complicated solution for battery charging.

3.2 CONTINUOUS CONDUCTION MODE (CCM) BASED PFC EV CHARGERS

3.2.1 NEED OF POWER FACTOR CORRECTION (PFC)

AC-DC converters are an essential part of the battery chargers in order to convert AC voltage to stiff DC. Diode rectification is one of the most common ways which is implemented in current chargers but draws discontinuous peaky current. Input current waveform and FFT analysis with diode rectification-based chargers are shown in Figure 3.5(a) and (b). It is seen that odd harmonics are injected into the grid leading to poor input current THD and reduced power quality. Such battery chargers require a bulky passive filter to filter out these harmonics leading to increased weight and size. As per IEC-6000-3-2 [13], the input current THD limit should be below 5% for automobile battery chargers.

The PFC control objectives can be defined as follows:

1. Sinusoidal input current is in-phase with the input voltage over a range of frequencies.
2. Low input current THD over the entire range of power.
3. Stiff Regulated DC output voltage.

Thus, an active PFC becomes an essential need for battery charging applications in order to meet the abovementioned control objectives. In order to reduce the cost and the weight single-stage, isolated, and non-isolated charger configurations have been developed [14]. Such chargers don't require a dedicated second stage for charging, which reduces cost and improves efficiency. Isolated single-stage topologies present poor efficiencies because of losses due to leakage inductance. Huge voltage spikes at turn-off reduce reliability and require an RCD snubber which reduces the efficiency drastically [15–27]. Secondary converters and control schemes are developed in order to improve the efficiencies of such topologies [28]. On the other hand, non-isolated topologies possess no such problems owing to the absence of high-frequency transformer, thus presenting higher efficiencies. But tackling the problem of an increased number of semi-conductor count and sensors increases the cost greatly. Semiconductor devices need to be over-rated to consider line voltage variations and guarantee safe operation. Moreover, non-isolated chargers have common ground with AC line. This leads to safety concerns in high-line voltage countries like India and China. Thus, isolation becomes necessary as per standards set by governments. Two-stage battery charging topologies have been popular in industries and have one stage dedicated to PFC regulation and the second stage for battery charging control [29]. As shown in Figure 3.6 such battery charger can have two configurations:

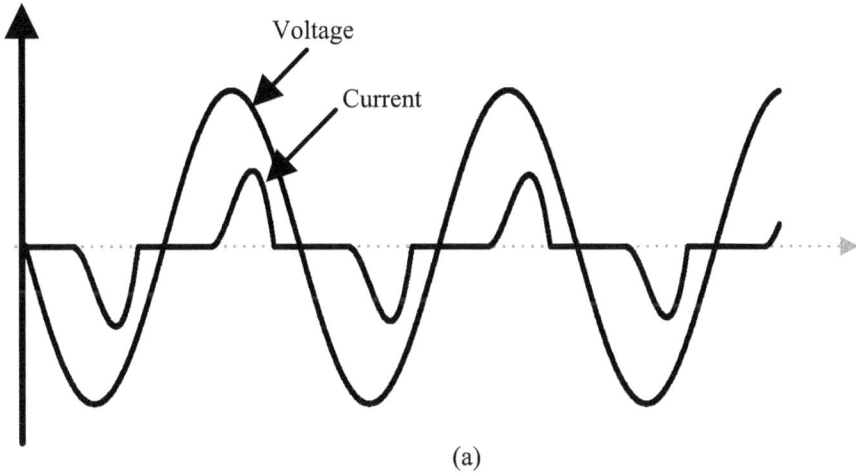

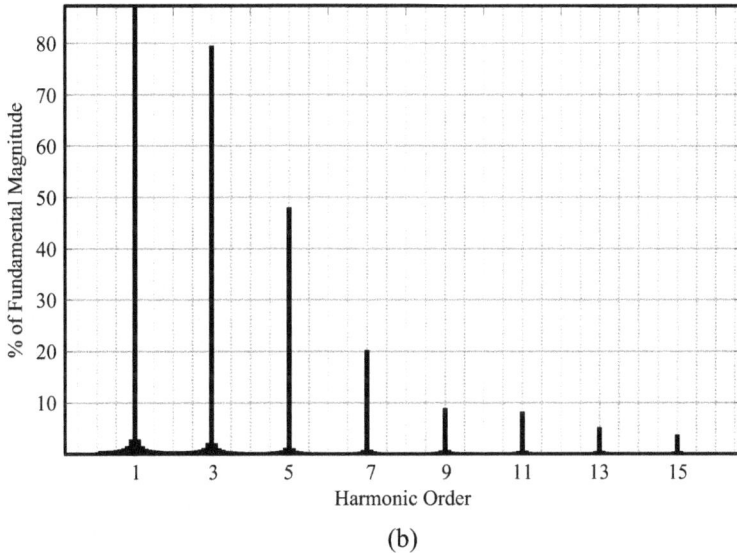

FIGURE 3.5   (a) Diode bridge voltage and current waveform. (b) Input current THD.

1. Post-regulator type
2. Pre-Regulator type

In a post-regulator configuration, an isolated converter is used to provide isolation from AC mains and a non-isolated converter to shape the input current and control the battery charging process. Such a battery charging configuration has been reported in [30] but presents comparatively less efficiency as the primary side switches have high conduction losses. Moreover, such a topology is not suited for charging low-voltage battery packs as the current wave-shaping unit has high

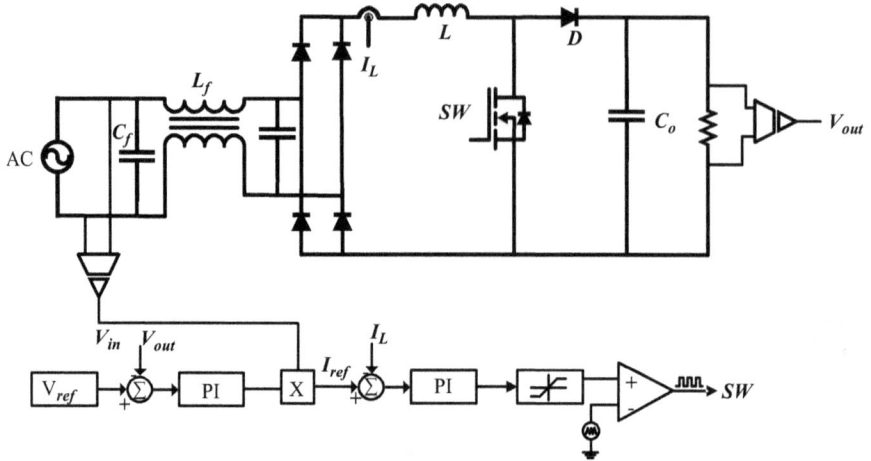

FIGURE 3.6 Boost converter as pre-regulator in CCM.

conduction and switching losses. On the other hand, a pre-regulator configuration [31–35] has a non-isolated converter used to shape the input current to achieve UPF and an isolated converter to control the battery charging process. This configuration is more reliable and is used in current battery chargers.

The most commonly used pre-regulator topology for such an implementation is the use of a boost converter because of its simple structure and grounded FET as shown in Figure 3.6 [35]. Boost inductors are operated in CCM and control is designed in order to shape this inductor current sinusoidal. Inductor current can be shaped in three ways mainly peak current control, average current control, and hysteresis control. These control techniques require sensing of input current for wave shaping and input voltage in order to synchronize with the grid. Figure 3.7 shows the circuit and control of a boost converter for active PFC. Implementing such control, it requires two control loops; the outer loop to regulate the output voltage and the inner loop to shape input current by maintaining UPF. Such control strategies have been extended to various topologies like SEPIC converter but deal with issues like increased component count leading to higher losses. This pre-regulator control scheme can also be applied to bridgeless topologies that are often used in the industry.

3.2.2 BRIDGELESS BOOST

Figure 3.7 shows the schematic of the bridgeless boost PFC converter. The bridgeless boost PFC topology eliminates the requirement of the diode rectifier at the input side, however, upholds the traditional boost topology features. Consequently, the loss associated with the diode rectifier bridge is reduced, making it suitable upto several kW where the need for high power density and efficiency is a major concern. The converter resolves the issues of heat management at the input side but raises the concern of high EMI [36]. The floating input line makes it

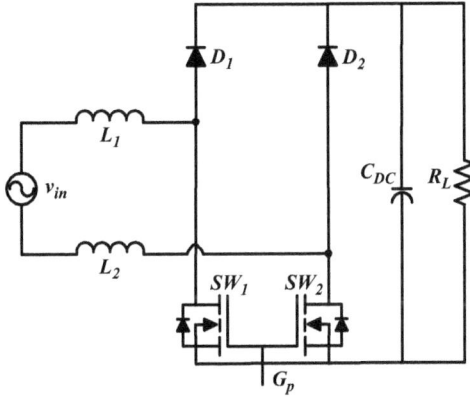

FIGURE 3.7 Bridgeless boost PFC converter.

impossible to sense the input voltage without a low-frequency transformer or an optical coupler. In order to sense the input current, a complex circuit is necessary to sense the current through the MOSFET and diode separately [37,38]. The topology also generates high common-mode noise than other bridgeless topologies.

3.2.3 SEMI-BRIDGELESS BOOST PFC

Figure 3.8 shows the schematic of the Semi-bridgeless boost PFC converter. The topology contains two slow diodes namely D_a and D_b. These diodes address the EMI issue at the input side as the current does not always return on this path and it also resolves the issue of floating ground. The conduction losses are very low in the converter. However, the converter control and current sensing are complex and expensive as it requires either three current transformers or the use of Hall Effect sensors and can also be measured by a differential amplifier. The efficiency is significantly improved at light load as compared to traditional bridgeless boost PFC

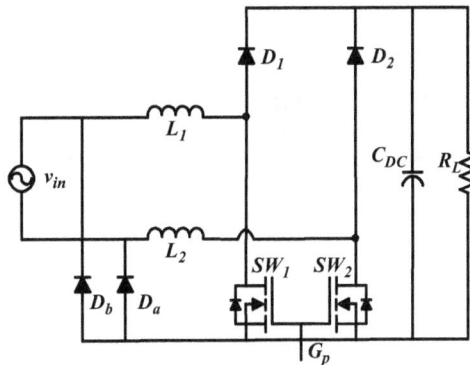

FIGURE 3.8 Semi-bridgeless boost PFC converter.

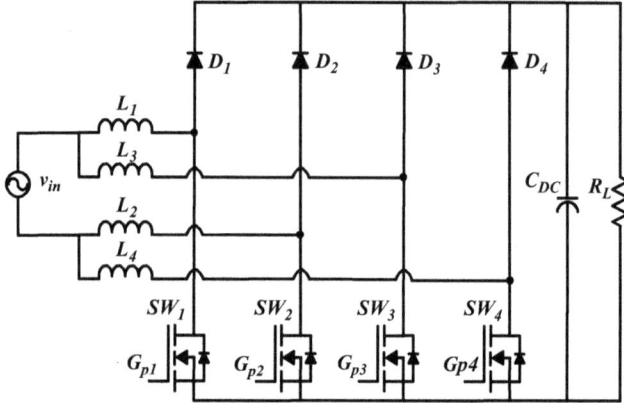

FIGURE 3.9 Bridgeless Interleaved Boost PFC converter.

topology. However, this topology does not achieve high full-load efficiency since there is high power loss in the MOSFETs due to high intrinsic body diode losses [39,40] and [41].

3.2.4 Bridgeless Interleaved Boost PFC Converter

Figure 3.9 illustrates the schematic of a bridgeless interleaved boost PFC topology. In comparison to the bridged interleaved boost PFC topology, it introduces two additional switches and trades bridge diodes with two fast diodes. The gating signals are 180° out of phase, like the interleaved boost PFC topology. The converter demonstrates a high input power factor, high efficiency, and low input current harmonics. The topology requires a small EMI filter at the input side and exhibits a low capacitor ripple. The converter consists of four diodes, four switches, and four inductors and is used for power levels above 3.3 kW. Hence, the topology has the highest number of components count than any other bridgeless PFC topology making it costly and bulky in size for practical usage with complex control strategies [42,43].

3.2.5 Bridgeless SEPIC and CUK PFC Converter

Figure 3.10(a) and (b) show the schematic of the bridgeless PFC circuits derived from the SEPIC and CUK topologies respectively. The topologies are formed by connecting two dc–dc SEPIC or CUK converters, one for each half-line cycle of the input voltage. The input ac line voltage is always connected to the output ground through the slow-recovery diodes D_p and D_n. Thus, the topologies do not suffer from the high common-mode EMI noise emission problem. Each topology utilizes two power switches (Q_1 and Q_2), two low-recovery diodes (D_p and D_n), and a fast diode (D_o). Passive components count increases due to the presence of the intermediate capacitor which leads to reduced power density. Also, there is one switch and two diodes in the current conduction path; hence, the conduction losses as well

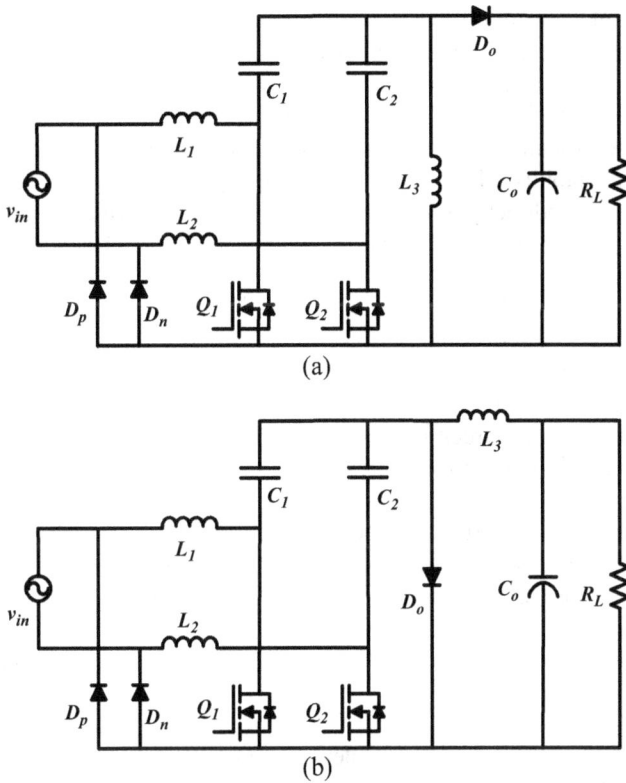

FIGURE 3.10 Converter configuration (a) Bridgeless SEPIC and (b) Bridgeless CUK.

as the thermal stress on the semiconductor devices are further increased. Moreover, the structure of the proposed topologies utilizes one additional inductor compared to the conventional topologies which consequently increases the size and cost of the converter. The converter operation is limited to low-power applications (<300 W). The bridgeless SEPIC converter demonstrates high output and input ripple current. The voltage stress in the switches of both converters is $V_{in} + V_o$ [44,45].

3.3 DISCONTINUOUS CONDUCTION MODE (DCM) BASED PFC EV CHARGERS

In the previous section, the front-end PFC converters discussed are CCM-based converters. Such converters require a bulky isolated transformer for input voltage and current sensing to implement phase-locked loop (PLL) for grid synchronization, which increases the cost and weight of the system significantly. It is observed that a total of three sensors or in some cases four are used to achieve control goals. Moreover, sophisticated microcontrollers are required to sample high-frequency data, leading to reduced reliability and converter robustness. In the prospect of control and cost, such converters present high complexity as it requires to determine the phase angle for implementation of PLL. Failure of PLL can lead to non-sinusoidal input currents, leading to poor THD and efficiency. As the front-end converter is operated in CCM, it presents right half plane zero (RHZ) during transient operations leading to instability. Thus, the controller designed for such implementation should be robust and reliable in order to meet the control objectives and maintain stability. This increases the computation burden on the microprocessor, which indirectly affects the converter performance and start-up.

Reduction in sensors and control complexity has several benefits and can be listed as follows:

1. System weight reduction.
2. Reduction in cost.
3. Improved system reliability.
4. Reduced control burden on the microcontroller.
5. High-frequency noise robustness.

DCM operation of converters can be used to simplify the control circuit and to reduce the number of sensors which consequently increases the converter reliability and robustness. Further, the design and performance of the converter are tested for input voltage change and load perturbations.

The following objectives can be achieved by operating converters in DCM mode:

1. Simplified control and easy implementation
2. Reduced number of sensors
3. Higher reliability
4. Improved power quality (THD < 5%)
5. Improved efficiency
6. UPF operation for input voltage variation

3.3.1 Interleaved DCM Buck-Boost Charger-Based PFC EV Chargers

CCM-based PFC converters have been in implementation, but the concept of DCM cannot be implemented in all the basic topologies. Boost converter and buck converter cannot be implemented in DCM for power factor correction for the following reason. The input current for a boost converter can be given as

$$i_{in}(t) = \begin{cases} \frac{v_{in}(t)}{L}t, & 0 < t \le t_{on} \\ \frac{(v_{in}(t) - V_o)}{L}t, & t_{on} < t < T_s \end{cases} \tag{3.1}$$

By performing FFT of the (3.1) using (3.2), one can get,

$$i_{in}(t) = \frac{a_0}{2} + \sum_{h=1}^{\infty} (a_h \cos(h\omega_{sw}t) + b_h \sin(h\omega_{sw}t)) \tag{3.2}$$

$$a_h = \frac{V_0}{h\omega_{sw}L}\left(D\sin(2\pi hD) + \frac{1}{2h\pi}\cos(2\pi hD) - \frac{1}{2h\pi}\right) \qquad (3.3)$$

$$b_h = \frac{V_0}{h\omega_{sw}L}\left(1 - D\cos(2\pi hD) + \frac{\sin(2\pi hD)}{2\pi h}\right) - \frac{V_{in}}{h\omega_{sw}L} \qquad (3.4)$$

From (3.3) and (3.4), it is observed that lower-order odd harmonics are present, which leads to poor input current THD. Figure 3.11 shows the input current waveform and FFT analysis for a boost PFC converter operating in DCM. It is seen that higher amplitude low-order harmonics are injected [(3.3), (3.4)] in input current leading to poor THD. Such a waveform is obtained due to its boost structure as the boost inductor is connected to the input throughout the switching cycle. Thus any non-linearity from the output is transferred to the input leading to a peaky input current waveform. As a consequence, DCM boost and boost-derived configurations require a large filter at the input side to filter out lower-order harmonics, which in

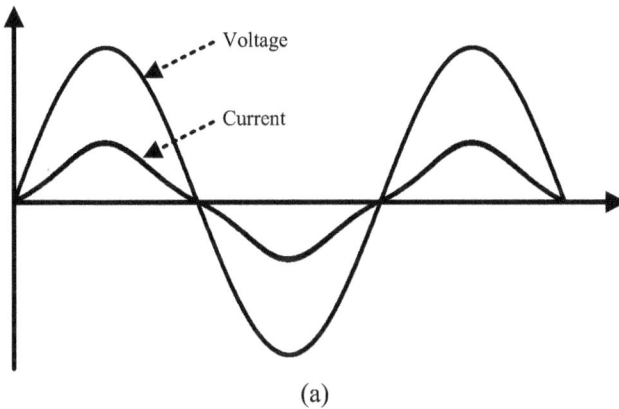

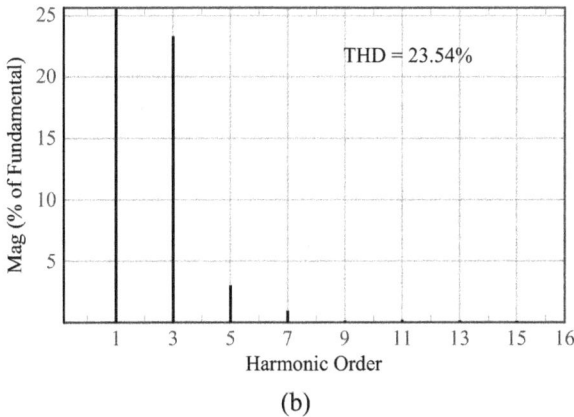

FIGURE 3.11 Boost PFC input current in DCM (a) waveform and (b) THD.

turn reduces the power density of the system. Several methodologies have been proposed for improving the current quality of DCM boost converter by reducing the lower-order harmonic content. However, such approaches are complex in nature and require extra sensing without having a significant impact on the harmonic content [46–50]. Therefore, DCM-based boost topologies are not feasible for such PFC applications. On the other hand, buck or buck-derived topologies cannot be implemented PFC, which can be understood by performing FFT analysis of input current expression expressed as

$$
i_{in}(t) = \begin{cases} \frac{(v_{in}(t) - V_o)}{L}t, & 0 < t \le t_{on} \\ 0, & t_{on} < t < T_s \end{cases} \tag{3.5}
$$

Using (3.1),

$$
a_h = \frac{(v_{in}(t) - V_o)}{h\omega_{sw}L}\left(D\sin(2\pi hD) + \frac{1}{2h\pi}\cos(2\pi hD) - \frac{1}{2h\pi}\right) \tag{3.6}
$$

$$
b_h = \frac{(v_{in}(t) - V_o)}{h\omega_{sw}L}\left(\frac{\sin(2\pi hD)}{2h\pi} - D\cos(2\pi hD)\right) \tag{3.7}
$$

Even though this topology does not present odd harmonics at the input as seen from (3.6) and (3.7), unlike the boost converter, power transfer doesn't take place for $v_{in}(t) < V_o$, as duty cycle saturates at its maximum limit, leading to poor power factor as seen from Figure 3.12.

A DCM buck-boost becomes advantageous as the input inductor is either connected to the input or the output, thus harmonics from the output are not reflected across the input side thus achieving a good THD and UPF operation. Moreover, it has fewer passive components count as compared to the conventional SEPIC or CUK converter making it a cost-effective option for charging.

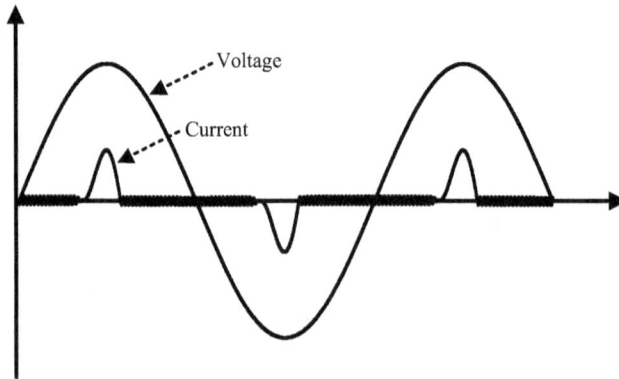

FIGURE 3.12 Input current and voltage waveform for buck converter in DCM.

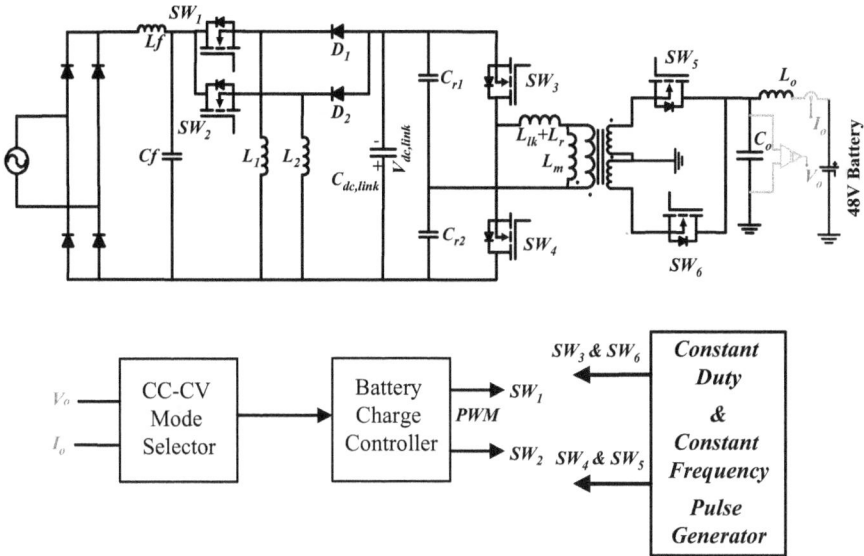

FIGURE 3.13 The proposed single-phase interleaved buck-boost derived PFC converter.

Figure 3.13 shows the proposed two-stage isolated charger configuration. First stage is an interleaved buck-boost DCM PFC converter which ensures UPF operation at all operating conditions. LC filter is connected to filter out high-frequency switching harmonics and allow pure sine wave current at the input. The interleaved PFC consists of two DCM buck-boost converters in parallel which operate at 180° out of phase. This reduces the input current ripple, resulting in the reduction of size and input filter. Since the input current is the sum of DCM inductor currents L_1 and L_2, the inductor peak current is lowered thus reducing the turn-off losses. Interleaving inherently reduces the conduction losses and by operating the buck-boot converter in boost mode it cuts down the current stresses on the switch and diodes. Even though voltage stress is high across semiconductor devices, low current stress on devices leads to smaller heatsink requirement and lower conduction losses, thus leading to an overall reduction of weight. Second stage is an isolated half-bridge LLC resonant converter with centre-tap transformer along with synchronous rectification on secondary.

This particular topology is selected because of the following reasons:

1. Less number of semiconductor device count.
2. By splitting the resonant capacitor into two capacitors half-bridge is formed in order to reduce resonant capacitor size thus reducing the number of primary turns.
3. LLC resonant tank when reflected to secondary appear as a current source [51], thus making it suitable for charging applications, along with the capability to achieve ZVS at light load.

Center-tap transformer and synchronous rectification on the secondary improves the overall efficiency and reduces the cost of the converter. Generally regulated LLC resonant converters with a wide operation range possess the problem of increased output voltage during hold-up time [8]. At the output rather than connecting an LC filter which induces duty cycle loss [52], a capacitive filter is used along with an output inductor to reduce the effect of a sudden change in output current during transient operation. The proposed configuration of the battery charger has easier control than conventional chargers that require the first stage for active PFC with three sensors and the second stage purely to control the charging voltage and current [14–28].

The second stage is an unregulated LLC resonant DC-DC converter to provide electrical isolation and voltage step-down. By operating this stage at constant duty and frequency, it not only reduces the control burden on the microcontroller in terms of voltage and current control but also the presence of magnetizing inductance large enough allows soft switching of half-bridge FETs [8]. Thus, conduction losses due to circulating currents are minimized. As the charger configuration is based on the pre-regulator concept and the second stage is a constant frequency, constant duty operated, the output of the buck-boost converter controls the total power transfer. In the absence of feedback control for DC-DC converter, it acts just as a voltage amplifier with a fixed gain, thus operating with minimum switching losses. As the stress on all semiconductor devices tends to increase or decrease based on dc-link voltage $V_{dc,link}$, it needs to be optimized to achieve a high overall efficiency of the converter. The method for dc-link voltage selection is explained in the next section. The output-to-input relationship of the proposed configuration is a product of individual gains of two stages and can be defined as

$$\frac{V_{dc,link}}{V_{in,pk}} \times \frac{V_{out}}{V_{dc,link}} = \frac{D}{1-D} \times n$$

$$\times \left(\sqrt{\left(1 + \lambda\left(1 - \left(\frac{\omega_o}{\omega}\right)^2\right)\right)^2 + \left(\frac{\pi^2}{8}Q\left(\frac{\omega}{\omega_o} - \frac{\omega_o}{\omega}\right)\right)^2} \right)^{-1}$$

$$(3.8)$$

where $Q = \frac{\omega_o L}{R_L}$, $\lambda = \frac{L_r}{L_m}$, $\omega_o = \frac{1}{\sqrt{L_r C_r}}$, $\omega = 2\pi f_{sw,HB}$. As the LLC converter switching frequency $f_{sw,HB}$ is selected the same as the resonant frequency, i.e., $\omega = \omega_o$ to obtain resonant tank gain as unity and to minimize circulating current losses, (3.1) can be simplified by

$$\frac{V_{out}}{V_{in,pk}} = \frac{V_{dc,link}}{V_{in,pk}} \times \frac{V_{out}}{V_{dc,link}} = \frac{D}{1-D} \times n \qquad (3.9)$$

3.3.1.1   Design of Interleaved DCM Buck-Boost Converter

The steady-state operation of the two-cell buck boost for one high-frequency switching cycle is shown in Figure 3. With the following assumptions [53,54] and the same analysis can be done for the single-cell converter.

a. All the components are ideal and lossless.
b. Within one switching cycle, input and output voltages are constant.
c. The output capacitor is bulky enough to maintain the output voltage constant.
d. Duty cycle is fixed for one power level.

Mode I. $0 < t < t_1'$

In this mode switch SW_1 is turned on with gating pulse V_{g_1}, L_1 charges and stores energy whereas capacitor C supplies the load. No energy is stored in L_2. Inductor current i_{L1} is given by

$$i_{L1}(t) = \frac{v_{in}(t)}{L_1} \tag{3.10}$$

Mode II. $t_1' < t < t_1$

Switch SW_1 is turned off and inductor current i_{L1} freewheels through diode D_1. Capacitor C filters out the ripple current and supplies pure DC current to the battery. Inductor current i_{L1} for this stage is given by

$$i_{L1}(t) = i_{L1,pk} - \frac{V_o}{L_1}t_1 \tag{3.11}$$

Mode III. $t_1 < t < T_s/2$

This is the DCM condition where energy stored in L_1 becomes zero and the capacitor supplies the load. At the end of this interval, half of the high-frequency cycle is over. This time duration is given by

$$t_{z1} = \frac{T_s}{2} - t_{on1} - t_{f1} \tag{3.12}$$

Mode IV. $T_s/2 < t < t_2'$

Switch SW_2 is turned ON in this mode, inductor L_2 charges, and stores energy. Capacitor C supplies the load and whereas energy stored in L_1 is zero. Inductor current $i_{L2}(t)$ is given by

$$i_{L2}(t) = \frac{v_{in}(t)}{L_2} \tag{3.13}$$

Mode V. $t_2' < t < t_2$

Gating pulse from switch SW_2 is removed. Inductor L_2 discharges through the load and diode D_2, thus supplies power to the load $i_{L2}(t)$ is given by

$$i_{L2}(t) = i_{L2,pk} - \frac{V_o}{L_2} \qquad (3.14)$$

Mode VI. $t_2 < t < T_s$

This is the same as Mode III, and the DCM period is given

$$t_{z2} = T_s - t_{on2} - t_{f2} \qquad (3.15)$$

The average load current in one switching cycle $i_{o,avg}$ is nothing but the average current of both diodes D_1 and D_2 i.e., the sum of area under the i_{L1} curve and i_{L2} curve during interval t_{f1} and t_{f2} respectively. Substituting $t = t_{f1}$ and $t = t_{f2}$ we get,

$$i_{o,avg} = \frac{1}{2} t_{f1} i_{L1,pk} + \frac{1}{2} t_{f2} i_{L2,pk} \qquad (3.16)$$

where,

$$t_{f1} = \frac{i_{L1,pk}}{V_o} L_1 \qquad (3.17)$$

$$t_{f2} = \frac{i_{L2,pk}}{V_o} L_2 \qquad (3.18)$$

$$i_{L1,pk} = \frac{v_{in}(t)}{L_1} D_{1,on} T_s \qquad (3.19)$$

$$i_{L2,pk} = \frac{v_{in}(t)}{L_2} D_{2,on} T_s \qquad (3.20)$$

Thus the average output over one switching cycle can be given as

$$i_{o,avg} = \frac{1}{2} \left(\frac{i_{L1,pk}^2}{V_o} L_1 + \frac{i_{L2,pk}^2}{V_o} L_2 \right) \qquad (3.21)$$

$$= \frac{v_{in}^2(t)T_s}{V_o}\left(\frac{D_{1,on}^2 L_2 + D_{2,on}^2 L_1}{L_1 L_2}\right) \tag{3.22}$$

The average output current $I_{o,avg}$ over a line period is calculated by the integration of the switching cycle average output current and can be given as

$$I_{o,avg} = \frac{1}{2\pi}\int_0^{2\pi} i_{o,avg}\,d\omega t \tag{3.23}$$

$$I_{o,avg} = \frac{V_{in,pk}^2 T_s}{4V_o}\left(\frac{D_{1,on}^2 L_2 + D_{2,on}^2 L_1}{L_1 L_2}\right) \tag{3.24}$$

where $D_{1,on} = D_{2,on} = D$, D = duty cycle.

Assuming the lossless circuit, i.e., 100% efficiency, the input power is equal to the output power

$$v_{in}i_{in} = v_o i_{o,avg} \tag{3.25}$$

Substituting (3.20) in (3.23)

$$v_{in}i_{in} = \frac{v_{in}^2(t)T_s}{2}\left(\frac{D_{1,on}^2 L_2 + D_{2,on}^2 L_1}{L_1 L_2}\right) \tag{3.26}$$

$$i_{in} = \frac{v_{in}(t)T_s}{2}\left(\frac{D_{1,on}^2 L_2 + D_{2,on}^2 L_1}{L_1 L_2}\right) = \frac{V_{in,pk}T_s}{2}\left(\frac{D_{1,on}^2 L_2 + D_{2,on}^2 L_1}{L_1 L_2}\right)\sin(\omega t) \tag{3.27}$$

$$i_{in} = I_{in,pk}\sin(\omega t) \tag{3.28}$$

where,

$$I_{in,pk} = \frac{V_{in,pk}T_s}{2}\left(\frac{D_{1,on}^2 L_2 + D_{2,on}^2 L_1}{L_1 L_2}\right) \tag{3.29}$$

Since the duty cycle of the converter is constant in DCM, the relation given by (3.26) indicates the UPF operation of the converter and the input current is sinusoidal with peak current $I_{in,pk}$ given by (3.29). The higher switching order harmonics present in the input currents will be filtered out by the input inductors and result in low input current THD.

Following inequalities must hold for DCM operation Figure 2.4

$$I_o < \Delta i_{L1} + \Delta i_{L2} \tag{3.30}$$

$$I_o < \frac{V_{in.pk} T_s}{2V_o} \left(\frac{D_{1,on} L_2 + D_{2,on} L_1}{L_1 L_2} \right) \tag{3.31}$$

where,

$$I_o = \frac{D V_{in,pk}}{(1 - D) * R} \tag{3.32}$$

$$K_{crit} > \frac{4 L_1 L_2}{R T_s (L_1 + L_2)} \tag{3.33}$$

To maintain DCM for all cases, the critical conduction parameter is given as K_{crit}

$$K_{crit}(D) = 1 - D \tag{3.34}$$

This can also be expressed in terms of critical load resistance R_{crit}. Thus, equation can be expressed as

$$R_{crit} > \frac{4 L_1 L_2}{D_{crit} T_s (L_1 + L_2)} \tag{3.35}$$

Critical duty cycle D_{crit} is calculated to maintain DCM for all cases. To maintain DCM at all times

$$t_{on1} + t_{f1} + t_{on2} + t_{f2} < T_s \tag{3.36}$$

$$D \left(1 + \frac{1}{M} \sin(\omega t) \right) < 1 \tag{3.37}$$

where $M = V_o / V_{in,pk}$
   Average output current is given by

$$I_{o,avg} = \frac{V_o}{R} \tag{3.36}$$

$$D = M \sqrt{K_{cond}} \tag{3.37}$$

where K_{cond} is the conduction parameter for interleaved buck-boost and is defined in equation (3.37). The critical value of K_{cond} to keep the converter in DCM is

$$K_{crit} = \frac{1}{2(M+1)^2} \tag{3.38}$$

To ensure PFC, the inductor needs to be in DCM for worst-case input voltage as input current will be maximum for that case. The inductor can be computed as follows.

$$L < \frac{V_{in,pk}^2 \, D^2 T_s}{2P_o} \tag{3.39}$$

A value below this critical inductor value should be selected in order to ensure DCM operation.

In PFC converters, the output capacitor is designed to filter the harmonic components occurring at twice the line frequency. Thus, the variation in the power (input and output) discharges through the output filter capacitor and is expressed as

$$P_c(t) = P_{ac}(t) - P_o(t) \tag{3.40}$$

$$V_o i_c = V_{in,pk} I_{in,pk} \cos 2\omega t - V_o I_o \tag{3.41}$$

Considering efficiency equal to 100%,

$$i_c(t) = \frac{V_{in,pk} I_{in,pk}}{V_o} \cos 2\omega t = I_o \cos 2\omega t \tag{3.42}$$

The output voltage ripple equation is given by,

$$V_{o,ripple}(t) \approx \frac{1}{C} \int i_c(t) dt \tag{3.43}$$

By putting (3.21) into equation (3.22),

$$V_{o,ripple}(t) \approx \frac{1}{C} \int i_c(t) dt \tag{3.44}$$

$$C = \frac{I_o}{2\omega V_{o,ripple}} \tag{3.45}$$

Input current in an interleaved DCM buck-boost converter is defined as

$$i_{in}(t) = \begin{cases} \frac{v_{in}(t)}{L}, & 0 < t \le t_{on} \\ 0, & t_{on} < t \le T_s \end{cases} \tag{3.46}$$

On performing FFT of input current using (3.2), we get

$$a_0 = \frac{v_{in}(t)}{L} D^2 T_s \tag{3.47}$$

$$a_h = \frac{v_{in}(t)}{2\pi h f_{sw} L} \left(D \sin(2\pi h D) + \frac{1}{2h\pi} \cos(2\pi h D) - \frac{1}{2h\pi} \right) \tag{3.48}$$

$$b_h = \frac{v_{in}(t)}{2\pi h f_{sw} L} \left(\frac{\sin(2\pi h D)}{2h\pi} - D \cos(2\pi h D) \right) \tag{3.49}$$

Equation (3.47) indicates the fundamental component of the input current and (3.48) and (3.49) indicate the switching order harmonics present, which need to be filtered out. Upon comparing (3.3) and (3.4) with (3.48) and (3.49), it is noted that, unlike conventional boost converter, the proposed converter does not inject higher lower order amplitude harmonics in the input and thus requires a small filter. By designing a low-pass LC filter with a cutoff frequency much lower than the switching frequency, the harmonic currents can be filtered out.

The criteria to design a low-pass LC Filter are as follows:

1. Selection of cut-off frequency f_c given by

$$f_c = \frac{1}{2\pi} \sqrt{\frac{1}{L_f C_f}} \tag{3.50}$$

2. Minimization of filter reactive power consumption for 60 Hz at 1.0 kW. The reactive power is minimum when the filter characteristic impedance is equal to the converter impedance i.e.,

$$Z_{ch} = \sqrt{\frac{L_f}{C_f}} = Z_{in} \tag{3.51}$$

where Z_{ch} is the characteristic impedance and Z_{in} is the input impedance at rated load and is given by

$$Z_{in} = \frac{2L}{D^2 T_s} \tag{3.52}$$

Using (3.51) and (3.52), the low-pass filter parameters L_f and C_f can be obtained as

$$L_f = \frac{Z_{ch}}{2\pi f_c} \qquad (3.53)$$

$$C_f = \frac{1}{2\pi Z_{ch} f_c} \qquad (3.54)$$

In order to select an optimal dc-link voltage, it is important to derive a relation between all semiconductor devices and dc-link voltage $V_{dc,link}$ to minimize losses. Detailed loss analysis needs to be performed to run the battery charger at maximum efficiency. Table 3.2 summarizes the loss equations for the proposed charger. The proposed converter is designed as per the specifications listed in Table 3.3. Based on the loss analysis equations derived from the above sections, and using the components and parameters listed in Table 3.4, optimal DC-Link is selected as shown in Figure 3.10. At a dc-link voltage of 402.7 V the total losses are 43.7 W. It is observed that the interleaved buck-boost losses tend to increase significantly from the obtained optimal point due to high voltage stresses on semiconductor devices even though other losses reduce drastically. A dc-link voltage of 400 V is selected to design the proposed charger and the passive components are designed accordingly.

To design the converter with the proper voltage and current limits, it is important to understand the battery dynamics. As a practical battery was not available for experimentation, it is important to understand the feasibility of the proposed charger for battery charging applications. The proposed battery charger is for a low-voltage battery charging application that uses 850 W motor powered by four 12 V series-connected, 120 Ah VRLA lead-acid battery packs [55]. By observing the C/7 discharge rate for 120 Ah battery [55], the allowable depth-of-discharge (DOD) is 20% for VRLA battery, as below this point the battery voltage drops sharply i.e., below 11 V point. Moreover, lead-acid battery packs when discharged, have higher internal resistance as lead sulfate is accumulated over the electrodes. With a rise in battery voltage, the concentration of sulphuric acid increases thus decreasing the effective internal resistance. So in general it can be said that for any battery, internal resistance is a function of the state-of-charge (SOC) of the battery [60]. Battery resistance of a lead-acid battery pack decreases exponentially to SOC [56]. With only 20% DOD allowed, internal resistance remains relatively constant for higher SOC and temperatures [57]. The battery equivalent model when connected to a battery charger is shown in Figure 3.11.

The battery equivalent circuit is represented by a parallel combination of self-discharge resistance R_p and storage capacitor C_p which describes the total energy stored in the battery. Equivalent internal resistance R_{int} is connected in series with a dc voltage source representing the battery voltage. Value of C_p can be calculated by

$$C_p = \frac{kWh * 3600 * 1000}{0.5(V_{batt,max}^2 - V_{batt,min}^2)} \qquad (3.55)$$

TABLE 3.2

Equations for Loss Analysis

Buck-Boost FET Conduction Losses	$I_{SW_{BB},rms} = \frac{V_{in,pk}D}{Lf_{SW,BB}}\sqrt{\frac{D}{6}}$	$P_{SW_1} = I_{SW_1,rms}^2 R_{DS,on}$ $P_{SW_2} = I_{SW_2,rms}^2 R_{DS,on}$

Buck-Boost
FET Switching Losses

$$P_{SW_{BB},off} = V_{sw,avg}I_{SW_{BB},rms}f_{SW,BB}\left(\frac{Q_{GD,BB}}{I_{g1,off}} + \frac{Q_{GS,BB}}{I_{g2,off}}\right)$$

$$P_{coss,BB} = \frac{1}{2} \times C_{oss,BB}f_{SW,BB}\int_0^{2\pi} C_{oss,BB}v_{sw}^2(\omega t)$$

DCM inductor Losses

$$I_{L,rms} = \frac{V_{in,pk}D}{Lf_{SW,BB}}\sqrt{\frac{D}{18\pi V_{dc,link}}(3\pi(V_{dc,link}-1)+8V_{in,pk})}$$

$$P_{L,losses} = 2[P_{c,limit} \times V_e + I_{L,rms}^2 \times R_{DCR}]$$

Buck-Boost Diode losses

$$P_{diode_{loss},BB} = 2[(I_{d,rms})^2 \times R_d + (I_{d,avg} \times V_f)]$$

$$I_{d,rms} = \frac{V_{in,pk}DT_{sw,BB}}{L}\sqrt{\frac{D}{3\pi V_{dc,link}}\left(\frac{4V_{in,pk}}{3} - \frac{\pi}{2}\right)}$$

$$I_{d,avg} = \frac{V_{in,pk}^2 D^2 T_{sw,BB}}{4LV_o}$$

DC-Link Capacitor
Losses

$$I_{C_{dc,link},rms} = \frac{V_{in,pk}DT_{sw,BB}}{L}\sqrt{\frac{V_{in,pk}D}{V_{dc,link}}\left(\frac{8}{9\pi} - \frac{V_{in,pk}D}{4V_{dc,link}}\right) - \frac{D}{3V_{dc,link}}}$$

$$P_{C_{dc,link}} = I_{c,rms}^2 \times R_{C_{dc,link}ESR}$$

Unregulated LLC FET
Conduction Losses

$$P_{HB,cond} = \left(\frac{nV_o}{8R_L}\sqrt{\frac{2R_L^2}{n^4L_m^2f_{sw,HB}^2} + 8\pi^2}\right)^2 \times R_{DS,HB}$$

Unregulated LLC FET
Switching Losses

$$P_{SW_{HB},off} = 0.1667 \times f_{sw,HB} \times V_{dc,link}\left(\frac{V_o}{4nL_mf_{sw}} - \frac{C_{oss,HB}V_{dc,link}}{\left(\frac{Q_{GS,HB}}{I_{g3,off}} + \frac{Q_{GD,HB}}{I_{g4,off}}\right)}\right)$$

$$\times \left(\frac{Q_{GS,HB}}{I_{g3,off,HB}} + \frac{Q_{GD,HB}}{I_{g4,off,HB}}\right)$$

Synchronous
Rectification
Conduction Losses

$$P_{cond,SR} = \left[\left(I_{SR,rms}\frac{\sqrt{3}V_o}{24\pi R_L}\sqrt{12\pi^4 + \frac{(5\pi^2-48)R_L^2T_{SW,HB}^2}{n^4L_m^2}}\right)^2 \times R_{DS,SR} + (2\right.$$

$$\left. \times T_{delay} \times I_{SR,rms} \times f_{SW,HB} \times V_{f,SR}\right]$$

In the above expression by substituting $V_{batt,max}$ and $V_{batt,min}$ with 56 V and 44 V, respectively, which are the charged/discharged voltages of a 48 V, 120 Ah VRLA battery pack [4], C_p can be obtained as $3.456 \times 10^4 F$. Self-discharge resistance is taken to be 10 kΩ [4], thus the effective impedance of the parallel network is very low, and hence the voltage drop is neglected across that network. By applying Kirchoff voltage law (KVL) and neglecting the voltage drop across the series resistance $R_{L_o,int}$ of the inductor L_o, we get

$$I_o R_{int} + V_{batt} = V_o \tag{3.56}$$

TABLE 3.3
Design Specifications

Parameters	Value
P_o	1.0 kW
$V_{in,rms}$	110 V
$f_{SW,BB}$	50 kHz
V_o	65 V
$f_{SW,HB}$	50 kHz

TABLE 3.4
Actual Parameters for Loss Analysis

Parameter	Value	Parameter	Value	Parameter	Value
$R_{g,switch}$	9 Ω	V_f	0.8 V	$V_{f,SR}$	1.6 V
$Q_{GD,BB}$	18 nC	$R_{DS,HB}$	80 mΩ	L_m	150 µH
$Q_{GS,BB}$	17 nC	$Q_{GD,HB}$	11 nC	$R_{C_o,ESR}$	31.8/18 mΩ
$R_{DS,on,BB}$	75 mΩ	$Q_{GS,HB}$	19 nC	R_L	4.225 Ω
R_{DCR}	15 mΩ	$C_{oss,HB}$	77 pF	V_{miller}	8 V
$R_{Cdc,link\,ESR}$	34.7/12 mΩ	$C_{oss,SR}$	640 pF	$R_{g,driver}$	7.5 Ω
R_D	98 mΩ	$R_{DS,SR}$	28 mΩ	$P_{e,limit}$	500 mW/cm^3
L_{para}	4 nH	V_e	0.559 cm^3		

Output voltage for the converter needs to be designed with the worst case which occurs when the converter operates with maximum power where the battery is fully charged and the charging current is high as shown in Figure 3.12. Considering a total internal resistance R_{int} of 0.16Ω for a battery voltage of 56 V [57], with a charging current of 15 A [55], the required output voltage can be computed to be 58.4 V. The entire process of battery charging is controlled by a battery management system (BMS) in sophisticated battery chargers in order to maintain battery health. Such BMS system controls the CC-CV mode of charging by ensuring battery cell voltage balancing, thus providing protection against overheating and overvoltage. Even though such systems are present, battery aging is one such issue that is more prominent in a lead-acid battery. Due to low DOD and repetitive charging of VRLA batteries, the battery gets heated up often, leading to increased internal resistance R_{int} with time. Due to an increase in R_{int} voltage drop across the resistance; thus higher charging voltage is required in order to charge the battery pack. An output voltage of 65 V is selected assuming these voltage margins and physical constraints such as temperature and wire resistance. It should be noted that any battery charger should be efficient over a range of power and should exhibit high efficiency from 80% of the rated power to 10%. High efficiency at light load becomes a crucial aspect of designing a battery charger.

Figure 3.13 shows the battery charging profile for a 48 V, 120 Ah battery. Charging starts when the output capacitor senses a voltage of 44 V or below. Current control is initiated by sensing the battery voltage (CC mode) which ensures a constant current flow of 15 A into the battery. CC mode is enabled until the battery voltage reaches up to 56.4 V with 14.1 V for each battery [58]. The converter operates with maximum power at this point as indicated in Figure 3.13. From [58] it is noted that the temperature rise is not significant for a battery voltage of 14.1 V as compared to other charging voltage levels where temperature increases significantly for every 10% rise in SOC. This voltage is known as the boost voltage of the battery and the charge mode is shifted to constant voltage (CV) mode where it holds the same battery voltage of 56.4 V. As the battery gets charged, its opposition to charge current is high in the absence of the current controller. Thus the charging current starts tailing off when a constant voltage is applied across the battery terminals. As the current reaches a value equal to 20% of the initial charging current, the charging process is terminated. At this point, the battery is 95% charged.

The small-signal model of the proposed converter is less complex as the second stage acts as a voltage amplifier with a gain proportional to the turns ratio, n, neglecting the dynamics offered by L_r and C_r. Thus, the secondary side state variables can be referred to as the primary as shown in Figure 3.14. The current injected equivalent circuit approach (CIECA) [59–61] is used to model the charger dynamic model. This modeling approach is simpler with respect to the state-space average model as it only accounts for the transfer characteristics of the converter. By considering the battery equal to R_L, and applying perturbations to (3.31) and neglecting the second order terms, we get

$$\hat{i}_{o,BB} = \frac{V_{in,pk}^2 DT_{sw,BB}}{LV_{dc,link}}\hat{d} + \frac{V_{in,pk}D^2T_{sw,BB}}{LV_{dc,ink}}\hat{v}_{in,pk} \tag{3.57}$$

From Figure 3.14 we know that

$$\hat{i}_{o,BB} = \left(s\left(C_{dc,link} + n^2C_o\right) + \frac{n^2}{sL_o + R_L}\right)\hat{v}_o \tag{3.58}$$

On equating (3.57) and (3.58) and substituting $\hat{v}_{in,pk} = 0$, CV mode transfer function to control the output voltage of the converter is given by

$$\frac{\hat{v}_o}{\hat{d}} = \frac{V_{in,pk}^2 DT_{sw,BB}(sL_o + R_L)}{LV_{dc,link}\left(s^2L_o\left(C_{dc,link} + n^2C_o\right) + sR_L\left(C_{dc,link} + n^2C_o\right) + n^2\right)} \tag{3.59}$$

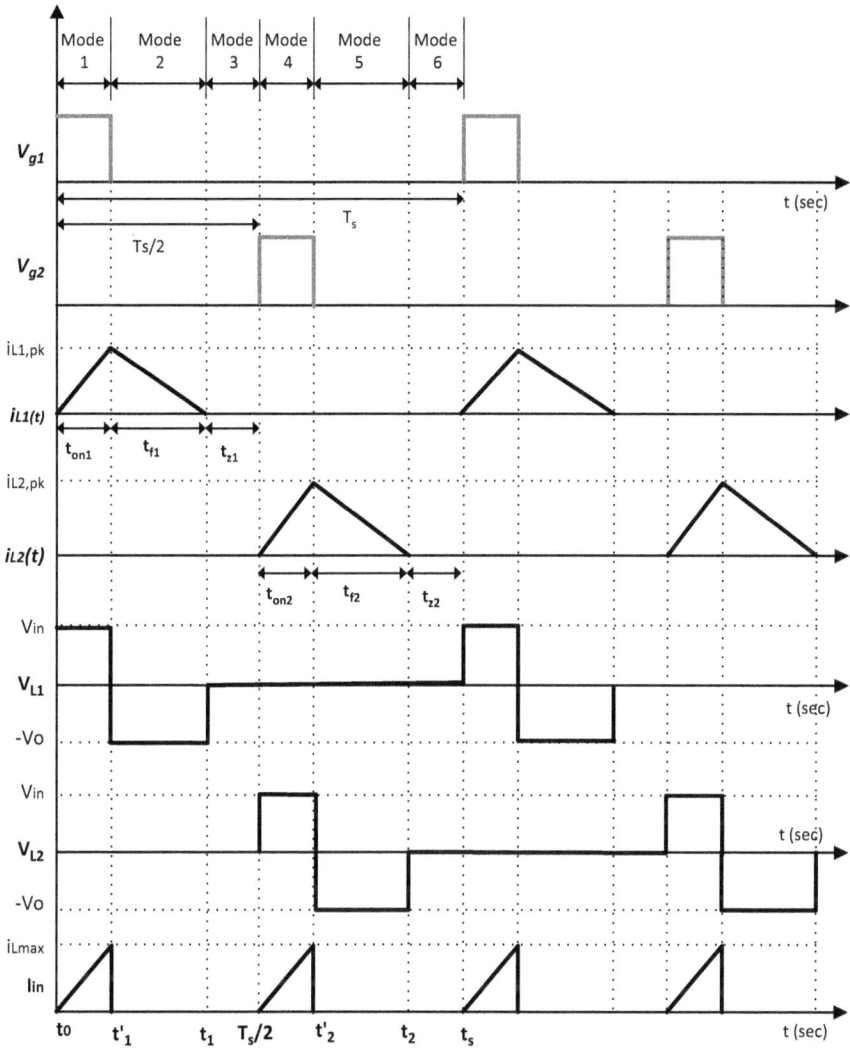

FIGURE 3.14 Steady-state waveforms of two-cell converter.

In order to control the charging current of the converter, L_o current is controlled. From Figure 3.14 the charging current expression can be given as

$$\hat{i}_o = \frac{n^2 \hat{v}_o}{sL_o + R_L} \tag{3.60}$$

On substituting $\hat{v}_o$ from (3.59), CC mode transfer function to control the charging current is given as

$$\frac{\hat{i}_o}{\hat{d}} = \frac{n^2 V_{in,pk}^2 DT_{sw,BB}}{LV_{dc,link}\left(s^2 L_o\left(C_{dc,link} + n^2 C_o\right) + sR_L\left(C_{dc,link} + n^2 C_o\right) + n^2\right)} \quad (3.61)$$

Figure 3.15 shows the control scheme for the proposed battery charger. Charging is initiated and follows the charge control algorithm as per Figure 3.16. As primary switches are directly controlling the charging voltage and current along with drawing sinusoidal input current, the bandwidth of the controller is selected to be less than 120 Hz. PI controllers are used to control the charging current and voltage with a phase margin of 60° and 70°, respectively.

This section presents the simulation results of the proposed battery charger using PSIM 11.1 software. The converter is designed with the specifications mentioned in Table 3.3, and the designed parameters are given in Table 3.4. The input LC low-pass filter is designed for a cut-off frequency of 10 kHz. The control-to-output

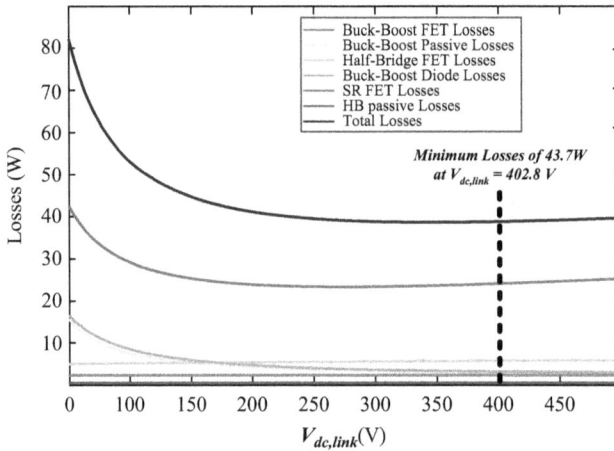

FIGURE 3.15   Calculated losses according to various DC-Link voltages.

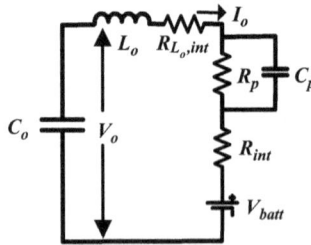

FIGURE 3.16   Battery equivalent circuit when connected to battery charger.

current and control-to-output voltage transfer function is obtained by and is given as (3.59) and (3.61)

$$I(s) = \frac{\hat{i}_o}{\hat{d}} = \frac{8.96}{9.64e^{-7}s^2 + 0.0109s + 1} \tag{3.62}$$

$$V(s) = \frac{\hat{v}_o}{\hat{d}} = \frac{0.0305s + 343.74}{9.64e^{-7}s^2 + 0.0109s + 1} \tag{3.63}$$

For a double pole system, a simple PI controller is sufficient to get the desired response. The controller transfer function is designed at a phase margin 60°, and gain crossover frequency 80 Hz. Figures 3.17 and 3.18 shows the frequency response of plant transfer function I(s) and G(s), controller transfer function H(s), and the open-loop transfer function I(s) * H(s). Both open-loop transfer function has

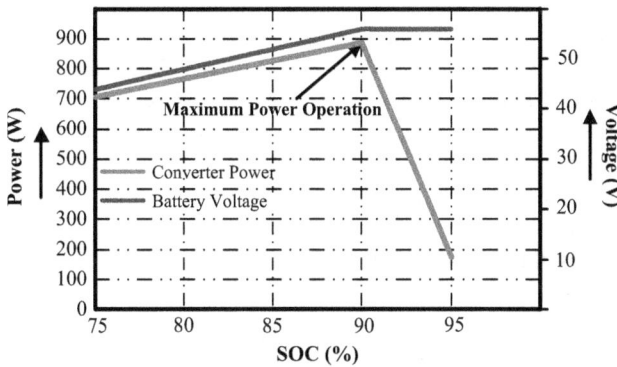

FIGURE 3.17 Variation of converter power and battery voltage with respect to SOC.

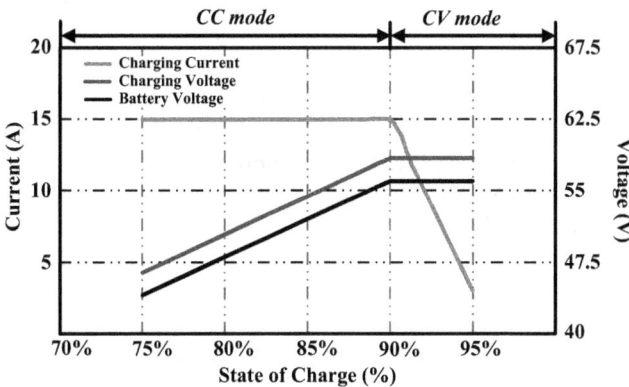

FIGURE 3.18 Charging profile of lead acid battery pack.

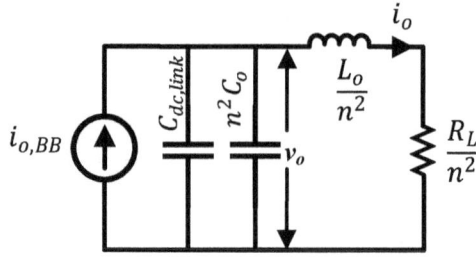

FIGURE 3.19 CIECA equivalent model.

an infinity dc gain which indicates the system reference tracking with zero steady-state error, and the system robustness for the input and load disturbances.

With the designed controller, a closed-loop simulation for the proposed converter is done using PSIM11 software, and the results are depicted in Figure 3.18. Figure 3.19(a) shows the input current and input voltage waveform during the CC mode of operation. It is observed that the input current and the voltage are sinusoidal and in-phase with each other confirming the UPF operation of the charger. Figure 3.19(b) shows the converter output voltage, and charging current waveforms at rated output power. The output current is constant and settled at a reference current of 15 A. Figure 3.19(c) and (d) shows the current controller response during input voltage variation from 110 V to 80 V and 110 V to 130 V. Output current is maintained at constant 15 A and the input current is closely tracking the input voltage both being in-phase and shape. The output current is stable and tracks the reference current with a settling time of 15 ms, which confirms the robustness of the designed current controller. Figure 3.19(e) shows the input inductor current waveforms at rated output power, the inductor currents are discontinuous thus validating the design. Figure 3.19(f) shows the converter response for 10% load perturbation from 1 kW to 100 W which is the turning point for the battery charger controller (CC mode to CV mode). The output voltage is stable and tracks the reference voltage with a settling time of 10 ms, which confirms the robustness of the voltage controller. Figure 3.19(g) shows the ZVS turn-on operation of the back-end DC-DC converter. The switch turns on with zero voltage, thus confirming the soft switching of the primary side half-bridge switches. Figure 3.19(h) shows dc-link voltage variation during the transient condition. It is observed that the dc-link voltage remains relatively the same during load change which verifies the assumption and the gain expression. DC-link tries to reduce huge load changes as the duty cycle reduces drastically and dc-link capacitor supplies power to the load. To verify the above-derived formulas in section 3.3 for dc-link selection, the average and RMS currents for an output power of 1.0 kW and input voltage of 110 V_{rms} are calculated and compared with the simulated values as listed in Table 3.5. The calculated values are very close to the simulated values, thus validating the dc-link voltage selection criteria.

TABLE 3.5

Comparison of Analytically Calculated and Simulated Average and RMS Current Values of Active and Passive Devices at Rated Condition of 1 kW. V_{in} = 110 V, D = 0.56, L = 75 μH

Parameters	Calculated	Simulated
I_{in}	9.16 A	9.11 A
$I_{SW_{BB},rms}$	7.07 A	7.01 A
$I_{L,rms}$	8.14 A	8.12 A
$I_{d,rms}$	4.04 A	4.06 A
$I_{d,avg}$	1.255 A	1.251 A
$I_{C_{dc,link}rms}$	5.13 A	5.19 A
$I_{sw_{HB},rms}$	5.70 A	5.88 A
$I_{SR,rms}$	12.08 A	12.2 A
$I_{C_O,rms}$	8.20 A	8.28 A

3.3.2 BRIDGELESS BUCK BOOST DERIVED FRONT-END PFC CONVERTER

One of the major limitations of the two-stage buck-boost charger proposed in the previous section is the higher losses of the front-end diode bridge. Thus, the charger cannot be scaled to high-power charger and thus a bridgeless topology needs to be derived for such applications.

Figure 3.20 shows the proposed single-phase switched-mode bridgeless AC-DC buck-boost derived PFC converter. The proposed converter is derived from the classical buck-boost converter. The diode rectifier is removed from the front end and integrated on the load side in the form of a voltage doubler configuration. The converter embodies two back-to-back connected MOSFETs, two diodes, one inductor, and two electrolytic capacitors. The voltage across the devices is reduced by the DC-split output configuration in the proposed converter, which reduces the switching losses. The back-to-back connected MOSFETs are connected to a common source configuration and as they are controlled by the same gate signal; it is considered as a single switch S in the analysis. The buck-boost inductor is designed for DCM operation to realize the inherent PFC at the AC power source. In the DCM operation of the converter, the value of the input voltage determines the amount of energy stored in the buck-boost inductor. Therefore, the average input current inherently trails the input voltage. Figure 3.21 illustrates the control circuit for the proposed converter. The converter output is controlled by only one control loop and a single sensor. As the controlled variable is DC output voltage, the proposed converter duty cycle is constant for rated output power and input voltage.

The proposed converter demonstrates several advantages like the reduced number of components as compared to the conventional boost and bridgeless boost converters. Moreover, only one semiconductor device conducts current at a time which significantly reduces the conduction losses benefiting high power conversion

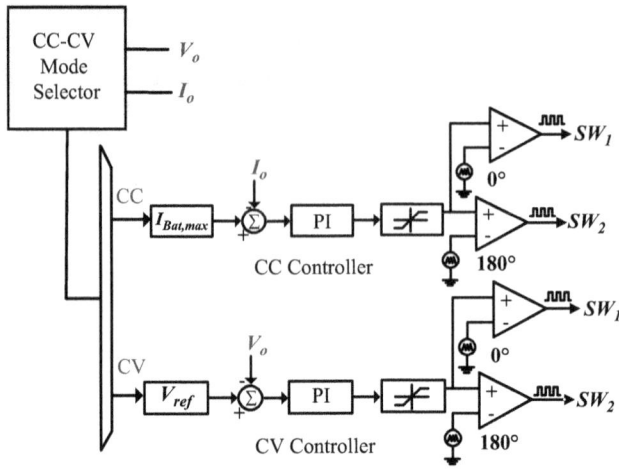

FIGURE 3.20   CC-CV charge controller.

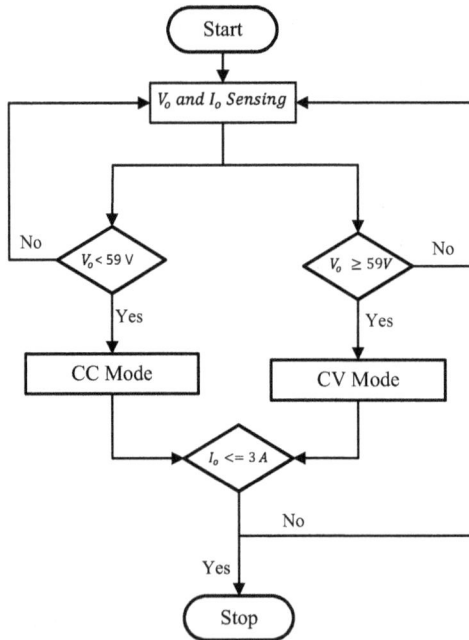

FIGURE 3.21   CC-CV mode selector algorithm.

efficiency and power density. To validate the performance of the proposed converter, a comparative evaluation with the state-of-the-art AC-DC PFC topologies converters is provided in Table 3.6. In all the above state-of-the-art converters it is observed that at a given point of time, more than one semiconductor is in the path of current conduction.

TABLE 3.6

Comparison of the Proposed Converter with the State-of-the Art Converters

Attributes	Bridgeless Boost	Semi-Bridgeless	Bridgeless Buck-Boost	Bridgeless SEPIC	Bridgeless CUK	Proposed
Output voltage	−ve	+ve	−ve	+ve	+ve	+ve
Line diodes requirement	−	2	2	2	2	−
Switching Devices in operation over one switching cycle	2sw+2D	1sw+2D	1sw+2D	1SW+1D	1SW+1D	2SW or 1D
HF inductors	2	2	2	3	3	1
HF Diode	2	2	2	2	2	2
Intermediate capacitor	No	No	No	Yes	Yes	No
Conduction Loss	High	Low	Medium	High	Low	Low
Switch Voltage Stress	High	Medium	High	High	High	Low

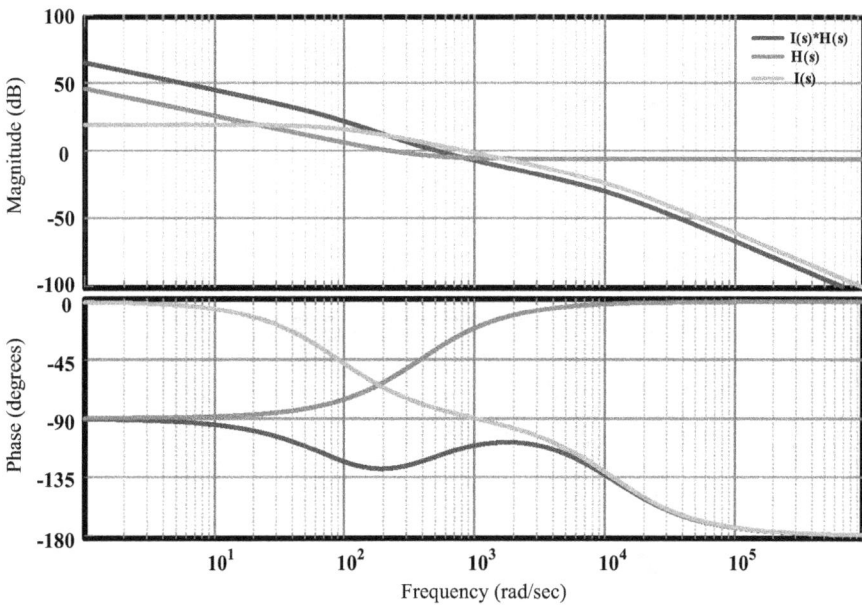

FIGURE 3.22 Frequency response of plant I(s), controller H(s) and open loop I(s) ∗ H(s).

The proposed converter also illustrates some additional benefits such as reduced voltage stress of $V_{pk} + \frac{V_o}{2}$ across all semiconductor devices compared to a traditional buck-boost converter. The proposed converter achieves PFC over the range of input voltage while maintaining a low THD below 5%. It also maintains a stiff regulated DC voltage. Due to DCM operation, the sensor requirement is reduced to one voltage sensor and avoids the sensing of input current, input voltage, and output current.

3.3.2.1 Design Bridgeless Buck Boost Derived Front-End PFC Converter

The proposed converter only operates in the boost mode in order to reverse bias the output diodes when the switch S is conducting. The equivalent circuit S of operation during positive and negative half-line cycles are shown in Figure 3.22(a) and (b), respectively. It must also be considered that either the switch S or only one diode (D_1 or D_2) is in the current flowing path, which consequently reduces the conduction losses. The converter is designed to be operated in DCM to achieve natural PFC at AC input. Figure 3.24 shows equivalent circuits of operation during a positive half-cycle of input voltage.

The steady-state waveform of the proposed converter for one switching cycle is shown in Figure 3.23 with the following assumptions. Figure 3.24

    a. All components are ideal and the input voltage and the output voltage are considered constant within one switching cycle.

    b. The output side filter capacitor is large enough to maintain the output voltage constant in one switching cycle.

    c. The output capacitors 'C_{01}' and 'C_{O2}' share half of the output voltage.

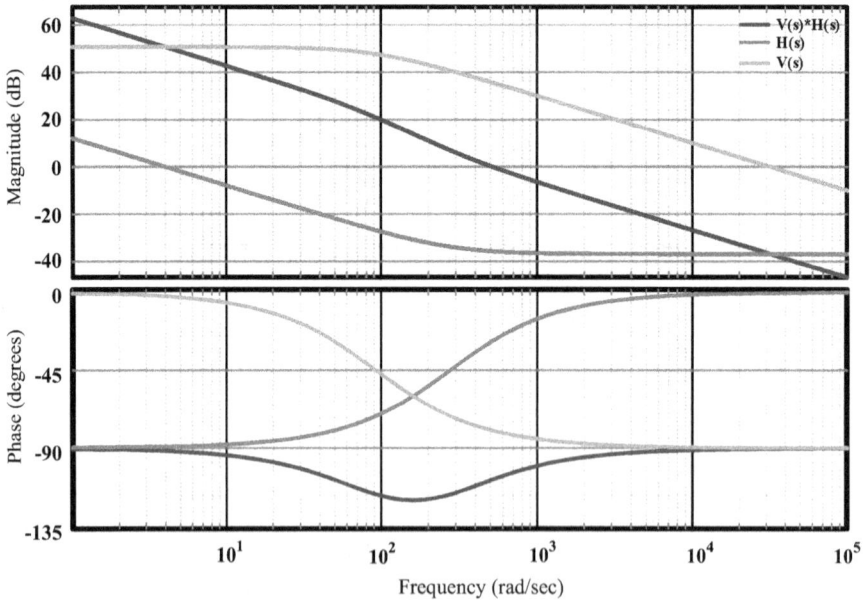

FIGURE 3.23 Frequency response of plant V(s), controller H(s) and open loop V(s) $*$ H(s).

Mode I. $(0 < t < t_1')$

In mode I, switch S is turned on with the gate signal V_g. Inductor L stores the energy and capacitors C_{01} and C_{O2} supply power to the load. The expression for the inductor current $i_L(t)$ is given as

$$i_L(t) = \frac{V_{in}}{L} * (\Delta t) \qquad (3.64)$$

where V_{in} = input voltage.

Mode II. $(t_1' < t < t_1)$

In this mode, the gate signal is removed and switch S is turned off. The inductor L starts demagnetizing by delivering the stored energy to the load while the capacitor C_{O2} gets charged. The expression for the inductor current $i_L(t)$ is given as

$$i_L(t) = i_{L,pk} - \frac{V_o}{2L} * (\Delta t) \qquad (3.65)$$

where $i_{L,pk}$ is the peak inductor current given by

$$i_{L,pk} = \frac{V_{in}}{L}(DT_s) \qquad (3.66)$$

where, DT_s = switch on-time.

This mode ends when the current through the diode D_2 is zero, that implies

$$D_1 T_s = \frac{2V_{in}}{V_o} DT_s \qquad (3.67)$$

where, $D_1 T_s$ = diode conduction time.

Mode III. $(t_1 < t < T_s)$

In this mode, all semiconductor devices are in off-state, and capacitors C_{o1} and C_{o2} supply power to the load.

The current supplied to the load is nothing but the average diode D_1 current $i_{D_2,avg}$ in the positive half-line cycle. Since the current is triangular in shape, its average can be given as

$$i_{D_2,avg} = \frac{i_{L,pk} D_1 T_s}{2T_s} \qquad (3.68)$$

Substituting (3.66) and (3.67) in (3.68), the average output current for one switching cycle can be given as

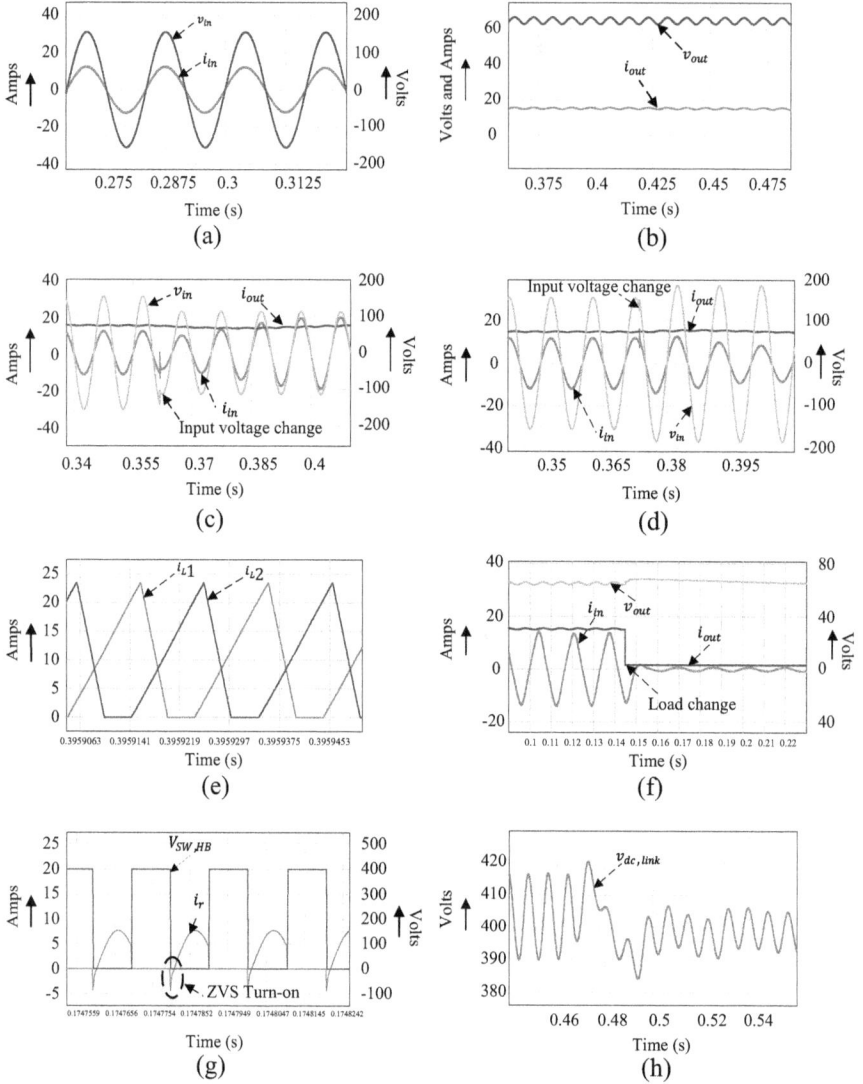

FIGURE 3.24 Simulation results (a) input voltage and current. (b) output voltage and current. (c) input voltage change from 110 V to 80 V. (d) input voltage change from 110 V to 130 V. (e) inductor current. (f) load change from 100% to 10%. (g) ZVS turn-on of half-bridge switch. (h) DC-link voltage variation during load change.

$$i_{o,avg} = \frac{i_{D_2,avg}}{2} = \frac{V_{in}^2 D^2 T_s}{2LV_o} \qquad (3.69)$$

Where D = duty cycle and V_o = output voltage.

Thus the average diode current $I_{D_2,avg}$ for half-line period can be defined as

$$I_{o,avg} = \frac{1}{2\pi} \int_0^\pi i_{D2,avg} \, d\omega t \tag{3.70}$$

$$I_{o,avg} = \frac{V_{pk}^2 D^2 T_s}{4LV_o} \tag{3.71}$$

The input current expression of the proposed converter for one switching cycle can be defined as

$$I_{in}(t) = \begin{cases} \frac{V_{in}}{L}t, & 0 < t \le t_{on} \\ 0, & t_{on} < t \le T_s \end{cases} \tag{3.72}$$

On performing FFT of (3.82)

$$a_0 = \frac{V_{in}}{L}D^2 T_s \tag{3.73}$$

$$a_h = \frac{V_{in}}{L}\left(Dsin\,(2\pi hD) + \frac{1}{2h\pi}\cos(2\pi hD) - \frac{1}{2h\pi}\right) \tag{3.74}$$

$$b_h = \frac{V_{in}}{2\pi hf_{sw}L}\left(\frac{sin\,(2\pi hD)}{2h\pi} - Dcos\,(2\pi hD)\right) \tag{3.75}$$

On substituting $V_{in} = V_{pk}\sin(\omega t)$ in (3.73) the fundamental component of the input current is obtained (3.72) and (3.73) shows the switching order harmonics, which needs to be filtered out. From (3.74) and (3.75), it is noted that unlike conventional boost converter, the proposed converter does not inject harmonics into the input and thus requires a relatively small filter. By designing a low-pass LC filter with a cut-off frequency much lower than the switching frequency, the harmonic currents can be filtered out. Therefore, the resulting input current contains only the fundamental current component, and it can be obtained by applying the power balance expression (3.76). The input current can be found by,

$$V_{in}I_{in} = V_o i_{o,avg} \tag{3.76}$$

Substituting (3.71) in (3.76)

$$V_{in}I_{in} = \frac{V_{in}^2 D^2 T_s}{2L} \tag{3.77}$$

$$I_{in} = \frac{V_{in}D^2T_s}{2L} = \frac{V_{pk}D^2T_s}{2L}\sin(\omega t) = I_{pk}\sin(\omega t) \qquad (3.78)$$

where, $I_{pk} = \frac{V_{pk}D^2T_s}{2L}$ peak input current.

Equation (3.78) shows that the filtered input current is sinusoidal and is in phase with the input voltage, which proves the UPF operation of the converter.

The following inequalities must hold for DCM operation which is given as

$$DT_s + D_1T_s < T_s \qquad (3.79)$$

On substituting (3.67) in (3.86)

$$D < \frac{M}{(M + 2\sin(\omega t))} \qquad (3.79)$$

where, $V_{in} = V_{pk}\sin(\omega t)$, $M = V_o/V_{pk}$.

In (3.79), the worst case occurs at $\omega t = 1$, thus by substituting $\omega t = \frac{\pi}{2}$ in (3.79) the condition to operate the converter in DCM is given as

$$D < \frac{M}{(M + 2)} \qquad (3.80)$$

From (3.80), the critical value of voltage conversion ratio M_{cric} for a given duty cycle can be defined and is given as

$$M_{cric} < \frac{2D}{(1 - D)} \qquad (3.81)$$

Output current is given as

$$I_o = \frac{V_o}{R} \qquad (3.82)$$

Substituting (3.71) in (3.82),

$$D = M\sqrt{2K_{cond}} \qquad (3.83)$$

where K_{cond} = conduction parameter of the converter

$$K_{cond} = \frac{2L}{RT_s} \qquad (3.84)$$

From (3.80) and (3.83), critical conduction parameter K_{cric} can be calculated with

$$K_{cric} = \frac{1}{2(M+2)^2} \tag{3.85}$$

To maintain PFC under all conditions, the inductor current needs to be in DCM for the worst-case input voltage. The DCM inductor can be computed by using (3.71) and (3.80) and is given by

$$L < \frac{V_{pk}^2 V_o^2 T_s}{4P_o \left(V_o + 2V_{pk} \right)^2} \tag{3.86}$$

In a single-phase PFC rectifier, the output capacitors are designed to filter out the second-order supply frequency oscillations present in the output voltage. The output ripple is caused by the unbalanced instantaneous power between input and output. Therefore, capacitors are designed to buffer this unbalanced power and filter out oscillations. Thus, by considering $C_{o1} = C_{o2} = C_o$, the low-frequency output voltage ripple $V_{o,ripple}$ is given as

$$\Delta V_{o,ripple} = \frac{1}{C_o} \left(\int i_{co1} dt + \int i_{co2} dt \right) \tag{3.87}$$

$$= \frac{1}{C_o} \int (i_{D1} - 2i_o) dt = \frac{2i_o}{\omega C_o} \tag{3.88}$$

$$C_o = \frac{2I_o}{\omega V_{o,ripple}} \tag{3.89}$$

The criteria to design a low-pass LC filter are as follows:

1. Selection of cut-off frequency f_c given by

$$f_c = \frac{1}{2\pi} \sqrt{\frac{1}{L_f C_f}} \tag{3.90}$$

2. Minimization of filter reactive power consumption for 60 Hz at 1.0 kW. The reactive power is minimum when the filter characteristic impedance is equal to the converter impedance i.e.,

$$Z_{ch} = \sqrt{\frac{L_f}{C_f}} = Z_{in} \tag{3.91}$$

where, Z_{ch} is the characteristic impedance and Z_{in} is the input impedance at rated load and is given by

$$Z_{in} = \frac{2L}{D^2 T_s} \tag{3.92}$$

Thus, low-pass filter parameters L_f and C_f can be obtained as

$$L_f = \frac{Z_{ch}}{2\pi f_c} \tag{3.93}$$

$$C_f = \frac{1}{2\pi Z_{ch} f_c} \tag{3.94}$$

Traditional front-end converters of battery chargers use complex control which requires input voltage and current sensing along with PLL. Such systems pose a higher burden on microcontrollers as more computation speed is required. The proposed converter mitigates these problems by eliminating the input sensing and just using one sensor to control the output as shown in Figure 3.7(b). The small signal model of the proposed converter is obtained by using the current injected equivalent circuit approach (CIECA) [59–61]. This approach is better than the conventional state-space averaging approach as it becomes cumbersome and complex in DCM. Such complex models are tough to derive even for simple DC-DC based DCM converters [62]. On the other hand, the CIECA approach is much easier as it only models the transfer properties of the converter [59]. In CIECA, the entire circuit can be scaled down as shown in Figure 3.25. The non-linear parameters of the circuit are linearized by injecting the average output current produced by the non- linear part. From Figure 3.25

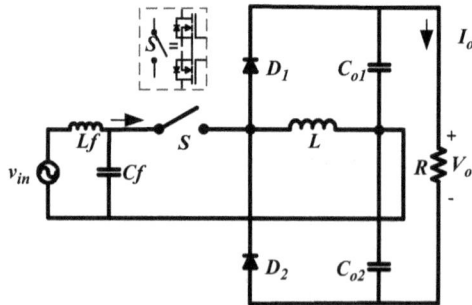

FIGURE 3.25 Proposed single-phase switched-mode bridgeless AC-DC buck-boost derived PFC converter.

$$\hat{i}_{o,avg} = \left(sC + \frac{1}{R}\right)\hat{v}_o \qquad (3.95)$$

On applying perturbations in (3.71) we get

$$\hat{i}_{o,avg} = \frac{V_{pk}^2 DT_s}{2LV_o}\hat{d} + \frac{V_{pk}D^2 T_S}{2LV_o}\hat{v}_{pk} - \frac{i_{o,avg}}{V_o}\hat{v}_o \qquad (3.96)$$

On equating (3.95) and (3.96) and substituting $\hat{v}_{pk} = 0$

$$\frac{v_o(s)}{d_o(s)} = \frac{V_{pk}D}{K_{cond}M(sRC + 2)} \qquad (3.97)$$

where $C = \frac{C_{o1}C_{o2}}{C_{o1} + C_{o2}}$, $M = \frac{V_o}{V_{pk}}$, and R = load resistance.

The converter control to output transfer function is obtained by substituting the design parameters in (3.97). As the transfer function is a single-pole system, a simple PI controller $\left(K_p + \frac{K_i}{s}\right)$ is used to control the output voltage as shown in Figure 3.7(b). As the output capacitor sees a voltage ripple of twice the line frequency, a PI controller with bandwidth lower than 120 Hz is selected with a Phase Margin (PM) of 60°. The controller is tuned using sisotool in MATLAB and the controller parameters are computed as $K_p = 0.00252$ and $K_i = 0.21$. The output voltage is sensed using a hall-effect-based LV-25P sensor. The sensed voltage is compared with the reference voltage and the error is fed into the PI controller. The PI controller generates the duty cycle to control switch S. A limiter is connected in order to limit the duty during start-up and overload conditions. The proposed converter is simulated in PSIM 11.1 software to confirm the converter analysis and the design. The converter design specifications are given in Table 3.7. The buck-boost inductance value is calculated from (3.86). The output filter capacitance values are calculated from (3.89).

TABLE 3.7
Converter Design Specifications

Parameter	Value
Line voltage, $V_{in} RMS$	110 $V_{RMS\,nominal}$
Input frequency, f	60 Hz
Output power, P_o	1.0 kW
Output voltage, V_o	400 V
Switching frequency, f_{sw}	50 kHz
Duty cycle, D	0.638
Buck-Boost Inductance, L	24.45 μH
Output capacitance, C_{o1}, C_{o2}	82.4 μF
Output voltage ripple, $V_{o,ripple}$	5% of output voltage (V_o)

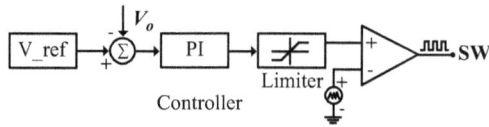

FIGURE 3.26 The control circuit for the proposed converter.

Using the designed parameters, the converter control-to-output transfer function is obtained from (3.97) and is given in (3.98). A PI-controller transfer function is designed for Phase Margin of 60° a bandwidth of 628.31 rad/sec. It determines $k_p = 0.00252$ and time constant $\tau = 0.012$ controller is designed and implemented.

$$\frac{v_o(s)}{d(s)} = \frac{1282.584}{1 + 0.032s} \tag{3.98}$$

With the designed parameters and the designed controller, the circuit is simulated, and the results have been presented for input frequency $f = 60$ Hz. The simulated input voltage and input current waveforms are shown in Figure 3.26(a). The output voltage and output current are shown in Figure 3.26(b). The output diode 'D1' current waveform is shown in Figure 3.26(c), which is discontinuous, and validates the converter design. The voltages across output capacitors are shown in Figure 3.26(d). Each output voltage capacitor is sharing half of the output voltage, which is in good agreement with the analysis. The inductor current waveform is shown in Figure 3.26(e). The output voltage and input current waveforms when the converter is subjected to a load disturbance from 50% to 100% of the rated power are shown in Figure 3.11(f). The controller responds immediately to the load change, and the output voltage is settled at the reference value 400 V.

3.3.3 EVALUATION OF PFC-BASED EV CHARGERS

In battery charging applications, AC-DC converters play a major role in PFC. This chapter compares various AC-DC topologies available for battery charger application with a detailed description comparing trade-offs of AC-DC topologies. Two AC-DC topologies for Plug-in chargers have been proposed. These charger topologies eliminate the use of input sensing and PLL which makes them more robust and reliable. A two-stage charger topology along with a detailed control design has also been described that enables readers to design DCM-based two-stage chargers. Proposed configurations have been compared with state-of-art AC-DC topologies that show various advantages of proposed converters Figures 3.27, 3.28, 3.29.

Interleaved buck-boost converter discussed in section 3.3.1 is compared with various CCM and DCM topologies. It is seen that the proposed charger configuration has a lower sensor count with very simple control and a high peak efficiency. Charger topologies reported in the literature suffer from issues like

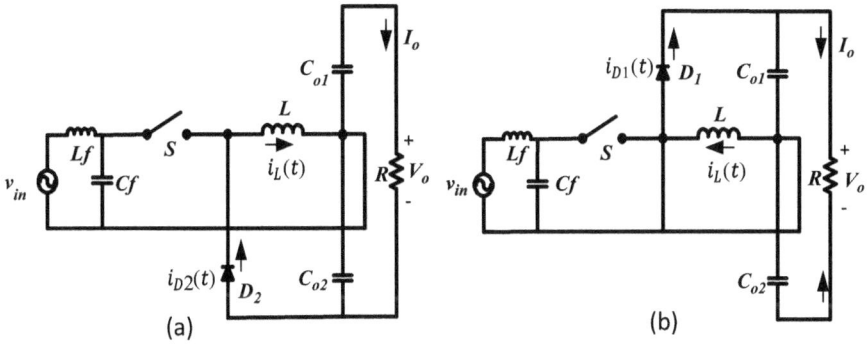

FIGURE 3.27 Proposed converter configurations during (a) positive half cycle, (b) negative half cycle.

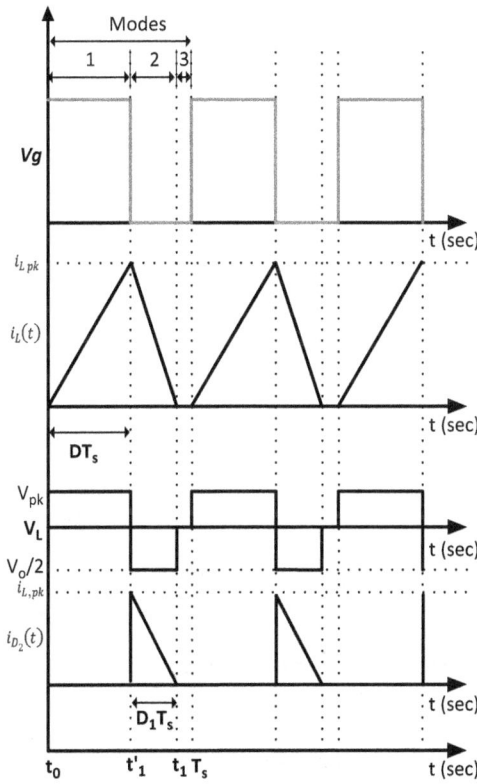

FIGURE 3.28 Waveforms for one switching cycle.

complex control, high device count, and lower efficiency. The bridgeless buck-boost derived converter discussed in section 3.3.2 eliminates the losses due to the diode bridge rectifier. As compared to other CCM-based bridges topologies that require isolation transformer or input sensing, the proposed topology eliminates

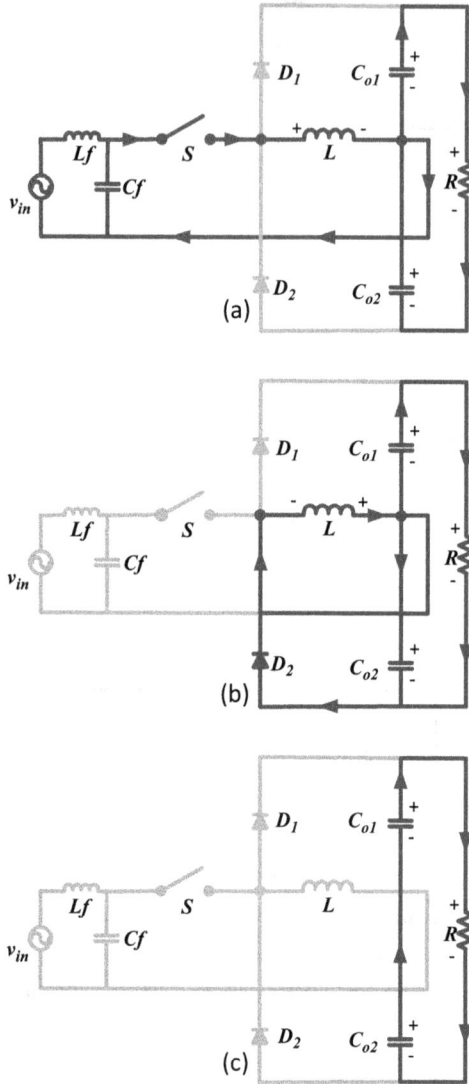

FIGURE 3.29 Equivalent circuits during positive half-cycle; (a) Mode-a; (b) Mode-b; (c) Mode-c.

sensing by operating in DCM operation. Moreover, by using split capacitors at the output, switch stress is reduced to $V_{pk} + \frac{V_o}{2}$ (Figures 3.30, 3.31, 3.32).

The proposed topology in section 3.3.2 can be extended to two-stage charger as shown in Figure 3.32. The front-end converter is a bridgeless converter analyzed in the previous section. It is derived from the traditional buck-boost converter. The converter is operating in DCM mode to achieve natural PFC at AC mains. The front-end converter consists of one bidirectional switch, two diodes, one inductor,

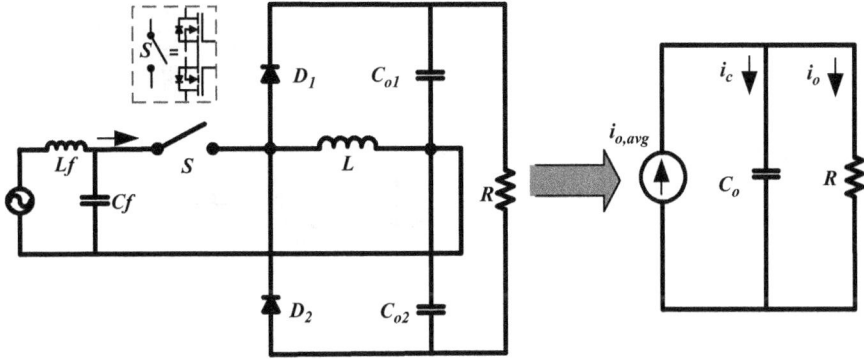

FIGURE 3.30 Equivalent circuit for small signal modeling.

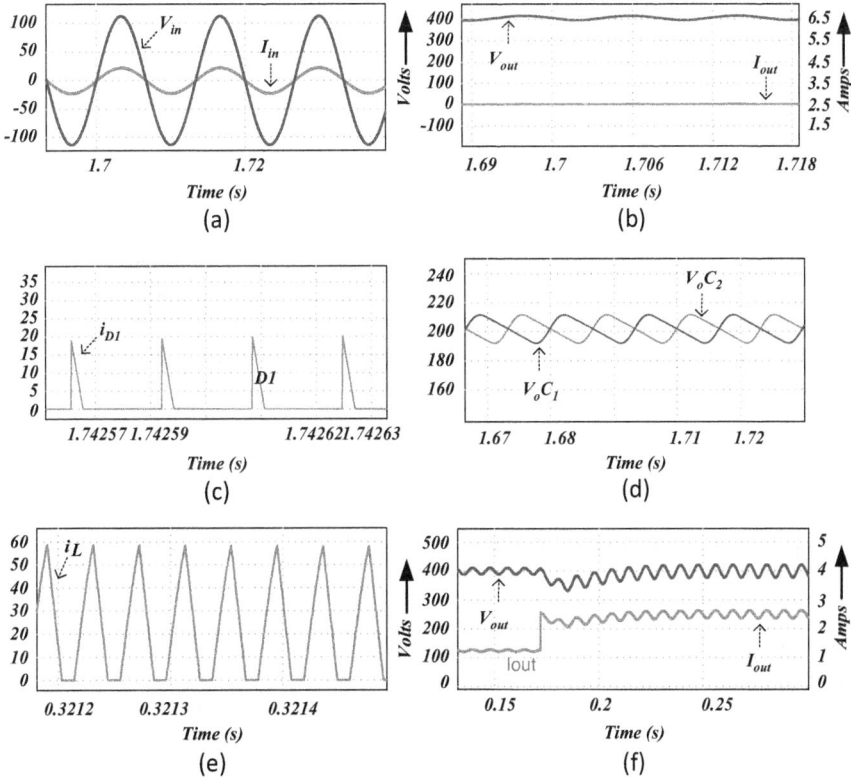

FIGURE 3.31 Simulation results (a) input voltage and input current; (b) output voltage and output current; (c) output diode $Do1$ current; (d) output capacitors voltages; (e) inductor current; (f) converter response for load change 50 % to 100 %.

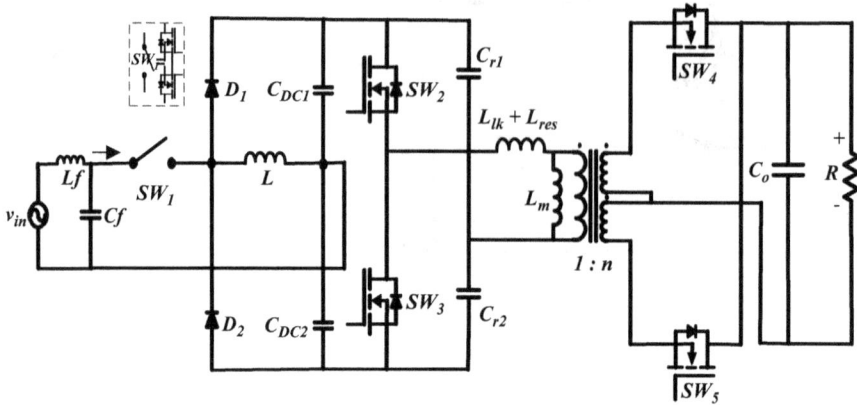

FIGURE 3.32 Proposed two-stage isolated EV charger configuration.

and two capacitors (voltage doubler). At the input mains, an LC filter is connected to filter out high-frequency switching harmonics and draw pure sine wave current from the source. As the diode bridge rectifier is eliminated and only one semiconductor device is in the current flowing path for one switching cycle, the overall conduction losses are reduced significantly. The diode has zero reverse recovery losses along with reduced voltage stress. This reduced voltage stress and conduction losses help in the reduction of the overall weight of the proposed charger The second stage is a half-bridge unregulated LLC resonant DC-DC converter to obtain the desired DC voltage of 65 V while offering isolation. Synchronous rectification on the secondary side is implemented to improve the overall efficiency of the converter. Merits of this topology over the conventional system. Additional merits (Table 3.8):

1. Less number of semiconductor devices count as the application requires unidirectional flow of high current and low output voltage.
2. LLC resonant tank when reflected to secondary appear as a current source, thus making it suitable for charging applications, along with the capability to achieve ZVS at low loads.

The converter utilizes a center tap transformer along with synchronous rectification on the transformer secondary to enhance the overall efficiency and decrease the cost of the charger. The operation of the second stage with a fixed duty cycle and switching frequency helps in the reduction of the control burden on the microcontroller. Moreover, soft switching of half-bridge MOSFETs is observed due to the resonant tank. The optimal design of the resonant tank resulted in reduced peak and circulating current in the back-end converter. In addition, owing to the sinusoidal current through the transformer results in improved transformer utilization.

TABLE 3.8

Comparative Analysis of Proposed Configuration with Other Topologies

Attributes		BL-BB	[14]	[17]	[16]	[8]	[33]	IL-BB
No. of Components	DBR	–	–	–	Yes	Yes	Yes	Yes
	Isolation	–	Yes	Yes	Yes	Yes	Yes	Yes
	S_w	2	3	3	3	6	6	6
	D	2	3	5	3	8	6	2
	L	1	2	4	4	4	2	3
	Int C	–	2	2	2	–	–	–
Mode of Operation		DCM	DCM	DCM	DCM	DCM	CCM	DCM
Switches with SS		2	3	3	3	6	4	6
Diode with Reverse Losses		0	0	0	0	4	6	0
Semiconductor Current Stress		High	High	High	Moderately high	Moderately Low	Low	Moderately Low
Semiconductor Voltage Stress		Moderate	High	High	High	Moderate	Moderate	High
Control and Modelling Approach		Very Simple	Complex	Complex	Complex	Very Complex	Very Complex	Very Simple
Efficiency at 100 % load		96%	90.8%	92%	91%	93.7%	95.3%	96.06%
Sensors		1	3	3	2	4	5	2
PLL requirement		No	No	No	No	Yes	Yes	No

*DBR – diode bridge rectifier, Int C- intermediate capacitor

BIBLIOGRAPHY

[1] Electric vehicle sales in 2020 para 1, line 2 Available [Online]. https://about.bnef.com/blog/electric-vehicle-sales-to-fall-18-in-2020-but-long-term-prospects-remain-undimmed/

[2] "Outlook for Electric Vehicles and Implications for the Oil Market" Available [Online]. https://www.bankofcanada.ca/wp-content/uploads/2019/06/san201919.pdf

[3] "Electric Mobility: Taking the Pulse in Times of Coronavirus" [Online]. https://www.itf-oecd.org/sites/default/files/electric-vehicles-covid-19.pdf

[4] "Global BEV & PHEV Sales for 2019" Available [Online]. https://www.evvolumes.com/

[5] C. Cochrane, T. Muneer, and B. Fraser, "Design of an electrically powered rickshaw, for use in India," Energies, vol. 12, no. 17, p. 3346, Aug. 2019.

[6] "Inside India's Messy Electric Vehicle Revolution" Available [online]. https://www.nytimes.com/2019/08/22/technology/india-electric-vehiclerickshaw.html.

[7] C. Oh, D. Kim, D. Woo, W. Sung, Y. Kim, and B. Lee, "A high-efficient non isolated single-stage on-board battery charger for electric vehicles," in IEEE Transactions on Power Electronics, vol. 28, no. 12, pp. 5746–5757, Dec. 2013, doi: 10.1109/TPEL.2013.2252200

[8] J. Lee, and H. Chae, "6.6-kW onboard charger design using DCM PFC converter with harmonic modulation technique and two-stage DC/DC converter," in IEEE Transactions on Industrial Electronics, vol. 61, no. 3, pp. 1243–1252, March 2014, doi: 10.1109/TIE.2013.2262749

[9] R. Kushwaha, and B. Singh, "UPF-isolated zeta converter-based battery charger for electric vehicle," in IET Electrical Systems in Transportation, vol. 9, no. 3, pp. 103–112, 9 2019, doi: 10.1049/iet-est.2018.5010

[10] R. Kushwaha, and B. Singh, "A modified luo converter-based electric vehicle battery charger with power quality improvement," in IEEE Transactions on Transportation Electrification, vol. 5, no. 4, pp. 1087–1096, Dec. 2019, doi: 10.1109/TTE.2019.2952089

[11] Rapidtron Electronika Private Limited" Available [online]. http://rapidtron.co.in/3-wheeler-chargers/

[12] S. Gangavarapu, A. K. Rathore, and D. M. Fulwani, "Three-phase single-stage-isolated cuk-based PFC converter," in IEEE Transactions on Power Electronics, vol. 34, no. 2, pp. 1798–1808, Feb. 2019, doi: 10.1109/TPEL.2018.2829080

[13] Limits for Harmonic Current Emissions (Equipment Input Current <16A per Phase), IEC/EN61000-3-2, 1995.

[14] R. Kushwaha, and B. Singh, "A bridgeless isolated half bridge converter based EV charger with power factor pre-regulation," 2019 IEEE Transportation Electrification Conference (ITEC-India), Bengaluru, India, pp. 1–6, 2019, doi: 10.1109/ITEC-India48457.2019.ITECINDIA2019-190

[15] R. Kushwaha, and B. Singh, "Bridgeless isolated zeta-luo converter based EV charger with PF pre-regulation," 2019 International Conference on Computing, Power and Communication Technologies (GUCON), NCR New Delhi, India, pp. 959–964, 2019.

[16] R. Kushwaha, and B. Singh, "A power quality improved EV charger with bridgeless cuk converter," 2018 IEEE International Conference on Power Electronics, Drives and Energy Systems (PEDES), Chennai, India, pp. 1–6, 2018, doi: 10.1109/PEDES.2018.8707701

[17] R. Kushwaha, and B. Singh, "An EV battery charger based on PFC Sheppard Taylor Converter," 2016 National Power Systems Conference (NPSC), Bhubaneswar, pp. 1–6, 2016, doi: 10.1109/NPSC.2016.7858944

[18] R. Kushwaha, and B. Singh, "An improved battery charger for electric vehicle with high power factor," 2018 IEEE Industry Applications Society Annual Meeting (IAS), Portland, OR, pp. 1–8, 2018, doi: 10.1109/IAS.2018.8544585

[19] B. Singh, and R. Kushwaha, "A PFC based EV battery charger using a bridgeless isolated SEPIC converter," in IEEE Transactions on Industry Applications, vol. 56, no. 1, pp. 477–487, Jan.-Feb. 2020, doi: 10.1109/TIA.2019.2951510

[20] R. Kushwaha, and B. Singh, "An improved PFC bridgeless SEPIC converter for electric vehicle charger," 2020 IEEE International Conference on Power Electronics, Smart Grid and Renewable Energy (PESGRE2020), Cochin, India, pp. 1–6, 2020, doi: 10.1109/PESGRE45664.2020.9070375

[21] B. Singh, and R. Kushwaha, "An EV battery charger with power factor corrected bridgeless zeta converter topology," 2016 7th India International Conference on Power Electronics (IICPE), Patiala, pp. 1–6, 2016, doi: 10.1109/IICPE.2016.8079359

[22] R. Kushwaha, and B. Singh, "An improved battery charger for electric vehicle with high power factor," 2018 IEEE Industry Applications Society Annual Meeting (IAS), Portland, OR, pp. 1–8, 2018, doi: 10.1109/IAS.2018.8544585

[23] R. Kushwaha, and B. Singh, "An electric vehicle battery charger with interleaved PFC cuk converter," 2018 8th IEEE India International Conference on Power Electronics (IICPE), Jaipur, India, pp. 1–6, 2018, doi: 10.1109/IICPE.2018.8709494

[24] J. Gupta, R. Kushwaha, and B. Singh, "Improved power quality charger based on bridgeless canonical switching cell converter for a light electric vehicle," 2020 IEEE 9th Power India International Conference (PIICON), Sonepat, India, pp. 1–6, 2020, doi: 10.1109/PIICON49524.2020.9112905

[25] J. Gupta, R. Kushwaha, and B. Singh, "Improved power quality charger based on bridgeless canonical switching cell converter for a light electric vehicle," 2020 IEEE 9th Power India International Conference (PIICON), Sonepat, India, pp. 1–6, 2020, doi: 10.1109/PIICON49524.2020.9112905

[26] R. Kushwaha, and B. Singh, "Power factor improvement in modified bridgeless landsman converter fed EV battery charger," in IEEE Transactions on Vehicular Technology, vol. 68, no. 4, pp. 3325–3336, April 2019, doi: 10.1109/TVT.2019.2897118

[27] D. Patil, and V. Agarwal, "Compact onboard single-phase EV battery charger with novel low-frequency ripple compensator and optimum filter design," in IEEE Transactions on Vehicular Technology, vol. 65, no. 4, pp. 1948–1956, April 2016, doi: 10.1109/TVT.2015.2424927

[28] J. Lee, Y. Jeong, and B. Han, "An isolated DC/DC converter using high-frequency unregulated LLC resonant converter for fuel cell applications," in IEEE Transactions on Industrial Electronics, vol. 58, no. 7, pp. 2926–2934, July 2011, doi: 10.1109/TIE.2010.2076311

[29] K. Yoo, K. Kim, and J. Lee, "Single- and three-phase PHEV onboard battery charger using small link capacitor," in IEEE Transactions on Industrial Electronics, vol. 60, no. 8, pp. 3136–3144, Aug. 2013, doi: 10.1109/TIE.2012.2202361

[30] H. J. Chae, W. Y. Kim, S. Y. Yun, Y. S. Jeong, J. Y. Lee, and H. T. Moon, "3.3 kW on board charger for electric vehicle," 8th International Conference on Power Electronics - ECCE Asia, Jeju, pp. 2717–2719, 2011, doi: 10.1109/ICPE.2011.5944762

[31] D. S. Gautam, F. Musavi, M. Edington, W. Eberle, and W. G. Dunford, "An automotive onboard 3.3-kW battery charger for PHEV application," in IEEE Transactions on Vehicular Technology, vol. 61, no. 8, pp. 3466–3474, October 2012, doi: 10.1109/TVT.2012.2210259

[32] H. Wang, S. Dusmez, and A. Khaligh, "Design and analysis of a full-bridge LLC-based PEV charger optimized for wide battery voltage range," in IEEE Transactions

on Vehicular Technology, vol. 63, no. 4, pp. 1603–1613, May 2014, doi: 10.1109/TVT.2013.2288772.

[33] H. V. Nguyen, and D. Lee, "Advanced single-phase onboard chargers with small DC-link capacitors," 2018 IEEE International Power Electronics and Application Conference and Exposition (PEAC), Shenzhen, pp. 1–6, 2018, doi: 10.1109/PEAC.2018.8590400.

[34] F. Musavi, M. Craciun, D. S. Gautam, W. Eberle and W. G. Dunford, "An LLC resonant DC–DC converter for wide output voltage range battery charging applications," in IEEE Transactions on Power Electronics, vol. 28, no. 12, pp. 5437–5445, December 2013, doi: 10.1109/TPEL.2013.2241792.

[35] P. Kong, S. Wang, and F. C. Lee, "Common mode EMI noise suppression for bridgeless PFC converters," in IEEE Transactions on Power Electronics, vol. 23, no. 1, pp. 291–297, January 2008, doi: 10.1109/TPEL.2007.911877.

[36] U. Moriconi, "A bridgeless PFC configuration based on L4981 PFC controller," STMicroelectronics Application Note AN1606, 2002.

[37] W. Frank, M. Reddig, and M. Schlenk, "New control methods for rectifier-less PFC-stages," Proceedings of the IEEE International Symposium on Industrial Electronics, 2005. ISIE 2005. Dubrovnik, Croatia, vol. 2, pp. 489–493, 2005, doi: 10.1109/ISIE.2005.1528966.

[38] S. F. Lim, and A. M. Khambadkone, "A simple digital DCM control scheme for boost PFC operating in both CCM and DCM," in IEEE Transactions on Industry Applications, vol. 47, no. 4, pp. 1802–1812, July-August 2011, doi: 10.1109/TIA.2011.2153815.

[39] K. Masumoto, and M. Shoyama, "Comparative study about efficiency and switching noise of bridgeless PFC circuits," 2012 International Conference on Renewable Energy Research and Applications (ICRERA), Nagasaki, pp. 1–5, 2012, doi: 10.1109/ICRERA.2012.6477267

[40] A. V. J. S. Praneeth, and S. S. Williamson, "A review of front end AC-DC topologies in universal battery charger for electric transportation," 2018 IEEE Transportation Electrification Conference and Expo (ITEC), Long Beach, CA, pp. 293–298, 2018, doi: 10.1109/ITEC.2018.8450186.

[41] M. Ancuti, M. Svoboda, S. Musuroi, A. Hedes, N. Olarescu, and M. Wienmann, "Boost interleaved PFC versus bridgeless boost interleaved PFC converter performance/efficiency analysis," 2014 International Conference on Applied and Theoretical Electricity (ICATE), Craiova, pp. 1–6, 2014, doi: 10.1109/ICATE.2014.6972651.

[42] V. Nithin, P. S. Priya, R. Seyezhai, K. Vigneshwar, and N. S. Sumanth, "Performance evaluation of Bridgeless and Phase Shifted Semi Bridgeless Interleaved Boost Converters (IBCS) for power factor correction," IET Chennai Fourth International Conference on Sustainable Energy and Intelligent Systems (SEISCON 2013), Chennai, pp. 119–124, 2013, doi: 10.1049/ic.2013.0303.

[43] P. R. K. Chetty, "Current injected equivalent circuit approach to modeling switching DC-DC converters," in IEEE Transactions on Aerospace and Electronic Systems, vol. AES-17, no. 6, pp. 802–808, November 1981, doi: 10.1109/TAES.1981.309131.

[44] Z. Li, K. W. E. Cheng, and J. Hu, "Modelling of basic DC-DC converters," 2017 7th International Conference on Power Electronics Systems and Applications - Smart Mobility, Power Transfer & Security (PESA), Hong Kong, pp. 1–8, 2017, doi: 10.1109/PESA.2017.8277782.

[45] A. J., Sabzali, E. H., Ismail, M. A., Al-Saffar, andA. A., Fardoun, "New Bridgeless DCM Sepic and Cuk PFC Rectifiers With Low Conduction and Switching Losses," IEEE Transactions on Industry Applications, vol. 47, pp. 873–88110.1109/tia.2010.2102996.

4 A Comprehensive Overview of Wide Band-Gap (WBG)-Based DC-DC Converters for Fast Plug-in EV Charging

Nil Patel, Luiz A. C. Lopes, and
Akshay Kumar Rathore

CONTENTS

4.1 INTRODUCTION

The utilization of fossil fuels extensively has led to an increase in pollution, and global warming effects and not only ending with that but has also led to the depletion of fossil fuel resources hence in an attempt to find an alternative source of energy with extensive research toward the electric vehicles as a medium for transportation [1]. EVs have thrived in recent decades because of their benefits in energy efficiency, environmental protection, and acceleration capability compared to regular internal combustion engine-based (ICE) vehicles. Aside from the financial issue, the distance-range concern is a significant factor for pure EV consumers [2]. Economical, effective, and convenient charging is the cardinal factor for further usage of electric vehicles. An on-board charger (OBC) is commonly seen in EVs. When the charger is plugged into a suitable 120-V socket in the USA, the recharging capacity is restricted to Level-1 power, as stated by SAE standard J1772 [3]. According to the Level 1 profile, the maximum charging power is 1.08 kW. The output power of the OBC may be raised to Level-2 for relatively fast charging when there is a design for a 208-V/240-V outlet, with the permissible limit charging power up to 14.4 kW. The size, space, and weight limits of a vehicle reduce the power density and efficiency of an OBC. The OBC range for light-duty vehicles (LDVs) is less than 14.4 kW. The Nissan Leaf EV includes a 7.7-kW OBC [4]. Tesla Model X and Model 3 OBC has a power range of 7.7 to 11.7 kW [5]. An EV's distance range may be recovered at a pace of 30–60 km/h with a level 2 OBC, which means it will take around 12 hours to get a driving range comparable to that of an ICE-based automobile. There is a pressing demand for charging power in the tens to hundreds of kilowatts range to replace a significant driving range in minutes, alleviating distance-range concerns. To fulfill the need, rapid-fast dc charging equipment has been created, which can deliver a DC straight to the battery pack. DC charging is classified as Level 1 and Level 2, with their power levels defined by SAE J1772. The peak power level of DC Level-1 is 36 kW whereas DC Level 2 has a peak power of 90 kW, whereas DC Level 3 has a peak power of 240 kW. CHAdeMO, Tesla supercharger, and GB/T 20234.3 are some of the other competing dc fast-charging standards. Tesla established 145-kW battery charger in 2019 and recently it developed 250 kW supercharger.

The recent number of EVs and charging stations is depicted in Figure 4.1. Plug-in hybrid electric vehicles (PHEVs) and battery electric vehicles (BEVs) have increased dramatically. By 2025, it is expected that the number of electric vehicles will have risen to 60–85 million, followed by it will further rise to 145–240 million by 2030, which is more than 21 times its current level [6,7]. The ever-increasing number of electric vehicles has created a tremendous need for advancements in recharging technologies. The efficiency is critical regardless of whether the charging unit is on the ground or on the EV, wired or wireless. Even a minor enhancement in converter efficiency can save substantial energy for the enormous number of vehicles on the ground. Hence, higher conversion efficiency and power density in power electronics applications lead to decreased volume and price of passive elements and cooling equipment [8]. The price and power capability are crucial factors on the roadside, while on the EV side, power density, efficiency,

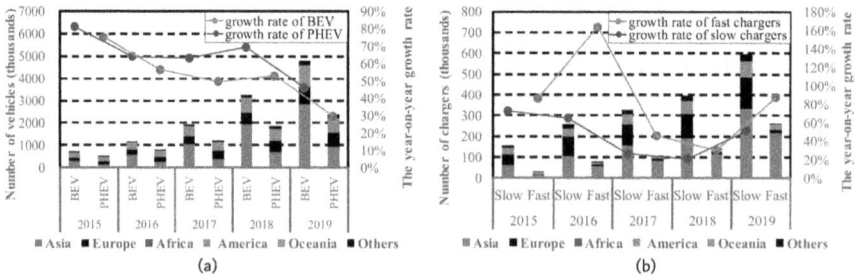

FIGURE 4.1 Development of EVs and charging technologies (a) PHEVs and BEV (b) Commercially chargers for public usage.

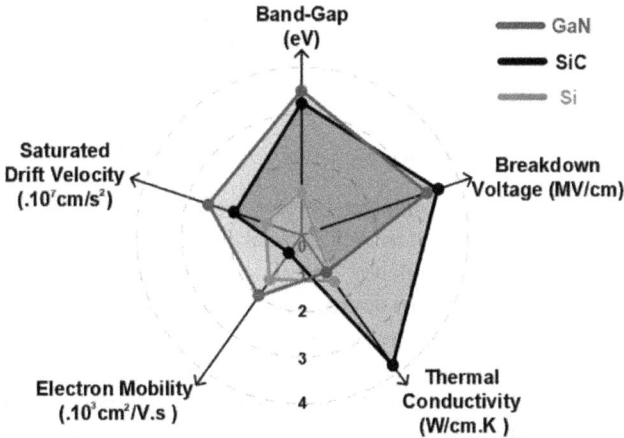

FIGURE 4.2 Comparison between Si, SiC, and GaN.

reliability, and price are paramount factors. The EV charging infrastructure and advanced technologies are continually changing to enhance efficiency, power capability, power density, and lowering prices. Power converters are fundamental to the aforesaid technologies, determining and influencing the progress of all other technologies. The power density and efficiency of EV recharging infrastructure are determined mainly by their operations and properties. After decades of progress, conventional silicon-based (Si) power devices have exceeded their limitations [9]. The WBG material, especially SiC and GaN, which has a more significant energy gap between the valence and conduction bands, is thought to be a new generation of materials that can significantly enhance the performance of power devices, as shown in Figure 4.2. The early research power device of SiC MOSFET and its capability has been beyond the hindrance of silicon devices [10,11]. WBG material-based devices have previously been used to produce a high-power-density battery charger with more than 98% efficiency [12]. WBG devices are being employed and launched in many research activities and large manufacturing.

In the recent decade, isolated dc-dc converters have received a lot of attention due to a reduction in the conduction losses, size, and cost by increasing the voltage gain and efficiency [13]. Most of the power converters can be unidirectional to transmit power to different applications such as UPS systems, advanced electric motor drives, on-board battery chargers, and renewable energy applications [14]. Nonetheless, bidirectional power conversion capabilities have become a requirement for fast-charging EVs [15]. One possibility is categorizing WBG DC-DC converter configurations into discrete and integrated stages. The usage of novel WBG designs and a topological overview of WBG semiconductor technology-based DC-DC converters for EV charging applications are discussed in the next section 4.2. High-frequency (HF) with high-efficiency-based SiC and GaN DC-DC converters for battery recharging of EVs are examined and analyzed. This chapter is organized as follows. Section 4.3 demonstrates the control strategies with tabular comparison incuding modeling approaches and stability issues. The critical assessment is presented in section 4.4. Section 4.5 discusses the commercialized WBG chargers which are available in the market. Future trends in EV industry are reported in section 4.6. Finally, the last section concludes this chapter.

4.2 REVIEW OF TOPOLOGIES

Wide band-gap (WBG) semiconductor devices are becoming more popular due to their unique characteristics, such as greater breakdown voltages, higher temperature handling capabilities, high-frequency operation, and higher switching transients sustain capability than silicon (Si) devices. The WBG-based DC-DC converters are categorized into SiC and GaN-based, as shown in Figure 4.3. In [16], B. Whitaker *et al.* introduced the SiC-based DC-DC converter with phase-shifted full-bridge (PSFB) circuit topology, as shown in Figure 4.4(a). This scheme has a high volumetric power density and gravimetric power density. However, Due to the forward voltage drop of the diodes, the diode bridge in this scheme is the lossiest in terms of conversion efficiency. Isolated Quasi-Switched Capacitor SiC-based DC-DC Converter is presented in [17] and shown in Figure 4.4(b). This converter decreases the voltage tension across the HV-side switches and the transformer voltage to two-third of the HV-dc-bus voltage, and also offer bidirectional power flow and easy control. However, the configuration can operate within the limited duty-ratio range and will suffer severe current stresses at the components level. GaN-based dual active bridge (DAB) DC-DC converter is exhibited in Figure 4.4(c), and it is good in performance and simple in circuit formation [18]. Nonetheless, to achieve a high instantaneous power balance between the mains and the batteries, the sinusoidal charging of the batteries needs a control strategy with bandwidth as high as possible, which reduces the reliability of the charger. In [19], Huang *et al.* proposed a GaN-based interleaved bidirectional buck-boost converter

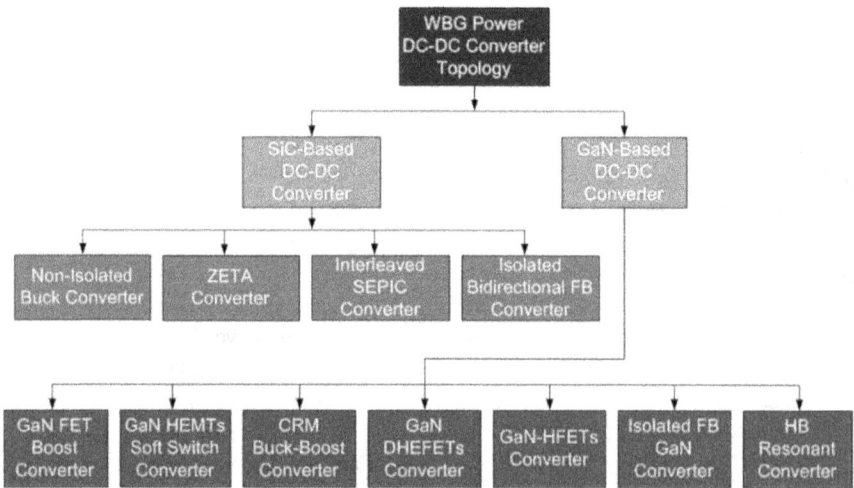

FIGURE 4.3   Generalized classification of WBG-based DC-DC converters.

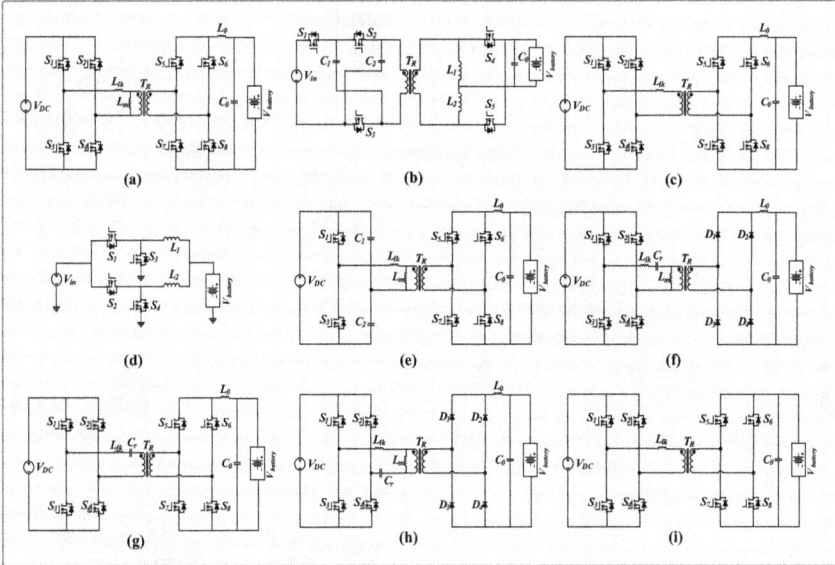

FIGURE 4.4 Configurations of WBG-based DC-DC converters: (a) Phase-shifted full-bridge isolated converter [16], (b) Isolated quasi-switched capacitor converter [17], (C) Dual active bridge converter [18], (d) Interleaved bi-directional buck-boost converter [19], (e) Bi-directional isolated LLC resonant converter [20], (f) Full-bridge LLC resonant converter [21], (g) Series resonant converter [22] (h) Full-bridge LLC resonant converter [23], and (i) Dual active bridge (DAB) converter [24].

with 98.5% efficiency as shown in Figure 4.4(d). In this configuration for an inverse coupled inductor, the resonant time in CRM is less than with a non-coupled inductor, which is advantageous for high-frequency operation, and both the soft-switching range and the circulating energy are increased utilizing an inverse coupled inductor in CRM. The demerit of this topology is the negative charge of the inductor current in the CRM operation, which does not move power to the output, is constant no matter its CRM operation, which does not move power to the output, is constant no matter its switching frequency. In [20], Half-bridge CLCL resonant converter is discussed with power-loss analysis as shown in Figure 4.4(e). The steady-state analysis of the bidirectional LLC resonant converter is processed and provides more detailed results regarding power loss and efficiency than others with this configuration. However, the design is complex and tricky. In Figure 4.4(f), the full-bridge (FB) LLC resonant-based converter achieves 97% efficiency and is explained in detail in [21]. This topology adopts the modulation of symmetric space vector pulse width, which, compared with other algorithms, essentially reduces the electrical stress of switches. However, in the complex resonant converter design, the transformer size is bulky, and the gate drive circuit loss is much higher, leading to lower power conversion efficiency. In [22], GaN-based series resonant topology, as shown in Figure 4.4(g), has more than 98% efficiency with higher reliability. This topology provides enhanced efficiency for series-resonant converters operating with an essential input and output voltage spectrum by significantly minimizing their switching frequency range. Nonetheless, the system is much more complex, and the cost is higher than other converters. The FB-LLC-based SiC converter is showcased in Figure 4.4(h) with a simple mode of operation [23]. The efficiency is higher, and the control algorithm is simplified. Power loss is slightly higher than other converters, and the design is tricky and complex. GaN-based DAB converter as shown in Figure 4.4(i) with effective modulation technique, is explained in [24]. Closed-loop control was carried out for the converter to decrease the output current ripple and align the power between three single-phase modules, which increases the power factor and improves THD. However, the transformer current is circulating, leading to conduction losses and efficiency will decrease.

In [25], Li *et al.* proposed a GaN-based CLLC resonant converter as shown in Figure 4.5(a) which has a higher operating temperature range. The converter has a high volumetric power density and efficiency as it adopts a two-stage control strategy. Due to the restriction of the helical structure and insensitive ac winding loss, magnetic integration is more difficult for PCB winding transformers. GaN-based DAB converter, as shown in Figure 4.5(b), is explained in [26]. The converter is much smaller, cheaper, and more efficient with improved thermal performance. The closed-loop control and design of parasitic components are tricky. In [27], SiC-based bidirectional LLC resonant converter is discussed, which is shown in Figure 4.5(c). The converter is under pulse frequency modulation power, and instead of a dc transformer with a set switching frequency, it has to remove the double-line frequency ripple in a low switching frequency range. However, the

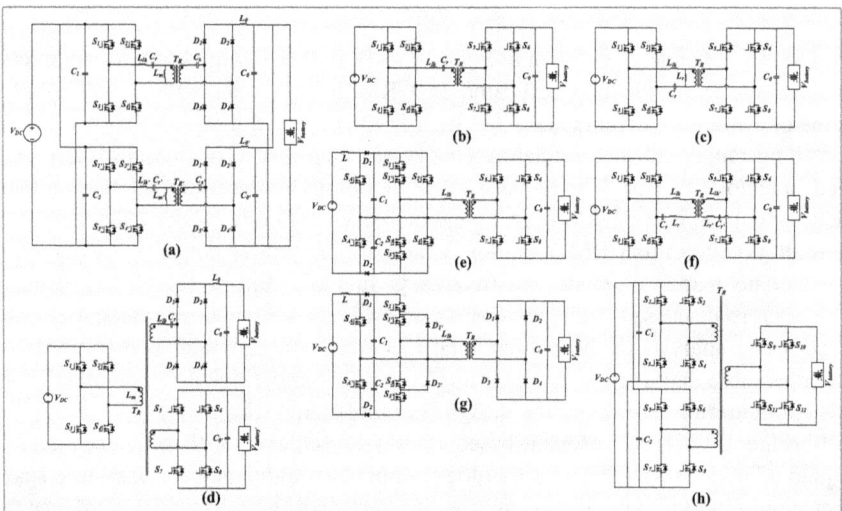

FIGURE 4.5 Configurations of WBG-based DC-DC converters: (a) CLLC resonant converter [25], (b) Dual active bridge (DAB) converter [26], (C) Bi-directional LLC resonant converter [27], (d) Isolated three-port converter [28], (e) Dual active bridge (DAB) converter [29], (f) Isolated bi-directional converter [30], (g) Multi-cell boost (MCB) with input-series output-parellel (ISOP) connected converter [31], and (h) Isolated three-port converter [32].

converter design is complex when the voltage gain is in a lower switching frequency range. Figure 4.5(d) represents the isolated three-port converter with a higher operational thermal and temperature range [28]. The dynamic performance and efficiency are much higher than other converters. At lower output voltages due to higher current operation, the efficiency deteriorates. SiC-based DAB converter as shown in Figure 4.5(e) is presented in [29]. This topology gives a considerable decrease in device footprint, which achieves higher efficiency. Nonetheless, this scheme increases the size of the converter and, due to the increased part count, significantly affects the reliability of fast chargers. A GaN-based isolated bidirectional converter is illustrated in Figure 4.5(f) [30]. It provides reliable high performance, with higher efficiency and rated capacity over a broad battery voltage range. Operation and design are complex, with limited information on the thermal insulation of this topology. In [31], Multi-Cell Boost (MCB) with Input-Series Output-Parellel (ISOP) Connected converter as shown in Figure 4.5(g) is discussed. This topology reduced the switching loss and magnetic loss with higher reliability. However, the transformer leakage inductance needs to be thoroughly balanced to ensure a soft switching of the switches for the desired load current set. In [32], Su *et al.* proposed a GaN-based isolated three-port converter, as shown in Figure 4.5(h). It has higher reliability and low cooling requirements. This converter is very compact, lightweight, and highly efficient. The improvement in the switching commutation loop, which increases the parasitic inductance in the loop and increases voltage ringing through the switches, is one potential disadvantage of putting the flying capacitors on the top side. Table 4.1 demonstrates the comparison

TABLE 4.1
Comparison Between SiC and GaN-Based DC-DC Converters

Parameters	GaN-based converter topologies	SiC-based converter topologies
Power loss	High	Low
Converter Size	Small	Medium
Switching Frequency	High	Low
Power Density	High	Medium
Operating Temperature	High	High
Switching Speed	High	Low
Voltage Stress	Low	Medium
Control Mechanism	Complex	Easy
Voltage-drop on devices	Low	Medium
Filter Component	Light	Reasonable
Discharge	High	Low
Control Variable	High	High
Necessity switching device	Low	Less
Complexity	High	Less
Efficiency	High	Low

analysis including size, power density, voltage stress, power loss, switching frequency, switching speed, control variables, efficiency, control complexity, and operating temperature for SiC and GaN-based DC-DC converters for EV chargers.

4.3 CONTROL STRATEGIES

B. Whitaker et al. introduced two converters in the cascaded form with a closed-loop system, but a detailed control system analysis is not reported [16]. In [17], the proposed control system can be implemented at a low cost. L. Xue et al. proposed closed-loop control is to directly control the dc-link voltage-ripple to compensate for the imbalance completely and at two times the line frequency, frequency while attaining zero steady-state error. However, closed-loop control is complex due to the more state variables in [18], as shown in Figure 4.6(a). In [19], a closed-loop control system is not revealed. In [20], the converter is operated in either the continuous current mode (CCM) or the discontinuous current mode (DCM),

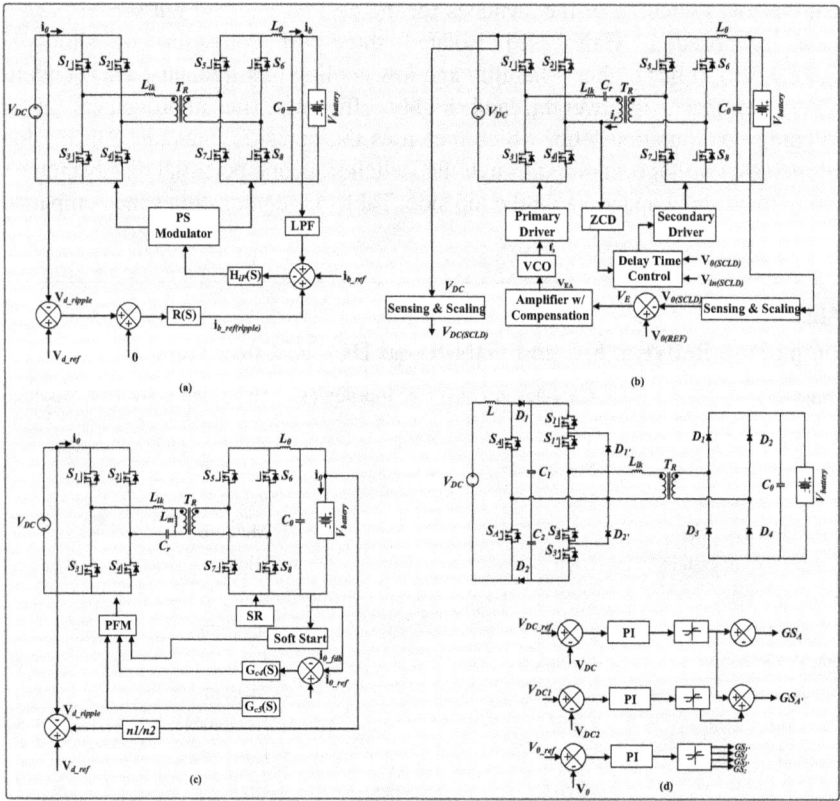

FIGURE 4.6 Control diagrams of WBG-based DC-DC converters: (a) Control diagram of DAB converter [18], (b) Control diagram of Series resonant converter [22], (c) Control diagram of bidirectional LLC converter [27], and (d) Control diagram of multi-cell boost (MCB) with input-series output-parallel (ISOP) connected converter [31].

depending on the load condition, the closed-loop control is much more effective with an easily calculated transfer function, and the stability issue is not present. In [21], the converter has an easy closed-loop control algorithm. Nevertheless, the system is not stable. G. Liu *et al.* proposed a control system in which the reduction within the switching frequency spectrum is accomplished by managing the output voltage with a combination of closed-loop frequency control of front-end switches and secondary-side turn delay-time monitoring. However, the system's power is complex because the number of PI loops is more in [22], as shown in Figure 4.6(b). The proposed converter is worked in continuous conduction mode (CCM), and the detailed small-signal modeling and transfer function are presented with the outer voltage loop and inner current loop in [23]. Besides, the closed-loop system's transfer function is very simple and the system is much more stable with two PI digital controllers used to stabilize the voltage and current control loop. In [24], the proposed closed-loop is mitigated the output current ripple while the grid is imbalanced. However, the system is complex because of the number of state variables and no stability issues. The configuration of control is simple, but the concept is novel. The dc-link voltage follows the battery voltage so that the gain of the stage stays unity, whereas the current loop of the dc-dc ensures that the no-second-order line frequency ripple is going through the filter from the dc-link to the battery [25]. The system is not stable and convincing, but the details analysis is not presented in [26].

In [27], the proposed control scheme is in stable condition and easy to implement, which is shown in Figure 4.6(c). Also, two PI control loop is present, making it a small and simple mathematical model matrix. In [28], the novel control strategy is used, and it is much helpful due to the simple transfer function of the closed-loop system, including voltage and current. Table 4.2 depicts the

TABLE 4.2
Comparative Analysis of Dontrol Strategies for WBG-based DC-DC Converters

Attributes	State variables	No. of loops	Stability level	Sensor count	Control complexity and modeling approach	PI controllers
[18]	High	3	High	3	Complex	2
[20]	Low	2	Low	2	Simple	2
[21]	Moderate	2	High	2	Effective	2
[22]	High	4	Low	2	Complex	3
[23]	Moderate	2	Moderate	4	Simple	2
[24]	Moderate	2	Moderate	1	Effective	2
[25]	Low	2	High	2	Easy	2
[27]	Low	2	High	2	Easy	2
[28]	Moderate	3	Low	3	Simple	1
[29]	High	3	Moderate	3	Complex	3
[31]	High	3	Low	3	Complex	3

comparative evaluation of control techniques for WBG-based DC-DC converters. The triple-phase shift control method is used to get voltage control with soft-switching, but the control scheme is very complex [29]. As shown in Figure 4.6(d), the PI controller has three control loops, and the control strategy is very complex [31]. In [32], the effectuation and design of the proposed control system are much more tricky and complex. The design and analysis of a closed-loop control system are tricky and challenging to implement [33]. In [34], the average current control technique is used, but the control complexity is higher due to a higher number of variables. Besides, this approach alone does not eradicate sensor noise, which may be a concern at low currents. The pragmatic approach to closed-loop is much more complex and costly [35]. In [36], the higher junction capacitance of diodes may damage the voltage gain of LLC converters in a higher frequency range, making the control system unstable. The closed-loop system is stable compared to the phase shift control mode and variable frequency control mode in [37]. The control system is bulky and very complex to analyze [38]. In [39], the cost of the prototype closed-loop control system is much higher than that of the conventional one.

4.4 CRITICAL ASSESSMENT

B. Whitaker *et al.* proposed a new DC-DC converter with higher efficiency and volumetric power density than the 2010 Toyota Prius Plug-in Hybrid battery charger [16]. In [17], the quasi-switched-capacitor dc-dc converter is introduced and has a tremendous advantage, like voltage stress is decreased by around one-third of the input voltage and transformer turns ratio is decreased. However, the converter is operated within its limited duty ratio range. In [18], the converter efficiency optimization can be possible through a novel developed GaN-based dc-dc converter via a small dc-link voltage ripple, but the closed-loop control is tricky. The critical current mode (CRM) method is used, and it is a simple and effective route to acquire the ZVS in the interleaved buck-boost converter, but it introduces a large ripple current [19]. The LLC DC-DC resonant converter is proposed, and a double pulse test (DPT) platform is used to give an accurate power-loss model [20]. Nevertheless, it is not examined thoroughly for bi-directional power flow. Table 4.3 presents the comparison of the State of Art Wide Band-Gap (WBG)-Based DC-DC Converters for Fast Plug-in EV Charging. In [21], the DC/DC converter requires to operate at a high switching frequency so that the iron loss of the transformer and switching loss is very high. G. Liu *et al.* proposed a control method that provides a better efficiency of series resonant converters that work with a wide input and output voltage range to minimize the switching frequency range [22]. In [23], the converter has the best-implemented control algorithm than others. The DC-DC converter achieved the utmost efficiency with compact size and reduced heat-sink requirement. In [24], the novel triple-phase shift control is employed to enhance the power factor and mitigate the output current ripple. The CLLC resonant dc/dc converter operates around the resonant frequency and removes the secondary order line-frequency ripple, but the intricate resonant tank design [25]. The DC-DC converter is cheaper and lighter, but the thermal performance is not better than others [26]. The proposed dc/dc converter has a compact size, more efficiency, and power density than wolfspeed converter [27]. Three-port dc/dc converter is one of the most reliable with lighter weight and more robust as per performance-wise, and it requires fewer switches and gate drivers [28,32]. In [29], the number of semiconductor devices and passive components is higher with difficult control complexity in the SiC-based DC-DC converter. F. Qi *et al.* proposed the converter, which gives consistently high efficiency at rated power over a wide battery voltage range, making the dual active bridge topology more capable [30]. J. Guo *et al.* proposed the optimization method for inductor design and maximized the efficiency of the converter, but the control complexity is higher among the others [34]. In [35], the introduced converter is safe, arc-free, and isolated in dynamic content with a novel solution for high-power electric vehicle fast charging. The proposed synchronous rectification technique is achieved to get reliable operation and higher efficiency at high frequency [36]. Compared to the converters, the SiC base DC-DC LLC resonant converter has a better and simple control system [37]. The converter has better power quality and less footprint size-wise [38].

4.5 COMMERCIALIZED CHARGERS

Many manufacturers are composed of WBG-based commercial battery chargers for commercial in the automotive market. Table 4.4 represents the WBG-based commercialized chargers in the market for fast EV charging. Wolfspeed developed a 22 kW SiC-based battery charger with more than 98% efficiency and a 140–250 kHz switching frequency range [40]. This topology has a flexible control scheme that enables high efficiency with bidirectional operation capability. France-based industry Brightloop Converters provides a GaN-based DC-DC-LP model with a higher operating temperature range [41]. SiC-based 22 kW OBC offered by Fraunhofer has a higher power density with a high input and output voltage range [42]. VisIC Technologies gives 6.6 kW OBC, simplifying the cooling systems, reducing the charging, and restricting size and cost [43]. Hella developed a GaN-based 22 kW fast EV charger with higher reliability and decreased vehicle weight, giving more excellent driving range and opening up new diversity in structure [44]. Continental-composed GaN-based OBC, which has a higher frequency operating range, leads to compact size, great air-cooled, and design flexibility for OBC incorporation [45]. US-based Canoo provides 7.2 kW OBC and has lower switching loss with efficient thermal management, leading to higher efficiency, higher power density, and lower weight [46]. On Semiconductor offers SiC-based 10 kW TND6318/D, which operates at 400 kHz switching frequency with controllable power components temperature [47]. ABB Terra54 gives a 50 kW SiC-based fast charger with higher efficiency and lower acoustic noise emissions [48]. Ingeteam RAPID50 developed a SiC-based battery charger that has dynamic power management [49]. SiC-based 50 kW Tritium chargers produced VEEFIL-RT model 50-kW fast chargers with higher efficiency, durability, and reliability.

TABLE 4.3

A Comparison of the State of Art Wide Band-Gap (WBG)-Based DC-DC Converters for Fast Plug-in EV Charging

Topology	Rated power (P_0), kW	Output voltage (V_0), V	Switching frequency (f_{sw}), kHz	No. of devices (Active switches/diodes/transformers/inductors/capacitors)	Control complexity	Cooling requirements	Volumetric power density, $\frac{kw}{dm^3}$	Efficiency at P_0 (%)
Ref. [16]	6.1	400	500	SiC/4/4/1/3/5	---	Low	5	96
Ref. [17]	1.016	300	500	SiC/5/0/1/2/4	---	Low	0.973	96
Ref. [18]	1	250	500	GaN/8/0/1/1/2	Complex	Moderate	---	95
Ref. [19]	1.2	150	1000	GaN/4/0/0/2/4	---	Low	---	98.5
Ref. [20]	10	400	120	SiC/6/2/1/0/1/3	Simple	High	---	96
Ref. [21]	10	350	185	SiC/4/4/1/2/2	Effective	Low	---	97
Ref. [22]	3.3	430	187	GaN/8/0/1/1/2	Complex	Moderate	0.97	98.1
Ref. [23]	1	420	200	SiC/12/12/3/6/6	Simple	Low	---	97.1
Ref. [24]	7.2	450	500	GaN/8/0/11/1/1	Effective	Moderate	3.3	>97
Ref. [25]	6.6	450	500	GaN/12/0/3/4/5	Easy	High	2.26	>96
Ref. [26]	7.2	400	500	GaN/8/0/1/1/2	---	Moderate	3.3	>97
Ref. [27]	6.6	350	300	SiC/8/0/1/2/2	Easy	High	1.93	96
Ref. [28]	5	400	220	SiC/8/4/1/2/4	Simple	High	2.74	98.2
Ref. [29]	350	850.6	25	SiC/12/0/1/3/4	Complex	High	1.6	98.1
Ref. [30]	3.3	300	100	GaN/8/0/1/4/4	---	Low	28.30	>97
Ref. [31]	50	400	25	SiC/6/8/1/2/2	Complex	High	0.81	96.6
Ref. [32]	6.6	400	100	GaN/8/0/1/0/2	---	High	10.5	96.6

TABLE 4.4
Wide Band Gap (WBG)-Based Commercialized Chargers in the Market for Fast EV Charging

Manufacturer	Semiconductor device	Output power (kW)	Weight (kg)	Power density $\left(\frac{kW}{dm^3}\right)$	Operating temperature range (°C)	Switching frequency range (kHz)	Efficiency (%)
Wolfspeed [40]	SiC	22	2.75	8	---	140–250	>98.5%
Brightloop Converters [41]	GaN	9.6	3.5	2.74	Up to 80	---	>94%
Fraunhofer [42]	SiC	22	2.8	13.8	Up to 85	100	>97%
VisIC Technologies [43]	GaN	6.6	4.5	3	---	---	>96%
Hella [44]	GaN	22	---	4	---	---	>96%
Continental [45]	GaN	3	---	---	---	900	>94%
Canoo [46]	GaN	7.2	---	---	---	---	>96%
ON Semiconductor [47]	SiC	10	---	---	---	400	>95%
ABB [48]	SiC	50	---	---	Up to 55	---	>95.5%
Ingeteam [49]	SiC	50	---	---	Up to 60	---	>94%
Tritium [50]	SiC	50	3	---	Up to 50	---	>92%

4.6 FUTURE TRENDS

This section covers futuristic research approaches in WBG-based power converter technology, soft-switching methods, and advanced control techniques for EVs. The subsequent sections discuss the outline some of the significant research developments.

4.6.1 POWER CONVERTER TOPOLOGY

Most plug-in EV chargers are projected to maintain two-stage structures within the next decade. More development is planned on resonant DAB, LLC, and CLLC converters for the DC-DC stage because of their simple constructions, control strategy, compact size, lower conduction losses, and higher power density. A three-phase system is necessary when the power level demand rises, and modular single-phase converters are a viable alternative for achieving 22 kW OBC. Because of its superior efficiency, a three-phase full-bridge DC-DC is also a good choice for bidirectional battery charging applications, particularly the three-level T-type converter, despite its higher device count and control complexity. Furthermore, bidirectional DC-DC converters would be chosen at higher power levels because they can attain high efficiency, high power density, and reduction in size, although posing specific hardware design and efficiency problems [51]. An enhanced control system and better devices might help to enhance the power conversion efficiency and dependability.

4.6.2 SOFT-SWITCHING TECHNIQUES FOR WBG-BASED POWER CONVERTERS

For EV applications, novel soft-switching approaches are being developed using unique power circuit topologies and digital control methodologies. The key

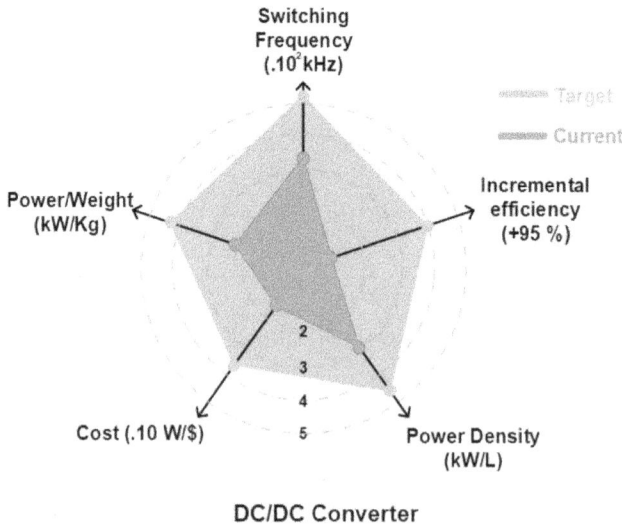

FIGURE 4.7 Generalized classification of WBG-based DC-DC converters.

performance indicators for the DC-DC converter are shown in Figure 4.7. This graph depicts the present DC-DC power converter's performance measurements and the performance goals that academics are aiming for in the near future within the next ten years. Therefore, soft-switching techniques play a cardinal role in reducing conduction losses. The goal of novel soft-switching approaches for EV power converters is to accomplish soft switching with minimum changes in power circuitry and to employ more complex modulation and control systems [52]. The study will primarily focus on employing easy and simple modulation and control techniques to improve the performance of power circuits utilizing power configuration.

4.6.3 ADVANCED CONTROL TECHNIQUES

Advanced control techniques can be applied to WBG-based DC-DC converters such as Model Predictive Control (MPC) which has a higher optimized performance level and precise tracking prediction with fast dynamics [53]. The backstepping control method has a higher signal and control stability [54]. Passivity-based control technique has more play and plug stability with accurate estimation without chattering problems [55]. The sliding mode control technique is simple in terms of implementation and higher robustness [56]. The observer technique has a precise tracking system against uncertainties and disturbances [57]. The intelligent control system has fast dynamic control and is less complicated in any system [58]. Artificial Intelligence (AI) control-based technique has higher accuracy and improves system performance [59]. Deep Learning (DL) and Machine Learning (ML)-based control methods reduce the control complexity [60].

Distance-range concerns, long-duration charging times, and significant financial pressure from off-board chargers will all be addressed in the future by increasing the size of the EV battery. The automotive industry's next step is to build a fast charger that enables high efficiency, power levels, and power density.

4.7 CONCLUSION

The WBG-based DC-DC converter topologies are proliferated in automotive and industrial usage, especially in EV charging applications. This chapter reviews WBG-based DC-DC converter topologies, control methodologies, and critical evaluation. WBG semiconductor devices are gaining popularity owing to their distinct advantages over silicon (Si) devices, such as larger breakdown voltages, greater temperature handling capabilities, high-frequency operation, and higher switching transients sustainability. SiC and GaN power devices also have greater switching frequencies, resulting in reduced switching losses, higher efficiency, cheaper costs, smaller size, and higher power density. As a result, WBG-based power converters might be a viable option for the future generation of rapid EV charging applications. All chargers are discussed for their advantages and disadvantages, and a thorough examination and comparison of commercially available chargers are depicted in this chapter. It can be concluded that advancement in WBG chargers is a promising solution for achieving higher power density, lower conduction losses, compact size, higher efficiency, and less cost.

REFERENCES

[1] M. Yilmaz and P. T. Krein, "Review of battery charger topologies, charging power levels, and infrastructure for plug-in electric and hybrid vehicles," *IEEE Trans. Power Electron.*, vol. 28, no. 5, pp. 2151–2169, 2013, doi: 10.1109/TPEL.2012. 2212917

[2] T. Franke and J. F. Krems, "What drives range preferences in electric vehicle users? Transport policy," vol. 30, p. 56, 2013, doi: 10.1016/j.tranpol.2013.07.005

[3] S. S. Williamson, A. K. Rathore and F. Musavi, "Industrial electronics for electric transportation: Current state-of-the-art and future challenges," *IEEE Trans. Ind. Electron.*, vol. 62, no. 5, pp. 3021–3032, May 2015, doi: 10.1109/TIE.2015.24 09052

[4] "Range & Charging & Saving | Nissan LEAF – Electric Family Car | Nissan." https://www.nissan.co.uk/vehicles/new-vehicles/leaf/range-charging.html (accessed May 05, 2022).

[5] "Tesla Model X and Model 3." https://www.tesla.com/compare (accessed May 05, 2022).

[6] "Global EV Outlook 2022 - Event — IEA." https://www.iea.org/events/global-ev-outlook-2022 (accessed May 05, 2022).

[7] S. Li, S. Lu and C. C. Mi, "Revolution of electric vehicle charging technologies accelerated by wide bandgap devices," *Proc. IEEE*, vol. 109, no. 6, pp. 985–1003, Jun. 2021, doi: 10.1109/JPROC.2021.3071977

[8] G. J. Su, "Comparison of Si, SiC, and GaN based isolation converters for onboard charger applications," *2018 IEEE Energy Convers. Congr. Expo. ECCE 2018*, pp. 1233–1239, Dec. 2018, doi: 10.1109/ECCE.2018.8558063

[9] T. V. Do, K. Li, J. P. Trovao and L. Boulon, "Reviewing of using wide-bandgap power semiconductor devices in electric vehicle systems: From component to system," *2020 IEEE Veh. Power Propuls. Conf. VPPC 2020 –Proc.*, Nov. 2020, doi: 10.1109/VPPC49601.2020.9330854

[10] M. Hoshi, "Electric vehicles and expectations for wide bandgap power devices," *Proc. Int. Symp. Power Semicond. Devices ICs*, vol. 2016, July, pp. 5–8, Jul. 2016, doi: 10.1109/ISPSD.2016.7520765

[11] Y. Berube, A. Ghazanfari, H. F. Blanchette, C. Perreault and K. Zaghib, "Recent advances in wide bandgap devices for automotive industry," *IECON Proc. (Industrial Electron. Conf.*, vol. 2020, October, pp. 2557–2564, Oct. 2020, doi: 10.1109/IECON43393.2020.9254478

[12] Y. Zhang *et al.*, "Development of a WBG-based transformerless electric vehicle charger with semiconductor isolation," *2018 IEEE 4th South. Power Electron. Conf. SPEC 2018*, Feb. 2019, doi: 10.1109/SPEC.2018.8636011

[13] S. Bhattacharya, "Wide-band Gap (WBG) WBG devices enabled MV power converters for utility applications - Opportunities and challenges," *2nd IEEE Work. Wide Bandgap Power Devices Appl. WiPDA 2014*, Nov. 2014, doi: 10.1109/ WIPDA.2014.6964611

[14] M. Parvez, A. T. Pereira, N. Ertugrul, N. H. E. Weste, D. Abbott and S. F. Al-Sarawi, "Wide bandgap DC-DC converter topologies for power applications," *Proc. IEEE*, vol. 109, no. 7, pp. 1253–1275, Jul. 2021, doi: 10.1109/JPROC.2021. 3072170

[15] X. Chen, Z. Li, H. Dong, Z. Hu, and C. Chris Mi, "Enabling extreme fast charging technology for electric vehicles," *IEEE Trans. Intell. Transp. Syst.*, vol. 22, no. 1, pp. 466–470, Jan. 2021, doi: 10.1109/TITS.2020.3045241

[16] B. Whitaker *et al.*, "A high-density, high-efficiency, isolated on-board vehicle battery charger utilizing silicon carbide power devices," *IEEE Trans. Power Electron.*, vol. 29, no. 5, pp. 2606–2617, May 2014, doi: 10.1109/TPEL.2013.2279950

[17] X. Zhang, C. Yao, C. Li, L. Fu, F. Guo and J. Wang, "A wide bandgap device-based isolated quasi-switched-capacitor DC/DC converter," *IEEE Trans. Power Electron.*, vol. 29, no. 5, pp. 2500–2510, May 2014, doi: 10.1109/TPEL.2013.2287501

[18] L. Xue, Z. Shen, D. Boroyevich, P. Mattavelli and D. Diaz, "Dual active bridge-based battery charger for plug-in hybrid electric vehicle with charging current containing low frequency ripple," *IEEE Trans. Power Electron.*, vol. 30, no. 12, pp. 7299–7307, Dec. 2015, doi: 10.1109/TPEL.2015.2413815

[19] X. Huang, F. C. Lee, Q. Li and W. Du, "High-frequency high-efficiency GaN-based interleaved CRM bidirectional buck/boost converter with inverse coupled inductor," *IEEE Trans. Power Electron.*, vol. 31, no. 6, pp. 4343–4352, Jun. 2016, doi: 10.1109/TPEL.2015.2476482

[20] X. Wang, C. Jiang, B. Lei, H. Teng, H. K. Bai and J. L. Kirtley, "Power-loss analysis and efficiency maximization of a silicon-carbide MOSFET-based three-phase 10-kW bidirectional EV charger using variable-DC-bus control," *IEEE J. Emerg. Sel. Top. Power Electron.*, vol. 4, no. 3, pp. 880–892, Sep. 2016, doi: 10.1109/JESTPE.2016.2575921

[21] H. Bai *et al.*, "Design of an 11 kW power factor correction and 10 kW ZVS DC/DC converter for a high-efficiency battery charger in electric vehicles," *IET Power Electron.*, vol. 5, no. 9, pp. 1714–1722, 2012, doi: 10.1049/iet-pel.2012.0261

[22] G. Liu, Y. Jang, M. M. Jovanović and J. Q. Zhang, "Implementation of a 3.3-kW DC-DC converter for EV on-board charger employing the series-resonant converter with reduced-frequency-range control," *IEEE Trans. Power Electron.*, vol. 32, no. 6, pp. 4168–4184, Jun. 2017, doi: 10.1109/TPEL.2016.2598173

[23] C. Shi, H. Wang, S. Dusmez, and A. Khaligh, "A SiC-based high-efficiency isolated onboard PEV charger with ultrawide DC-link voltage range," *IEEE Trans. Ind. Appl.*, vol. 53, no. 1, pp. 501–511, Jan. 2017, doi: 10.1109/TIA.2016.2605063

[24] J. Lu *et al.*, "A modular-designed three-phase high-efficiency high-power-density EV battery charger using dual/triple-phase-shift control," *IEEE Trans. Power Electron.*, vol. 33, no. 9, pp. 8091–8100, Sep. 2018, doi: 10.1109/TPEL.2017.2769661

[25] B. Li, Q. Li, F. C. Lee, Z. Liu and Y. Yang, "A high-efficiency high-density wide-bandgap device-based bidirectional on-board charger," *IEEE J. Emerg. Sel. Top. Power Electron.*, vol. 6, no. 3, pp. 1627–1636, Sep. 2018, doi: 10.1109/JESTPE.2018.2845846

[26] A. Taylor, J. Lu, L. Zhu, K. Bai, M. McAmmond, and A. Brown, "Comparison of SiC MOSFET-based and GaN HEMT-based high-efficiency high-powerdensity 7.2 kW EV battery chargers," *IET Power Electron.*, vol. 11, no. 11, pp. 1–9, Sep. 2018, doi: 10.1049/iet-pel.2017.0467

[27] H. Li, Z. Zhang, S. Wang, J. Tang, X. Ren and Q. Chen, "A 300-kHz 6.6-kW SiC bidirectional LLC onboard charger," *IEEE Trans. Ind. Electron.*, vol. 67, no. 2, pp. 1435–1445, Feb. 2020, doi: 10.1109/TIE.2019.2910048

[28] N. D. Dao, D. C. Lee and Q. D. Phan, "High-efficiency SiC-based isolated three-port DC/DC converters for hybrid charging stations," *IEEE Trans. Power Electron.*, vol. 35, no. 10, pp. 10455–10465, Oct. 2020, doi: 10.1109/TPEL.2020.2975124

[29] X. Liang, S. Srdic, J. Won, E. Aponte, K. Booth and S. Lukic, "A 12.47 kV medium voltage input 350 kW EV fast charger using 10 kV SiC MOSFET," in *Conference Proceedings – IEEE Applied Power Electronics Conference and Exposition –APEC*, vol. 2019, March, pp. 581–587, May 2019, doi: 10.1109/APEC.2019.8722239

[30] F. Qi, Z. Wang and Y. Wu, "650V GaN based 3.3kW Bi-directional DC-DC converter for high efficiency battery charger with wide battery voltage range," in *Conference Proceedings – IEEE Applied Power Electronics Conference and Exposition –APEC*, vol. 2019, March, pp. 359–364, May 2019, doi: 10.1109/APEC.2019.8721948

[31] S. Srdic, C. Zhang, X. Liang, W. Yu and S. Lukic, "A SiC-based power converter module for medium-voltage fast charger for plug-in electric vehicles," in *Conference Proceedings – IEEE Applied Power Electronics Conference and Exposition –APEC*, vol. 2016, May, pp. 2714–2719, May 2016, doi: 10.1109/APEC.2016.7468247

[32] G. J. Su, C. White and Z. Liang, "Design and evaluation of a 6.6 kW GaN converter for onboard charger applications," in *2017 IEEE 18th Workshop on Control and Modeling for Power Electronics, COMPEL 2017*, Aug. 2017, doi: 10.1109/COMPEL.2017.8013335

[33] Y. Guan, Y. Wang, D. Xu and W. Wang, "A 1 MHz half-bridge resonant DC/DC converter based on GaN FETs and planar magnetics," *IEEE Trans. Power Electron.*, vol. 32, no. 4, pp. 2876–2891, Apr. 2017, doi: 10.1109/TPEL.2016.2579660

[34] J. Guo *et al.*, "A comprehensive analysis for high-power density, high-efficiency 60 kW interleaved boost converter design for electrified powertrains," *IEEE Trans. Veh. Technol.*, vol. 69, no. 7, pp. 7131–7145, Jul. 2020, doi: 10.1109/TVT.2020.2991395

[35] R. B. Beddingfield, S. Samanta, M. S. Nations, I. Wong, P. R. Ohodnicki and S. Bhattacharya, "Analysis and design considerations of a contactless magnetic plug for charging electric vehicles directly from the medium-voltage DC grid with Arc flash mitigation," *IEEE J. Emerg. Sel. Top. Ind. Electron.*, vol. 1, no. 1, pp. 3–13, Jun. 2020, doi: 10.1109/jestie.2020.2999589

[36] Z. Zhang *et al.*, "1-kV input 1-MHz GaN stacked bridge LLC converters," *IEEE Trans. Ind. Electron.*, vol. 67, no. 11, pp. 9227–9237, Nov. 2020, doi: 10.1109/TIE.2019.2952806

[37] X. Ren *et al.*, "A 1-kV input SiC LLC converter with split resonant tanks and matrix transformers," *IEEE Trans. Power Electron.*, vol. 34, no. 11, pp. 10446–10457, Nov. 2019, doi: 10.1109/TPEL.2019.2896099

[38] A. Stillwell, M. E. Blackwell and R. C. N. Pilawa-Podgurski, "Design of a 1 kV bidirectional DC-DC converter with 650 V GaN transistors," in *Conference Proceedings – IEEE Applied Power Electronics Conference and Exposition –APEC*, vol. 2018, March, pp. 1155–1162, Apr. 2018, doi: 10.1109/APEC.2018.8341162

[39] R. Ramachandran and M. Nymand, "Experimental demonstration of a 98.8% efficient isolated DC-DC GaN converter," *IEEE Trans. Ind. Electron.*, vol. 64, no. 11, pp. 9104–9113, Nov. 2017, doi: 10.1109/TIE.2016.2613930

[40] P. Applications, "Design challenges and considerations of wolfspeed 22kW High efficiency Bi-directional DCDC converter," 2020. https://assets.wolfspeed.com/uploads/2022/05/Wolfspeed_Design_Challenges_and_Considerations_of_30kW_LLC_Converter.pdf

[41] D. Lp, "OUTLINE DRAWING EXAMPLE Dimensions in mm May vary depending on number of outputs ORDERING CODE EXAMPLE PRODUCT TYPE." [Online]. Available: www.brightloop.fr

[42] "22 kW 900 V bidirectional DC-DC converter." [Online]. Available: www.iisb. fraunhofer.de

[43] "VisIC's smallest 6.7 kW on-board-charger reference design best power density of 3kW/L and lightweight of 4.5kg.".

[44] "HELLA | GaN systems." https://gansystems.com/gan-applications/hella-22-kw-ev-onboard-charger/ (accessed Dec. 18, 2020).

[45] "CONTINENTAL | GaN systems." https://gansystems.com/gan-applications/ continental-3-kw-ev-onboard-charger/ (accessed Dec. 18, 2020).

[46] "CANOO | GaN systems." https://gansystems.com/gan-applications/canoo/ (accessed Dec. 18, 2020).

[47] O. Semiconductor, "TND6318 – On Board Charger (OBC) LLC converter." [Online]. Available: www.onsemi.com

[48] "Electric Vehicle Infrastructure Terra 54 HV UL 50 kW high-voltage DC fast charging station for HV battery electric vehicles," 2020. https://library.e.abb.com/public/ 729567b95c6d489b8b59cfaeb00340e3/Terra54HV_UL_Data-Sheet_C.pdf?x-sign= r2oxfLzrwielx57smlZ3nmfM5StGt89yPXFoL+mPPX/JXeQutSPtKNYCq5evfdgB

[49] "INGEREV® RAPID 50." https://www.ingeteam.com/en-us/sectors/electric-mobility/p15_58_164/ingerev-rapid.aspx (accessed Feb. 14, 2021).

[50] "Tritium VEEFIL-RT DC Charger | 50 kW - EVSE Australia." https://evse.com.au/ product/tritium-veefil-rt-dc-charger-50kw/ (accessed Feb. 15, 2021).

[51] J. Yuan, L. Dorn-Gomba, A. D. Callegaro, J. Reimers and A. Emadi, "A review of bidirectional on-board chargers for electric vehicles," *IEEE Access*, vol. 9, pp. 51501–51518, 2021, doi: 10.1109/ACCESS.2021.3069448

[52] M. Tissieres, I. Askarian, M. Pahlevani, A. Rotzetta, A. Knight and I. Preda, "A digital robust control scheme for dual Half-Bridge DC-DC converters," *Conf. Proc. – IEEE Appl. Power Electron. Conf. Expo. –APEC*, vol. 2018, March, pp. 311–315, Apr. 2018, doi: 10.1109/APEC.2018.8341028

[53] L. Chen, S. Shao, Q. Xiao, L. Tarisciotti, P. W. Wheeler and T. Dragičević, "Model predictive control for dual-active-bridge converters supplying pulsed power loads in naval DC micro-grids," *IEEE Trans. Power Electron.*, vol. 35, no. 2, pp. 1957–1966, Feb. 2020, doi: 10.1109/TPEL.2019.2917450

[54] A. Hernández-Méndez, J. Linares-Flores, H. Sira-Ramírez, J. F. Guerrero-Castellanos and G. Mino-Aguilar, "A backstepping approach to decentralized active disturbance rejection control of interacting boost converters," *IEEE Trans. Ind. Appl.*, vol. 53, no. 4, pp. 4063–4072, Jul. 2017, doi: 10.1109/TIA.2017.2683441

[55] M. A. Hassan, E. P. Li, X. Li, T. Li, C. Duan and S. Chi, "Adaptive passivity-based control of DC-DC buck power converter with constant power load in DC microgrid systems," *IEEE J. Emerg. Sel. Top. Power Electron.*, vol. 7, no. 3, pp. 2029–2040, Sep. 2019, doi: 10.1109/JESTPE.2018.2874449

[56] S. Dutta, S. Hazra and S. Bhattacharya, "A digital predictive current-mode controller for a single-phase high-frequency transformer-isolated dual-active bridge DC-to-DC converter," *IEEE Trans. Ind. Electron.*, vol. 63, no. 9, pp. 5943–5952, Sep. 2016, doi: 10.1109/TIE.2016.2551201

[57] H. Yuan and Y. Kim, "Equivalent input disturbance observer-based ripple-free deadbeat control for voltage regulation of a DC–DC buck converter," *IET Power Electron.*, vol. 12, no. 12, pp. 3272–3279, Oct. 2019, doi: 10.1049/IET-PEL.201 9.0652

[58] J. Wang, W. Luo, J. Liu and L. Wu, "Adaptive type-2 FNN-based dynamic sliding mode control of DC-DC boost converters," *IEEE Trans. Syst. Man, Cybern. Syst.*, vol. 51, no. 4, pp. 2246–2257, Apr. 2021, doi: 10.1109/TSMC.2019.2911721

[59] X. Shi, N. Chen, T. Wei, J. Wu and P. Xiao, "A reinforcement learning-based online-training AI controller for DC-DC switching converters," *2021 6th Int. Conf. Integr. Circuits Microsystems, ICICM 2021*, pp. 435–438, 2021, doi: 10.1109/ ICICM54364.2021.9660319

[60] M. S. M. Gardezi and A. Hasan, "Machine learning based adaptive prediction horizon in finite control set model predictive control," *IEEE Access*, vol. 6, pp. 32392–32400, May 2018, doi: 10.1109/ACCESS.2018.2839519

5 Advance Current-Fed DC/DC Converters

Pan Xuewei and Akshay Kumar Rathore

CONTENTS

5.1 GENERAL INTRODUCTION

Current-fed isolated converters are generally derived from the non-isolated boost converters by inserting one or several high-frequency transformers into specific places. Current-fed converters have been justified to be suitable for low voltage high current applications. Renewable energy sources such as solar photovoltaic and fuel cells generate low voltage and high current output, thus current-fed converters are good candidates to interface them with high voltage dc bus. In distributed generation system, the energy storage devices such as battery and supercapacitor are key elements that are able to absorb the reverse power flow and also output power as auxiliary sources. Therefore, current-fed converters with bidirectional power transfer capability are required. Current-fed converters have drawn a lot of research attention over the past decades due to the following merits:

1. Smaller input current ripple, which is beneficial to extract maximum power from renewable systems stably and extend lifespan. In addition, the

DOI: 10.1201/9781003330134-5

inductor at the input side of the current-fed converter is more reliable and has a longer lifetime compared with the electrolytic capacitor employed in the voltage-fed converter.

2. Lower transformer turns-ratio: Current-fed converters are boost-derived converters and have built-in boost function. Therefore current-fed converter results into lower transformer turns ratio, which can simplify the design and reduce losses.
3. Negligible diode ringing and free from duty cycle loss due to purely capacitive output filter.
4. Easier current control ability. The input current can be directly and precisely controlled.
5. Inherent short circuit protection. Current limiting through the components and short circuit protection become significant for high current applications.
6. No flux-imbalance problem.

Current-fed converters have inherent defects as well. The main drawbacks of current-fed converters are the issue of charging the inductor at start-up and the voltage spike across the turning-off switches caused by current mismatch between the input boost inductor and the leakage inductance of the HF transformer. Lots of literature have been focusing on overcoming these problems through the innovation and modification of topologies with new modulation and control methods and soft-switching techniques. Novel modulation and auxiliary components or circuits are often investigated to solve the intrinsical problems associated with the topologies and enhance the converter performance for specific applications.

5.2 TRADITIONAL CURRENT-FED CONVERTERS

5.2.1 TRADITIONAL UNIDIRECTIONAL CURRENT-FED CONVERTERS

Different topologies of unidirectional current-fed DC/DC converters have been researched. As illustrated in Figure 5.1, full-bridge, half-bridge, and push-pull topologies have been proposed for the low-voltage side inverter stage. For the high-voltage side rectifier stage, there exist some popular topologies like full-bridge diodes rectifier, half-bridge voltage doubler, and center-tapped rectifier. However, the major drawback of current-fed converter is high voltage spike across the device at turn-off owing to the energy stored in the leakage inductance [1]. Snubbers are generally required to limit the voltage spike to prevent the switching device from a permanent breakdown. Different snubber circuits such as dissipative snubbers, regenerative snubbers, active snubbers, etc. have been proposed as shown in Figure 5.2. These snubbers are employed to accommodate the whole boost inductor current until the HF transformer current is fully built up to the level of the boost inductor current [1].

Figure 5.3 illustrates current-fed full-bridge converters using conventional dissipative snubbers like RC or RCD snubbers. Dissipative snubbers lead to low efficiency owing to the energy dissipated in the snubber resistor. To improve the

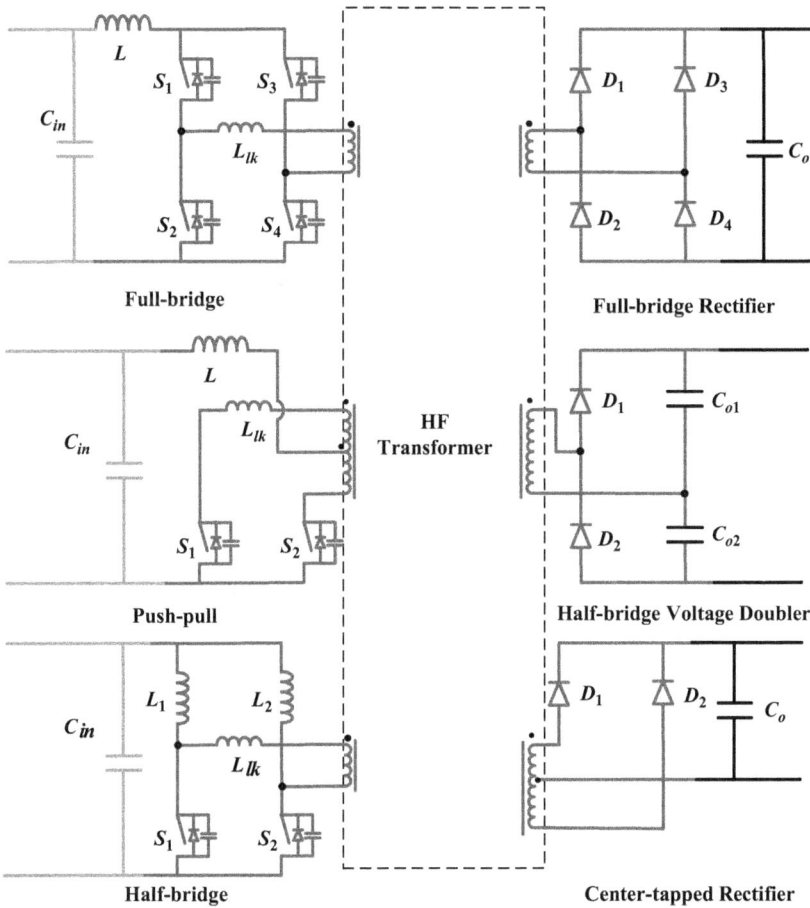

FIGURE 5.1 Conventional unidirectional current-fed DC/DC converter topologies.

efficiency, energy recovery LC snubbers have been proposed as shown in Figure 5.4. The LC snubber stores the surge energy in the capacitor during device turn-off. Once the switch is turned on, the capacitor is reset and energy stored in the inductor is fed back to the input instead of being dissipated. However, the conventional LC snubber has several problems like complex structure and difficult optimal design and does not assist in soft-switching. In [2], an auxiliary flyback snubber including a capacitor, a diode, and a flyback converter was introduced to recycle the absorbed energy as illustrated in Figure 5.5. The flyback snubber alleviates the voltage spike and transfers the trapped energy to the load. The circuit operates with hard switching. A similar auxiliary snubber was proposed in [3] to clamp the voltage across primary switches, recycle the absorbed energy to the load, and assist in achieving the ZVS of the primary switches. However, both of those two auxiliary snubbers are too complex.

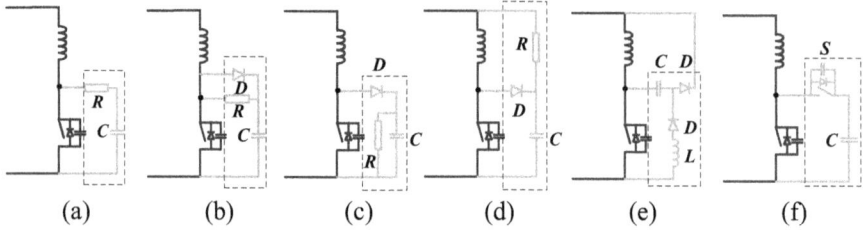

FIGURE 5.2 Snubbers for suppressing the voltage spike: (a) Dissipative RC snubber, (b–d) Dissipative RCD snubber, (e) Non-dissipative energy recovery LC snubber, and (f) Active-clamping snubber.

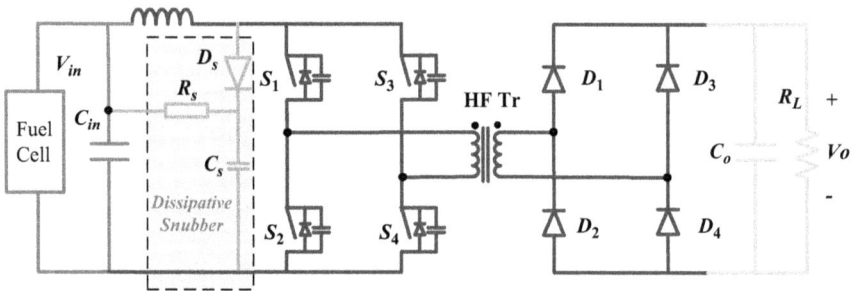

FIGURE 5.3 Current-fed full-bridge DC/DC converter employing dissipative snubbers.

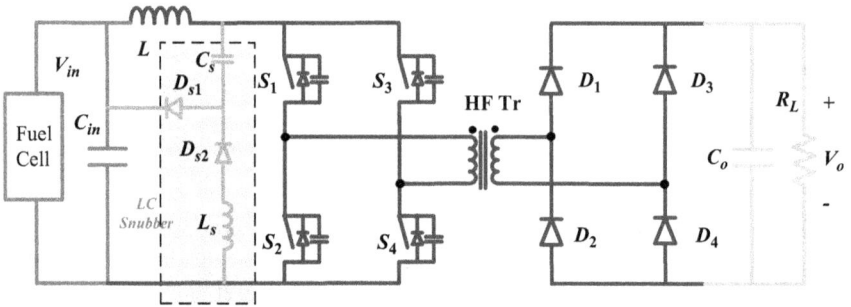

FIGURE 5.4 Current-fed full-bridge DC/DC converter employing non-dissipative snubbers.

Active-clamping snubber circuit [4], which consists of a switch and a capacitor, is proposed to clamp the device voltage. The energy stored in the leakage inductor can be recycled, thus solving the energy loss problem. With a proper parametric design, ZVS of primary switches can be achieved. Several examples of typical current-fed L-type half-bridge topologies employing active-clamping techniques are displayed in Figure 5.6 [4]. The active-clamping circuit proposed in Figure 5.6(a) can achieve ZVS both at turn-on and turn-off, but the voltages across the auxiliary switches are twice as those of the main switches. The active-clamping snubber

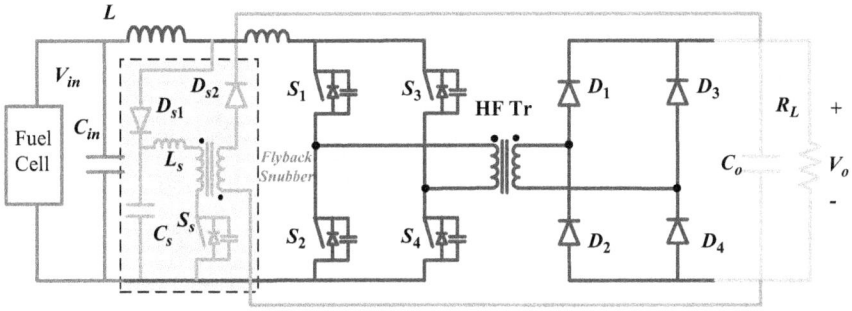

FIGURE 5.5   Current-fed full-bridge DC/DC converter with an auxiliary flyback snubber.

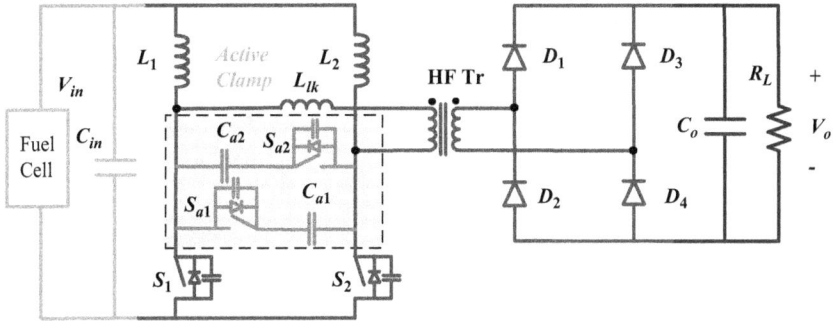

(a)

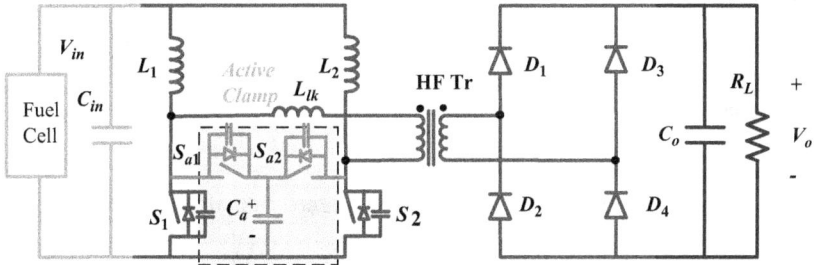

(b)

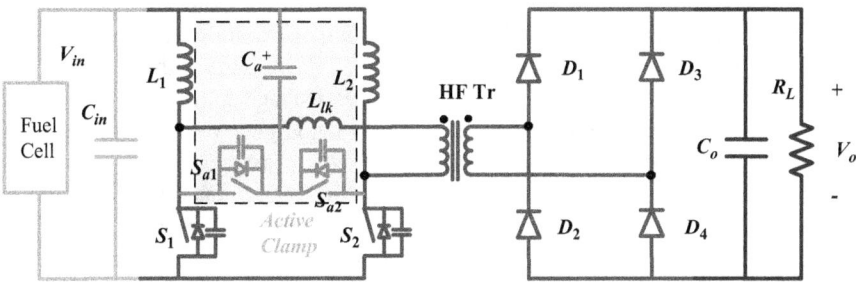

(c)

FIGURE 5.6   Current-fed L-type half-bridge DC/DC converters with three different types of active-clamping circuits.

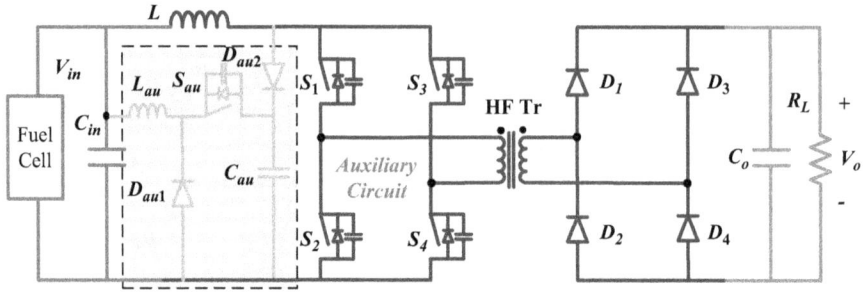

FIGURE 5.7 Current-fed ZCS full-bridge DC/DC converter with parallel auxiliary circuit.

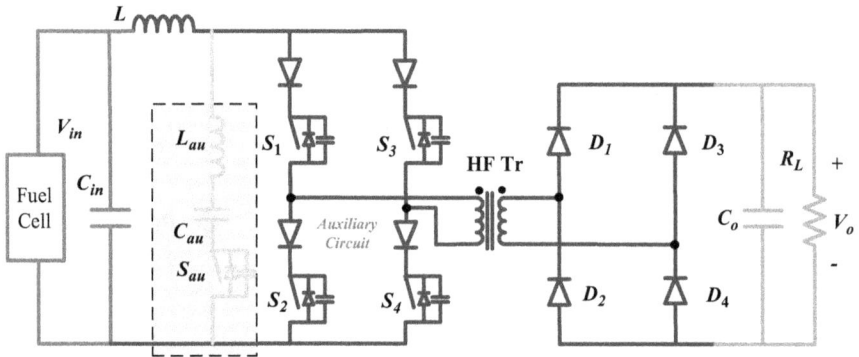

FIGURE 5.8 Current-fed ZCS full-bridge DC/DC converter with current-blocking diodes.

circuits need a floating active device(s) and a high-value of HF clamp capacitor for accurate and effective clamping.

Another approach is to apply ZCS to divert the current away from the switches before turning them off. External auxiliary circuits are utilized to achieve ZCS and reduce the circulating current for current-fed full-bridge topology as illustrated in Figure 5.7 [5]. For the topology shown in Figure 5.8, the transformer leakage inductance and device output capacitance are used to create a quasi-resonant path to facilitate ZCS [6]. The shortcoming is that four extra diodes are connected in series with the main switches to reduce the circulating loss. This brings more conduction losses and increases the cost and circuit footprints. For the topology proposed in [7] as shown in Figure 5.9, an external circuit consisting of two uni-directional switches and a resonant capacitor is employed to obtain ZCS turn-on/off of the switches. The ringing due to the interaction of the transformer leakage inductance and switch capacitance is the major limitation of this technique. Higher voltage rating devices or dissipative snubbers are needed. An external boost converter is added to provide a path for the boost inductor current and send the trapped energy to the load as shown in Figure 5.10 [8]. A small inductor is needed to achieve ZCS turn-on/off. Although the trapped energy can be recycled, the auxiliary circuits still contribute to a significant amount of loss. The higher cost and circuit complexity are two major concerns.

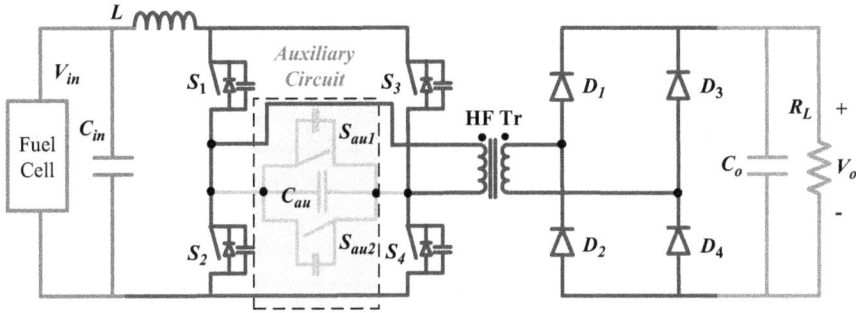

FIGURE 5.9 Current-fed ZCS full-bridge DC/DC converter with snubber energy.

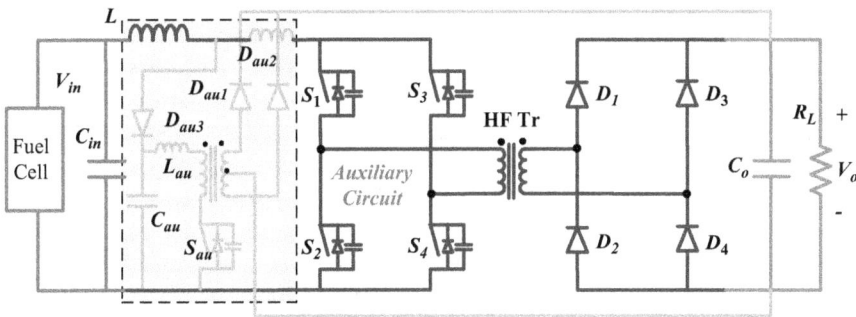

FIGURE 5.10 Current-fed ZCS full-bridge DC/DC converter with an external boost converter.

5.2.2 TRADITIONAL BIDIRECTIONAL CURRENT-FED CONVERTERS

5.2.2.1 Current-Fed Converters with Snubbers

In bidirectional current-fed circuits, hard commutation is also a major issue as it results in device voltage overshoot, which is caused by the current mismatch between the input boost inductor and the leakage inductance of HF transformer. Snubbers are generally required to limit the voltage spike to prevent the switching device from a permanent breakdown. Similar to the unidirectional current-fed converter, different snubber circuits such as dissipative snubbers and regenerative snubbers have been utilized as shown in Figure 5.2. These snubbers are employed to accommodate the whole boost inductor current until the HF transformer current is fully built up to the level of the boost inductor current.

Some work utilizes the auxiliary circuits to clamp the voltage spike during the switching transition. The flyback snubber was used to recycle the absorbed energy in the clamping capacitor [9]. The topology of the current-fed converter with a flyback snubber is shown in Figure 5.11 [9]. Since the flyback snubber can be operated independently, it can be freely controlled according to the requirements of the converter. Therefore, it can clamp the voltage of low-voltage switches to a level slightly higher than the voltage across the primary side of the transformer. Due to the

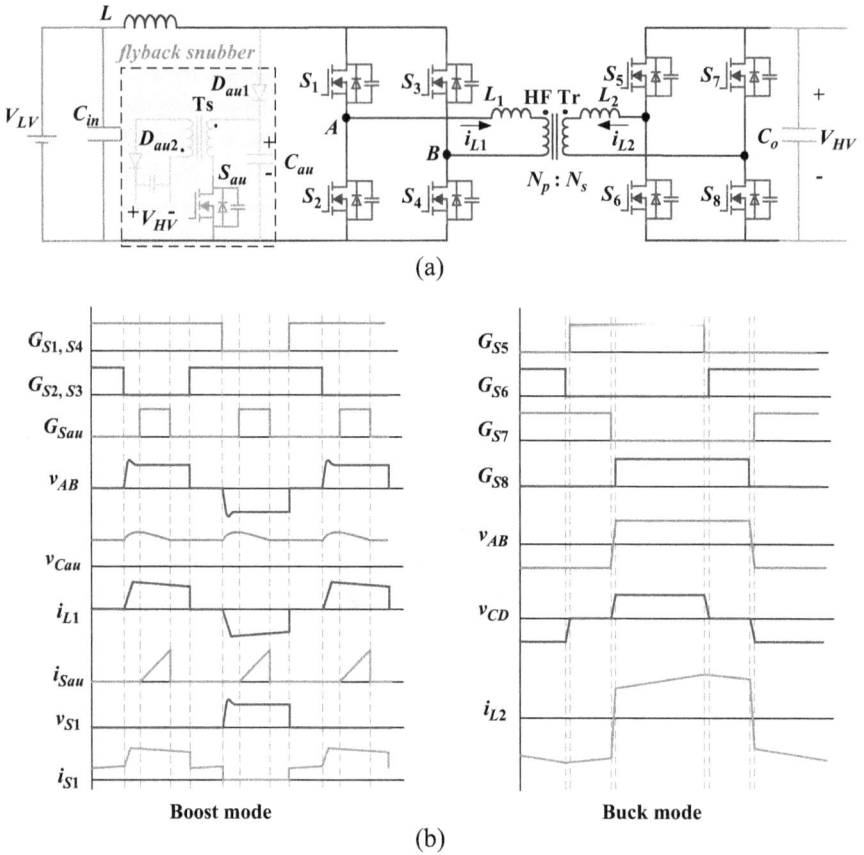

FIGURE 5.11   (a) Topology with flyback snubber. (b) Key operating waveforms.

existence of the clamping branch (D_{au1} and C_{au}) and the flyback circuit, the circulating current does not flow through the main switches, and the voltage spike is avoided during switching commutation, consequently improving the reliability of the converter. The clamping branch and the flyback snubber are activated in the start-up process and regular boost mode operation. The circuit can be feasible by pre-charging the high voltage side capacitor through the flyback snubber at start-up thus realizing soft start and suppressing inrush current. In boost mode, low-voltage side switches are controlled and the high-voltage side is synchronous rectifying. While in buck mode, the low-voltage side is operated as a rectifier and the topology is a phase-shift full-bridge converter without the participation of the flyback snubber. In [10], external auxiliary buck circuits are utilized to achieve ZCS and reduce the circulating current for current-fed full-bridge topology as illustrated in Figure 5.12.

5.2.2.2   Current-Fed Converters with Active Clamp Circuit

A simple way to suppress the voltage spike inherited in current-fed converter is to utilize power semiconductor devices and a large HF film capacitor to provide an

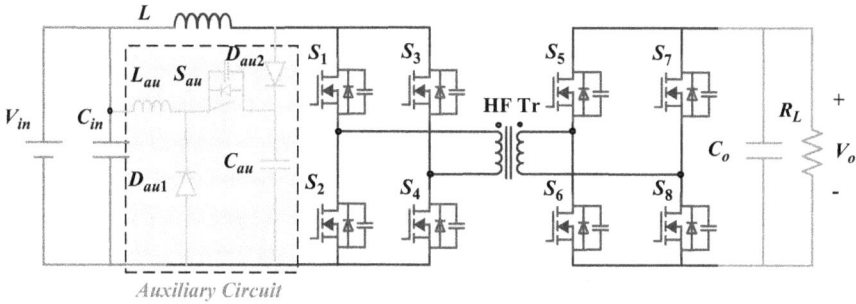

FIGURE 5.12 Topology with auxiliary circuit.

additional commutation path. It allows the current flowing through the leakage inductance to rise to the input inductor current in a controllable manner. The voltage spike can be considered to be clamped by the HF film capacitor. Thus this is called active clamp technique.

Active clamp has been implemented in many fundamental topologies, such as full-bridge current-fed topology as shown in Figure 5.13(a) [11]. A path consisting of a capacitor C_a and switch S_a in series is used to clamp the voltage spike when the main switches turn off and store the energy in auxiliary clamp capacitor C_a through the antiparallel body diodes D_{ax} of the auxiliary switch S_a. The corresponding operating waveforms are shown in Figure 5.13(b). In boost mode, the duty cycle of the main switches is larger than 0.5 and the gating signals of switches in the same leg are phase shifted by 180° with an overlap. The auxiliary switch S_a is on state during the non-conduction interval of the main switches. S_a will achieve ZVS if it is turned on before the clamp capacitor current i_{Ca} decreases to zero. When the current i_{Ca} is reversed, the energy stored in the capacitor will be released back to the main circuit. Similarly, the main switches can also be gated for ZVS turn on. Thus ZVS soft switching operation is achieved for all switches and high efficiency can be achieved. In buck mode, synchronous rectification is implemented in the current-fed side to reduce the conduction loss. While the phase shift control strategy is applied in the voltage-fed side. The free-wheeling current is reset by activating the auxiliary switch S_a and at the same time, ZCS of S_7 and S_8 is realized. But current-fed side switches are turned on twice in one switching cycle along with different duty ratios, increasing the control complexity. The problem of this fundamental active clamped topology is that the soft-switching can't be maintained for light load conditions.

A L-L type active-clamped current-fed converter is proposed as shown in Figure 5.14 with wide/extended ZVS range design [12]. As shown in Figure 5.14, the active clamp circuit is composed of two switches and a capacitor, and the switches provide access for the clamping capacitor C_a to absorb and release the energy from the current mismatch during commutation. Similar operating process to the aforementioned topology in Figure 5.13 is seen in this converter. The difference is that two auxiliary switches are gated on once in turns during one switching cycle while the auxiliary switch in Figure 5.13 is gated on twice in one cycle. Compared with active clamped full-bridge topology, L-L type half-bridge has less number of

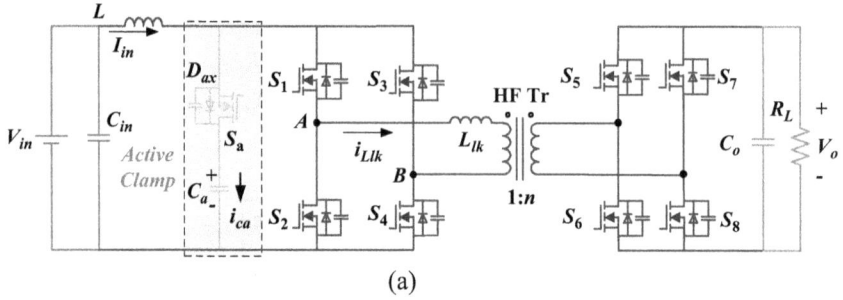

(a)

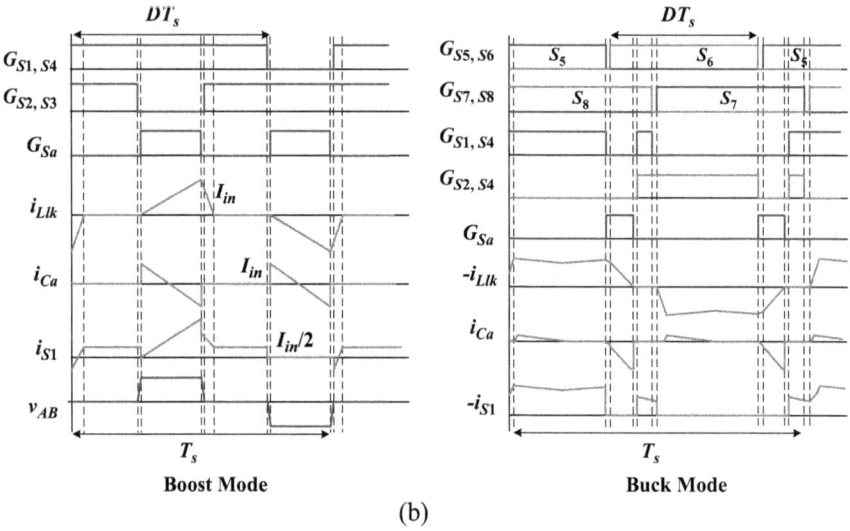

(b)

FIGURE 5.13   (a) Topology with the simplest active clamp. (b) Key operating waveforms.

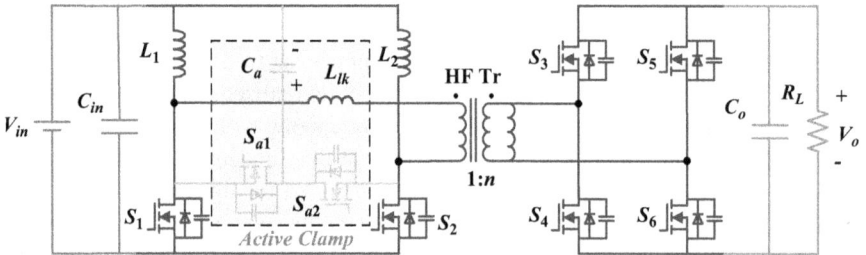

FIGURE 5.14   Active clamped current-fed topology with wide/extended ZVS range.

active switches and potentially lower input current ripple with interleaving effect. The clamping capacitor C_a can also be connected to the ground of the primary side [13]. It is intensively researched in the literature. The L-L type active-clamped converter can achieve a higher boost ratio on account of equivalent dual boost converters on the primary side. Moreover, the magnetizing inductance of HF transformer has been used to extend the soft-switching range. It maintains soft-switching over a wide voltage range and wide load range. All the switches realize soft switching even at light load conditions. It is worthy of mentioning that a tradeoff should be considered between the conduction loss and the range of soft switching for the primary switches.

Another novel active clamp circuit as given in Figure 5.15 can achieve ZVS both at turn-on and turn-off, but the voltages across the auxiliary switches are twice as those of the main switches [14]. The active clamp circuits need floating active device(s) and high value of HF clamp capacitor for accurate and effective clamping. In addition to the above converters, clamp circuit have also been introduced in other current-fed topologies [15,16]. A new topology with active clamped push-pull current-fed circuit at the low-voltage side and LC resonant hybrid full bridge circuit at high-voltage side is proposed in [15], as shown in Figure 5.16. Two novel control methods are presented to achieve the ZVS soft-switching and higher power handling capacity in both power transferring directions. In buck mode, commutation overlap period is reduced to expand the max duty cycle. In boost mode, active commutation is accelerated by utilizing the reflected voltage from the high-voltage side. The power transferring capacity is increased 1.5 times in buck mode operation

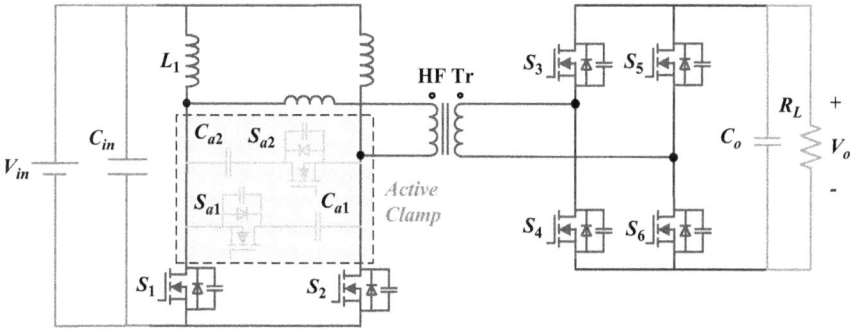

FIGURE 5.15 Current-fed topology with floating active clamp circuit.

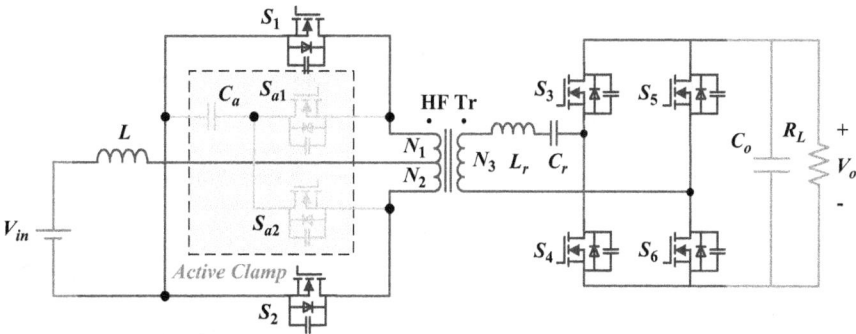

FIGURE 5.16 Active clamped push-pull current-fed converter.

TABLE 5.1
Comparison of Three Active Clamped Topologies

Topologies	Full-bridge	L-L type half-bridge	Push-pull
Numbers of low voltage switch	5	4	4
Voltage stress of low voltage switch	$\frac{V_{in}}{2(1-D)}$	$\frac{V_{in}}{(1-D)}$	$\frac{V_{in}}{2(1-D)}$
Current stress of low voltage switch	I_{in}	I_{in}	I_{in}
Transformer windows utilization	Good	Good	Poor
Input current ripple	Moderate	Good	Moderate

and 3.6 times in boost mode operation. In [16,17], similar control method has been studied for active-clamped current-fed full-bridge topology at low voltage side. At the high-voltage side, a fast recovery DC-link diode with a parallel switch is introduced to allow the use of high-voltage MOSFETs and achieve low switching loss. In boost mode, the added parallel switch is off-state and the DC-link diode reduces large reverse current from the high voltage terminal. In buck mode, the parallel switch is turned on and the high-voltage side full bridge works as an inverter.

In summary, the active clamp technique can be applied to three basic double-ended topologies namely push-pull, L-L type half-bridge, and L-L type full-bridge. All of them can achieve ZVS soft-switching in bidirectional energy transferring and have higher efficiency compared to dissipative/passive snubbers. They all suffer from the demerits of additional floating devices, drivers, and large HF capacitors adding to the volume. The main features of these three topologies are compared in Table 5.1, and they all have advantages and disadvantages.

5.2.2.3 Dual Half-Bridge Current-Fed Converter

F. Z. Peng *et al.* proposed a novel type of current-fed converter with boost half-bridge as shown in Figure 5.17(a) [18]. The converter is configured with dual half-bridge structure placed at both sides of the HF transformer and an inductor at the input of the low-voltage side. This topology is a boost-type circuit and minimizes the devices count for the same power rating compared to full-bridge topology, which allows a compact packaging. Each switch and corresponding capacitor will serve as the active clamp for the complementary one. In addition, the start-up problem of voltage-fed converter is inexistent, which eliminates the need for additional devices and reduces the complexity of the converter.

Figure 5.17(b) shows the idealized operating waveforms of the converter under different cases according to the phase shift Φ_1 and voltage relationships of V_1 & V_3 and V_2 & V_4. The optimum case happens when the duty cycle $D = 0.5$, $V_1 = V_3$ and $V_2 = V_4$, which minimizes the peak value of the leakage inductor current. S_1 and S_3 are controlled with the same duty cycle, which is named as symmetrical duty cycle (SDC) control. The duty cycle D can be regulated to match the voltages across both sides of the transformer. The power transfer and its direction are decided by the phase shift Φ_1 between two voltages across the transformer primary and secondary. When V_{r1} is

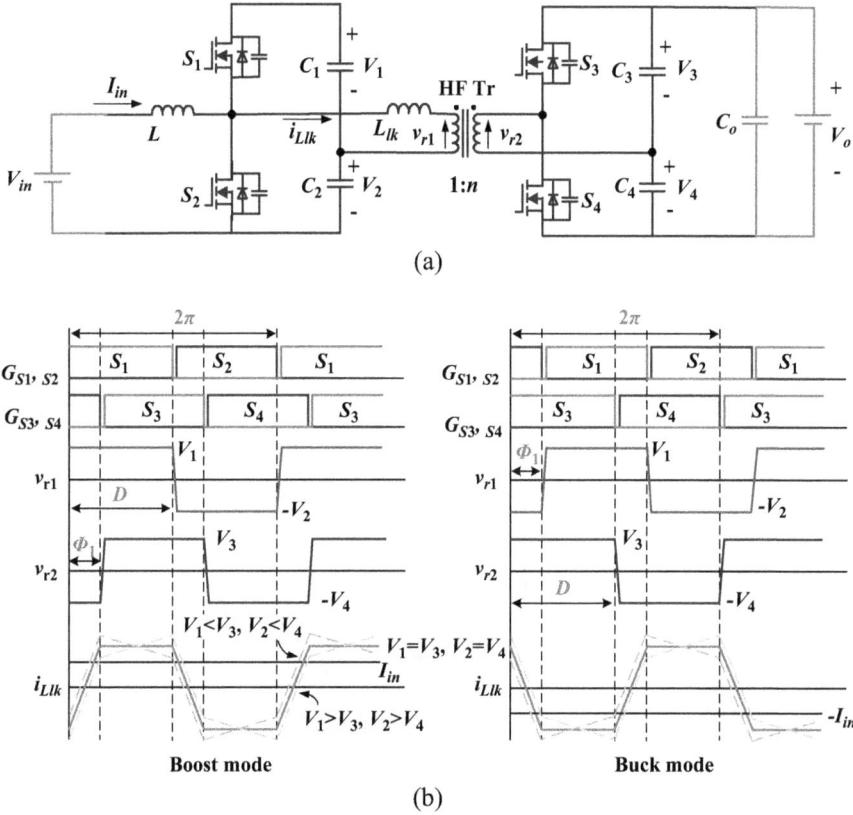

(a)

(b)

FIGURE 5.17 (a) Conventional current-fed dual half-bridge topology. (b) Key operating waveforms.

leading V_{r2}, the converter is working at boost mode. Otherwise, the converter is working in buck mode. The current stresses of switches are also related with Φ_1. Decreased value of Φ_1 brings less current stress. ZVS of all switches is achieved in either direction of power flow. However, the current stresses of low-voltage side switches are asymmetric (shown in Figure 5.18) and it is difficult to accommodate a wide range of voltage boost ratios as well. Practically, the ZVS soft-switching can't be maintained through the full range of load. Another drawback of half-bridge topology is the requirement of split capacitors which have to deal with the full load current.

The secondary side of the above topology in Figure 5.17 has been modified in [19] as given in Figure 5.19(a). The converter is bilaterally symmetrical and is also a variant of the boost circuit. In boost mode, the low voltage side can be considered as a combination of a boost converter and a half-bridge converter as shown in Figure 5.19(b). And for the high voltage side, the working principle of the converter can be regarded as a hybrid of a buck converter and a half-bridge converter as shown in Figure 5.19(c). Similarly, S_4 is the key switch. In order to avoid the unbalance voltage of C_1 and C_2 and the unsymmetrical voltage of the primary

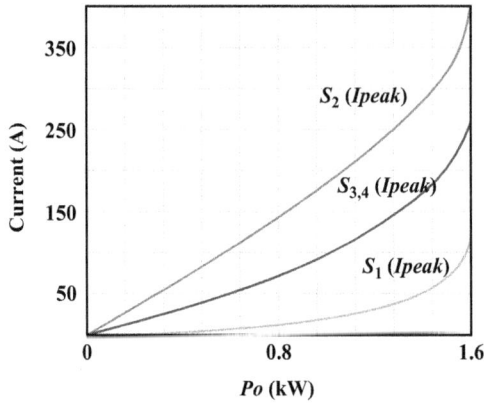

FIGURE 5.18 Current stresses of switches versus output power.

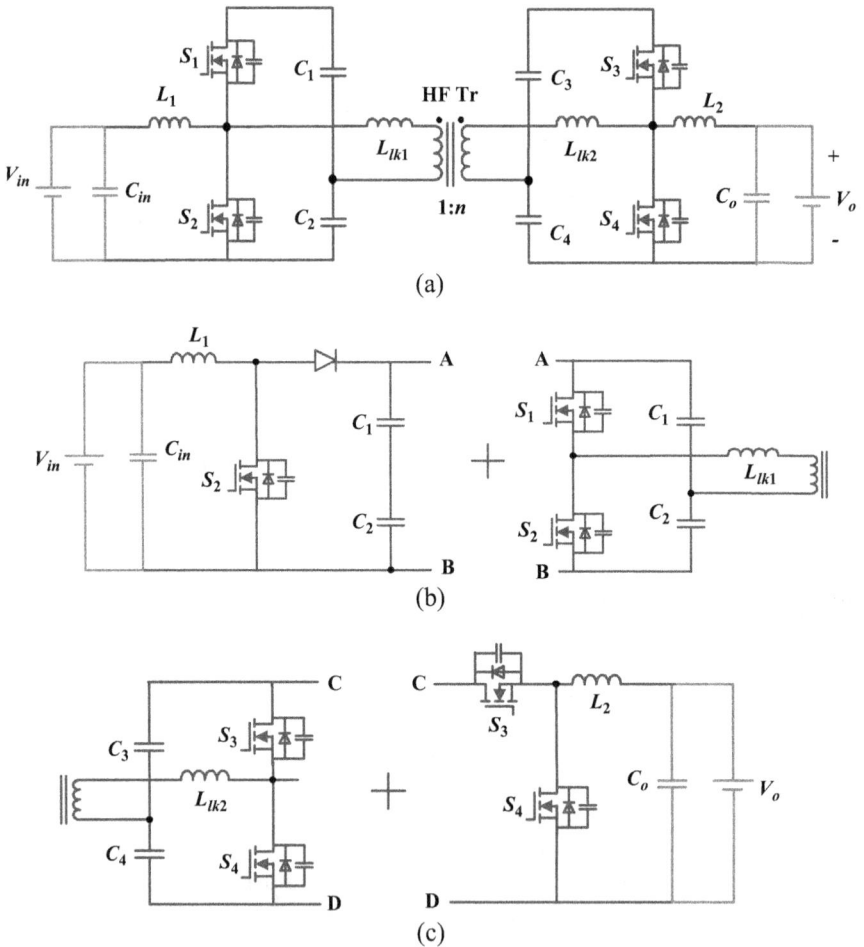

FIGURE 5.19 (a) Bilaterally symmetrical current-fed dual half-bridge topology. (b) and (c) Equivalent circuits.

winding, the duty cycle of all switches is fixed at 50%. Similar to [18], this topology is operating at phase shift control mode. Power flow, the output voltage and current, and soft-switching are all controlled by the phase-shift angle between the transformer's primary and secondary voltages with a fixed 50% duty cycle. Compared to the half-bridge topology in Figure 5.17, this modified topology achieves ZVS for all switches and zero current commutation for the rectifier diode over a wide load range. The control method, i.e., using a fixed duty cycle with phase shift to control power flow, is simple and easy to implement. But the fixed duty cycle (FDC) control method is not suitable for wide input applications. The current stress of switches stays asymmetric like topology in [18].

Based on the conventional dual half bridge current-fed converter [18,20] adds one more identical secondary-side configuration and combines the two capacitor legs into one as shown in Figure 5.20(a). The immediate effect of this change is that the secondary-side output voltage becomes three level from two level, and thus the converter can be optimized for low conduction loss design. PWM plus hybrid phase-shift modulation (PSM) is applied to this converter. As shown in Figure 5.20(b), there are three control variables, namely the duty cycle D of S_1, S_3, and S_5, the phase-shift angle φ_{ps} between the primary and secondary switching units and the secondary side internal phase shift angle φ_s. Three different operating modes of the converter are identified depending on the control of these variables. Boost operation occurs in modes I and II, and buck operation exists in modes II and III. The direction of power flow is controlled by φ_{ps} instead of φ_s. In order to minimize the RMS current, the duty cycles of the three half bridges are the same (SDC control). The duty cycle D can be regulated to match the voltages across both sides of the transformer. In addition, the current stresses, RMS current, and conduction losses of the converter are reduced with PWM control. Single phase-shift (SPS) modulation (modulation of φ_{ps}) and dual phase-shift (DPS) modulation (modulation of φ_s and φ_{ps}) are combined to form a hybrid PSM. The adoption of hybrid PSM can achieve practical ZVS operation of high voltage side switches within wide input voltage and full power range, resulting in lower switching losses and higher efficiency. Compared with the topologies and control scheme in [18,19], the proposed control strategy performs better under all modalities than SPS control and can better match the output voltage. However, this converter is more complex in terms of structure and control.

5.3   ADVANCE CURRENT-FED CONVERTERS

5.3.1   L-L Type Dual Active Bridge Current-Fed Converter

A novel type of current-fed converter has been introduced in [22] based on F. Z. Peng's single half-bridge topology, as shown in Figure 5.21. It can also be considered as one special type of active clamped L-L type current-fed converter. This L-L type dual active bridge current-fed converter has drawn a lot of research efforts. In [22], the converter is controlled by phase-shift plus low voltage side PWM scheme and the high voltage side pulse width is fixed at 0.5. ZVS soft-switching over a wide range of loads under a wide input voltage variation condition

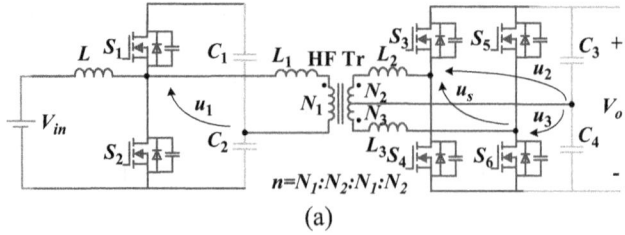

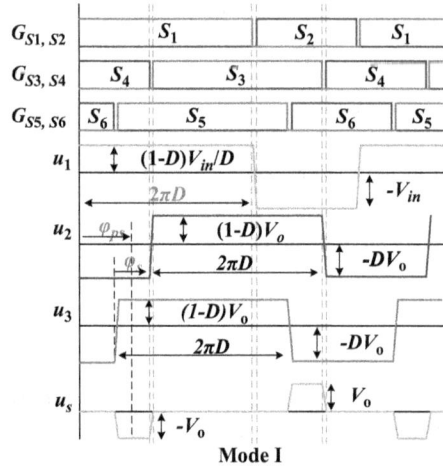

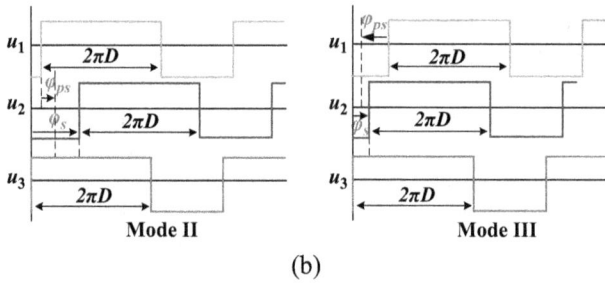

FIGURE 5.20 (a) Topology proposed in [20]. (b) Three operating modes.

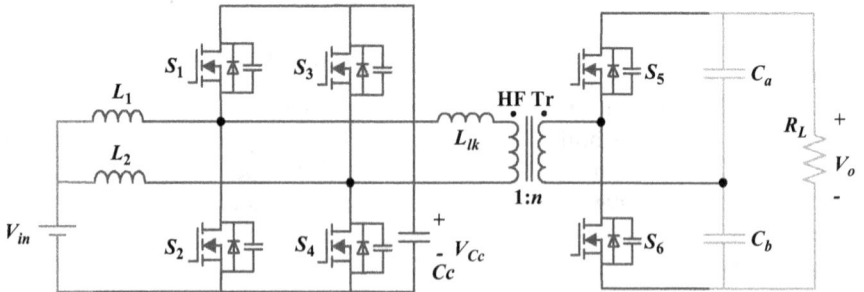

FIGURE 5.21 L-L type dual active bridge current-fed converter with voltage doubler.

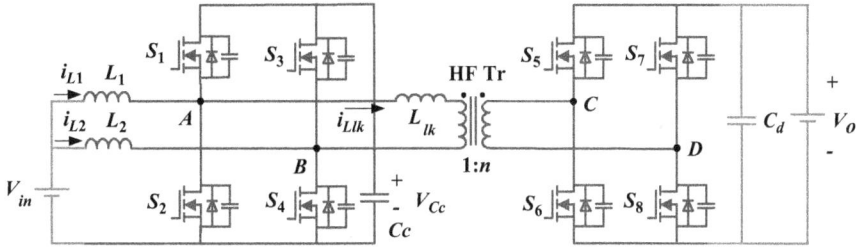

FIGURE 5.22 L-L type dual active bridge current-fed converter with full-bridge at high voltage side.

is achieved. The circulating current is also reduced considerably. However, the circulating energy is still relatively large with fixed high voltage side pulse width in boost mode and the realization of the wide range of output voltage changing in buck mode is difficult. [23] modifies Figure 5.21 into three-level topology (full-bridge at high voltage side), as shown in Figure 5.22. Therefore the pulse width of the high-voltage side can be modulated to obtain more operation modes. PWM plus SPS is adopted, and the duty cycle of both the low-voltage side and high-voltage side is kept the same (SDC). Voltage mismatching control is employed, where the voltage across clamp capacitor is not controlled to be equal to reflected high voltage ($V_{Cc} \neq V_o/n$). The studied converter minimizes the RMS value of the transformer current. But soft-switching through a full range of loads can't be achieved, and the current stress is still relatively high.

D. Sha *et al.* [23] studied the control of this converter further. Voltage matching control is implemented to avoid the high current stress caused by unmatched voltage control. All the possible and practical switching patterns on the premise of wide ZVS range, power controllability, and voltage matching control have been analyzed and compared with respect to three degrees of control freedom, i.e., the low-voltage side duty cycle D_p, the high-voltage side duty cycle D_s, and the phase shift angle φ_E. The practical operating patterns are depicted in Figure 5.23. To achieve ZVS for a wide load range, $D_p > D_s$ should be maintained. A new control strategy is proposed so that D_p is always larger than D_s by a fixed value and the bidirectional power is monotonously controlled by φ_E in the range of $-\pi/2$ to $\sim\pi/2$. The fixed value is designed as small as possible on the premise of ZVS achievement in order to reduce the circulation loss. Therefore, the independent control variables have been reduced from three to two. The converter is optimally operating at different modes according to the load conditions and low voltage variations. The closed-loop control is simple while maintaining the advantages of minimized circulating current. The proposed control strategy allows the converter to achieve soft switching over the full load range including no load condition.

Based on the operating modes shown in Figure 5.23, corresponding modified modes are presented in Figure 5.24 with the same three control freedoms D_p, D_s, and φ_E [24]. It is clearly demonstrated that the modified modes in Figure 5.24 are the boundary modes of that in Figure 5.23 which reduces the peak current and RMS current in the transformer. Compared to the previous control methods, lower

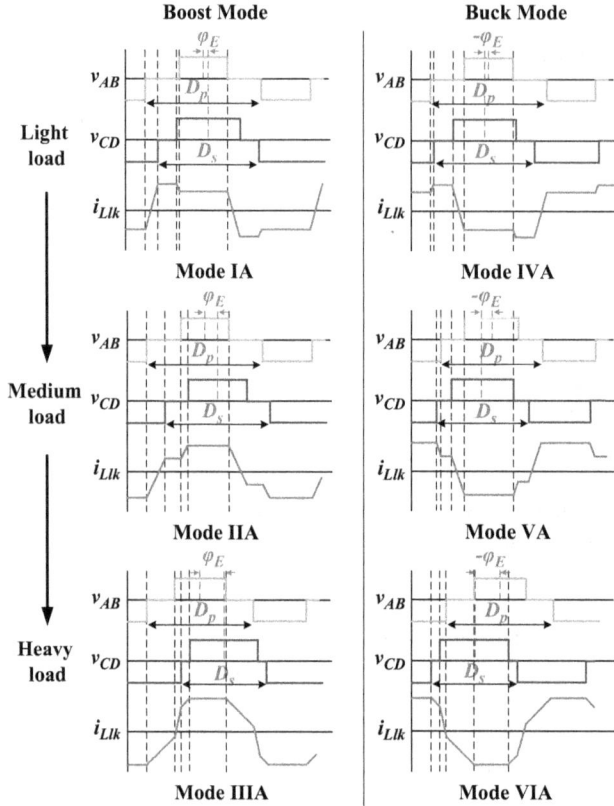

FIGURE 5.23   Operating patterns for L-L type dual active bridge current-fed converter with full-bridge at high voltage side.

conduction loss and core loss occur with this modified modulation strategy. However, the requirement of control precision is greatly increased.

[25] replaces the two input inductors in Figure 5.22 with magnetic-integrated inductors, thus improving the power density and the power conversion efficiency at light-load conditions. The duty cycle of low-voltage side bottom switches is controlled to be lower than 50% to reduce the conduction loss. Meanwhile, the voltage mismatching control with adaptive clamp voltage regulation helps achieve ZVS for all switches throughout the full load range. However, proper dead time to achieve ZVS has to be designed precisely which brings difficulty in control. To obtain both low conduction losses and circulating current, [26] proposed a new control method with double PWM plus double phase-shift (DPDPS). The gating signal of both sides of HF transformer is controlled with double phase-shift to achieve ZVS of low voltage side switches and ZVS/ZCS of high voltage side switches. Voltage matching control is also implemented. The advantage of this control method is that the soft-switching is maintained for a wide variation of low input voltage. Thus the conversion efficiency is higher than the PWM plus SPS control. A similar DPDPS

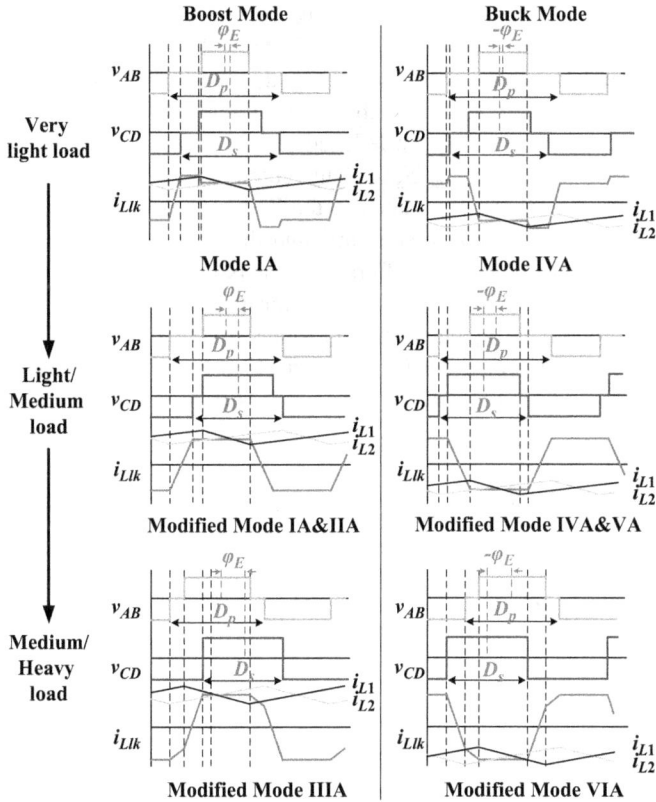

FIGURE 5.24 Modified operating modes for L-L type dual active bridge current-fed converter with full-bridge at high voltage side.

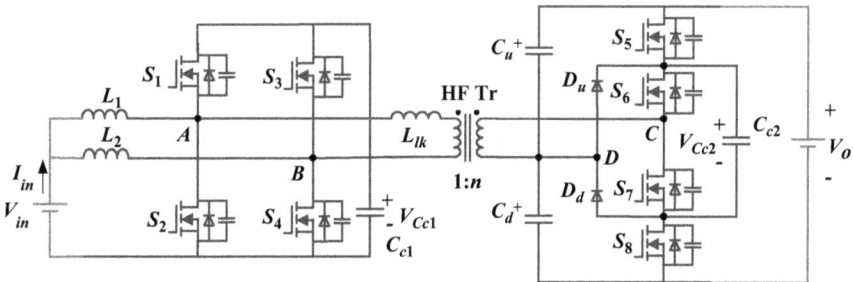

FIGURE 5.25 L-L type dual active bridge current-fed converter with a three-level neutral point clamped circuit.

technique has been implemented for the topology with a three-level neutral point clamped circuit at the high voltage side, as shown in Figure 5.25.

A novel L-L type hybrid dual active bridge converter is proposed in [27], as shown in Figure 5.26(a). The high-voltage side is a full bridge with an auxiliary half-bridge circuit, which allows five-level high-voltage operation at most. The theoretical operating waveforms are shown in Figure 5.26(b). The low voltage switches are conducted complementarily with 50% fixed duty cycles regardless of the input voltage and load variation. Therefore, the current ripples of L_1 and L_2 offset each other and the input current ripple can be reduced to zero theoretically. For high-voltage switches, the duty cycle of S_7 and S_8 is D and complementary to that of the auxiliary switches. The selection of D is related to the ZVS range for all switches, which is determined by the voltage conversion ratio and the phase shift angle φ. To achieve zero current ripple, this converter suffers from the issue of the high number of active switches and high current stress with voltage mismatching control.

A multiport topology suitable for hybrid battery and supercapacitor applications is presented in [28]. The converter is symmetrical on both sides of the transformer, shown in Figure 5.27(a). As indicated in Figure 5.27(b), the duty cycle D_1 controls the

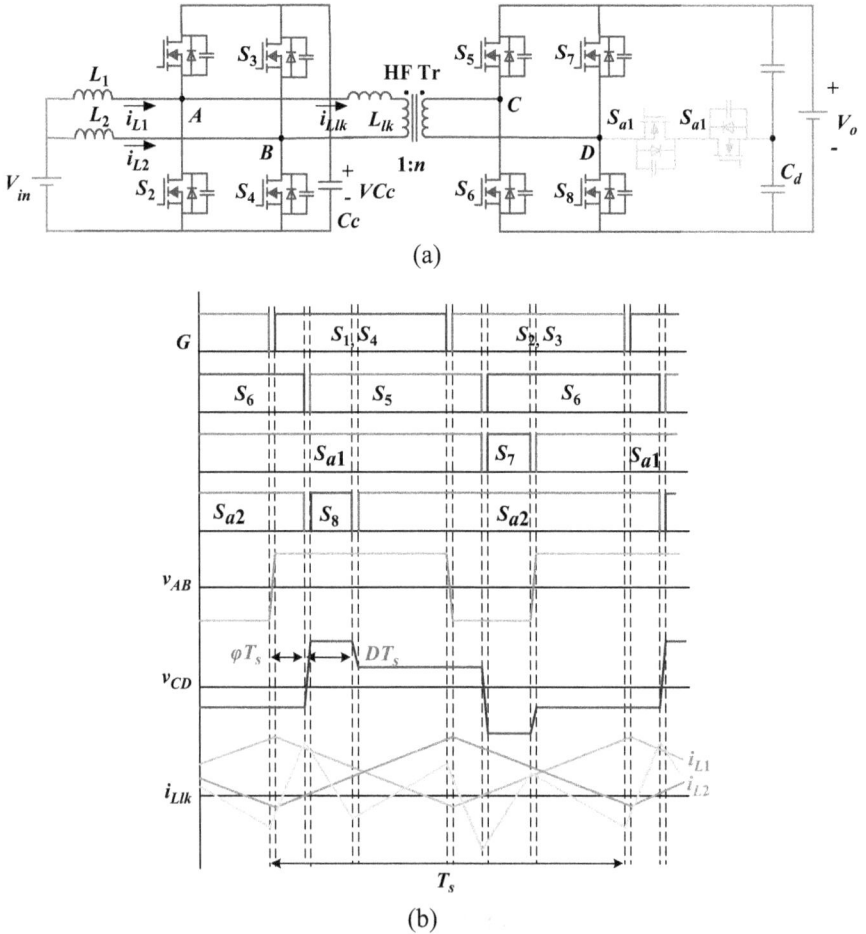

(a)

(b)

FIGURE 5.26 (a) L-L type dual active bridge current-fed converter with hybrid high voltage side topology. (b) Key operating waveforms.

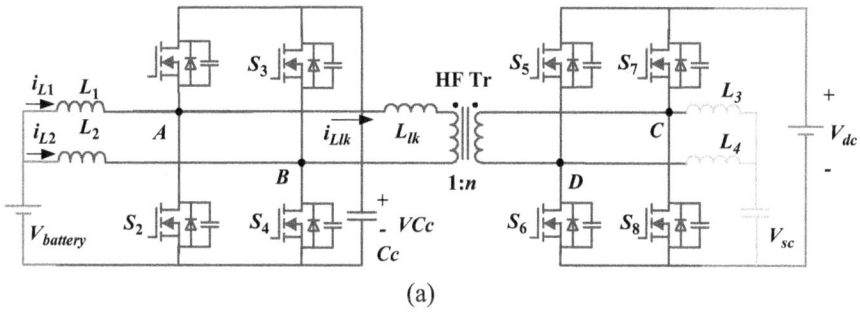

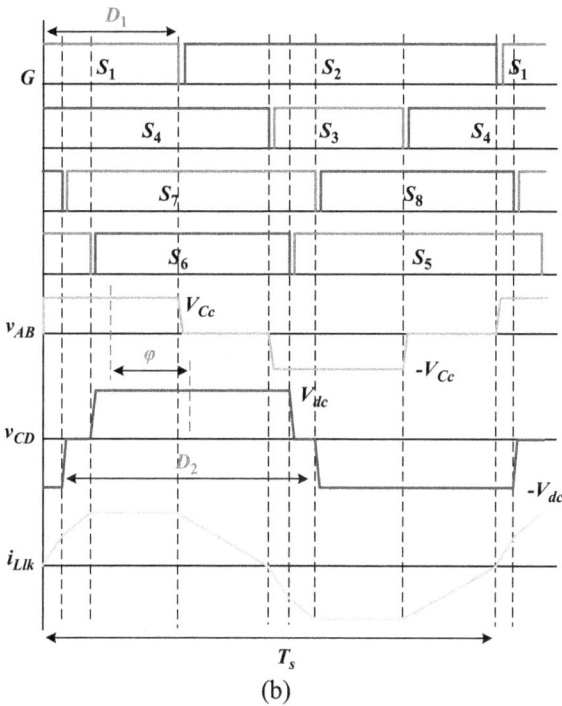

FIGURE 5.27 (a) L-L type dual active bridge current-fed converter with multiport interface. (b) Key operating waveforms.

voltage across the clamp capacitor Cc, and D_2 adjusts the voltage of dc-bus. Similarly, the phase angle φ controls the power flow between the battery and dc bus. For ZVS achievement of switches on the battery side in full load range, D_1 is set less than 0.5, while for switches on the dc-bus side, the inductors L_3 and L_4 should be designed properly to achieve the whole range of ZVS operation. Compared to the topologies without multiport, lower reactive power and circulating current are realized in this topology and it has the potential to attain the port extension. The operating patterns of the above research work in boost mode and buck mode are symmetrical and the power-transferring magnitude and direction can be generally controlled by the phase shift angle since they are piecewise related.

5.3.2 RESONANT TYPE CURRENT-FED CONVERTER

Resonant type current-fed converters make use of the transformer parasites (leakage inductor, parasitic capacitor) as resonant circuit elements and solve the problem of voltage/current spikes and high switching losses. Resonant converters are the earliest soft-switched converters. According to different types of resonances, resonant converters can be divided into series-resonant, parallel-resonant, and series-parallel resonant converters. Dozens of research works have applied these resonant techniques over current-fed converters. Many types of resonating current-fed converters such as LC-type, CLC-type, LLC-type, LCC-type, CLLC-type, etc. have been reported, which were distinguished based on the components of the resonant circuit. Normally the output voltage is regulated through varying frequency control, and switches can be soft-switched.

A new current-fed resonant topology for bidirectional wireless power transmission power electronics system is presented in [29]. A series-parallel CLC resonant tank is formed by connecting a designed capacitor in series with the transmitter coil. The topology is shown in Figure 5.28. Compared to the conventional current-fed topology with LC resonant tank, the voltage stress on the converter is much smaller. Variable frequency modulation is used to control the power flow. Regardless of the load current, the devices can obtain soft switching. The receiver-side voltage doubler acts as an uncontrolled rectifier in the forward direction. To compensate for the reactive power consumed by the receiver coil, a capacitor is connected to it in series. In the backward direction, the voltage doubler circuit operates as voltage-fed inverter with a fixed duty cycle 0.5. Varying frequency control brings difficulty to the optimized control and design of the converter.

To solve the issue of varying frequency control of resonating converters, some scholars focus on the study of fixed-frequency or limited-range frequency control. Figure 5.29 shows a novel fixed-frequency PWM-controlled LC series-resonant tank-based current-fed converter [30]. The primary side is an L-L type active bridge topology, while the secondary side is an active full-bridge circuit. The resonant tank, which is composed of a series inductor L_r, a capacitor C_r, and an HF transformer, is

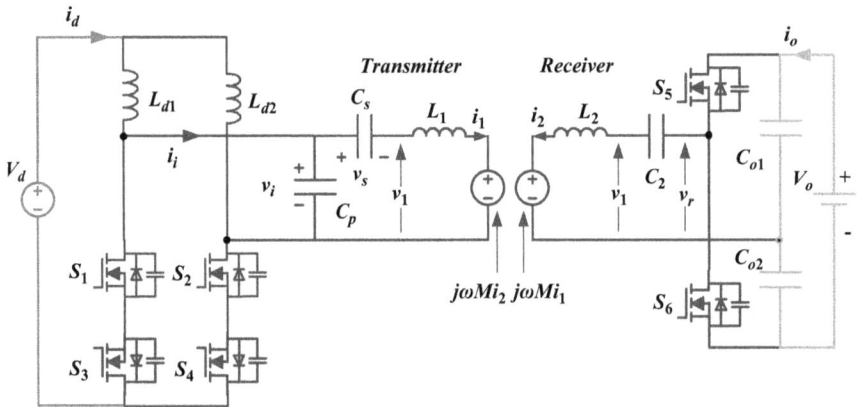

FIGURE 5.28   Current-fed converter with CLC resonant tank.

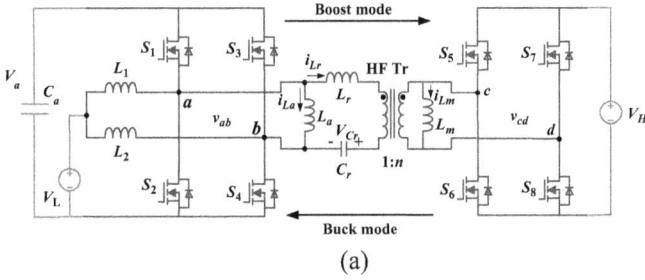

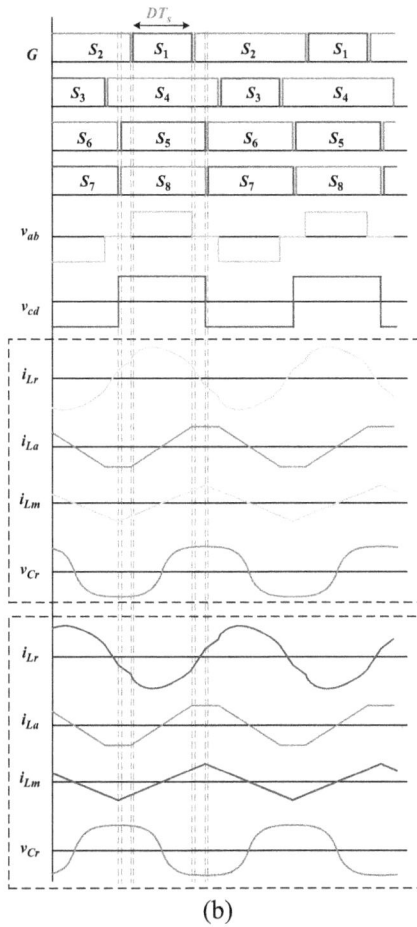

FIGURE 5.29 (a) Current-fed converter with LC series-resonant tank. (b) Key operating waveforms.

used to produce a nearly sinusoidal current at both sides of the transformer. This results in a lower switching loss of the switching devices. The auxiliary inductor L_a and magnetizing inductor L_m are employed to achieve ZVS for all the switches. Figure 5.29(b) shows the operating waveforms in boost and buck modes. It can be seen that the gating signals of switches in boost and buck modes are exactly the same, as well as the inductor currents i_{La} and i_{Lm}. The only difference is the direction of i_{Lr} and v_{Cr}. The most attractive feature of this proposed series-resonant converter is that the voltage gain characteristic is very similar to the well-known PWM converters, whose voltage gain is only determined by the duty cycle D of switches S_1 and S_3. D is the only control variable and varies around 0.5 to accommodate the wide voltage applications. The amplitude and direction of the transferred power are regulated easily and smoothly through simple PWM control.

Figure 5.30 shows a traditional LLC resonant converter with optimal modulation strategy analyzed in [31]. Compared to the conventional pulse-frequency modulation (PFM) which is not suitable for bidirectional operation within a wide voltage range due to the requirement of a wide frequency regulation range, this new control method employs a narrow operation frequency range which is beneficial for the design of magnetic components and achieves high efficiency. In boost mode, PWM control is employed. The duty cycle D of S_1 and S_3 is changed to regulate the voltage of capacitor C_a and the high voltage side switches are synchronous rectified.

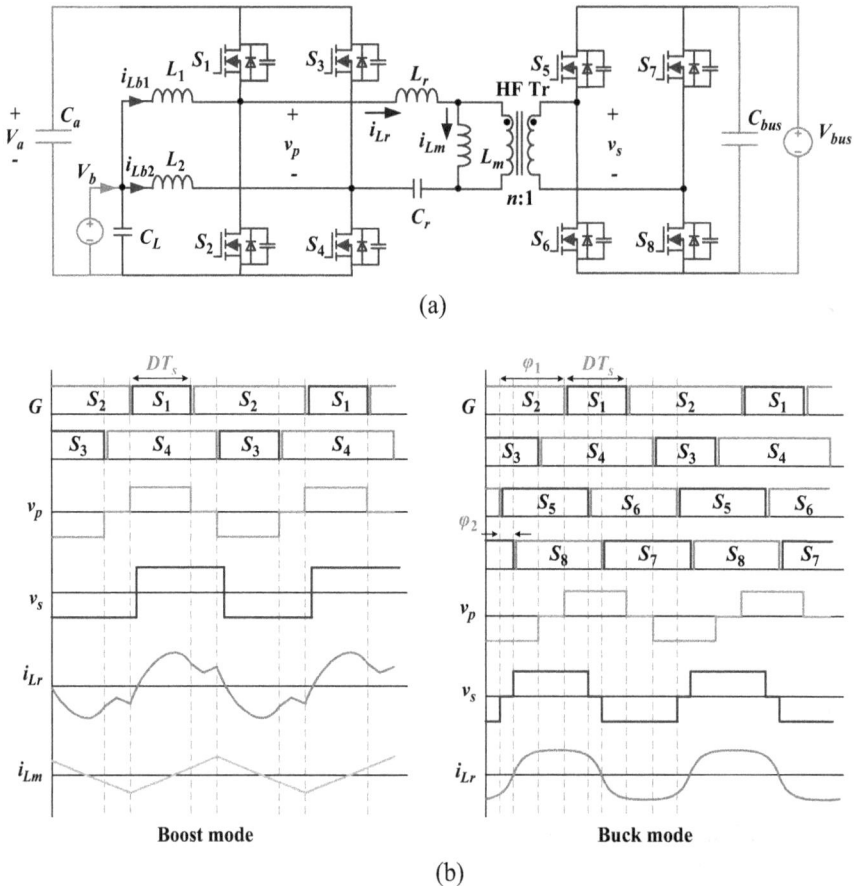

(a)

(b)

FIGURE 5.30 (a) Current-fed converter with LLC resonant tank. (b) Key operating waveforms.

The dc-bus voltage is controlled by PFM. While in buck mode, PWM plus PSM is adopted. The phase shift angle φ_1 between S_5 and S_1 and duty cycle D regulate V_b and V_a respectively. And the constant phase shift φ_2 between S_5 and S_8 helps to realize ZVS of all the switches. It should be noted that the converter parameters and control strategy need to be well-designed and tradeoffs must be made to ensure ZVS over a full operating range and better efficiency performance.

The disadvantages of the resonant type current-fed converter are that the resonant tanks need extra inductors and capacitors, and the resonating voltage and current stress of these passive components are relatively large.

In addition to the aforementioned converters with resonance in the whole cycle, there exist other types of resonant converters of which the resonance happens only for a part of the switching period. Instead of full resonance, partial resonance may obtain the advantages of soft commutation, limited device voltage, and current stress. Some of them are named as quasi-resonant converter [32,33]. The resonant components are the leakage inductor of HF transformer and the parasitic capacitor of the switches. The topology is shown in Figure 5.31 and its corresponding boost mode operating waveforms are shown in Figure 5.33. The resonance frequency should be selected several times the maximum switching frequency. The voltage gain is mainly affected by the resonant components and switching frequency, instead of the high-voltage side duty cycle. In buck mode operation, low-voltage side switches are operated the same as that in boost mode. Although the proposed converter is operated under ZVS at both power directions, the voltage stress of low-voltage side switches is still too high.

Similar techniques called impulse commutated converters have been reported recently [34]. Impulse commutation utilizes transformer leakage inductance and inter-turn winding capacitance as the resonance tank, which allows the device voltage to rise gradually and therefore solves the issue of device turn-off voltage spike without increasing complexity and degrading the performance of converters. The resonance happens only for a short period of time depending upon the overlap time and produces a resonance impulse. ZCS turn-off operation of low voltage switches and reduced peak current compared to full resonant converters are achieved and the voltage gain is immune to load variation. However, frequency modulation is still required to regulate the load voltage. The impulse commutation

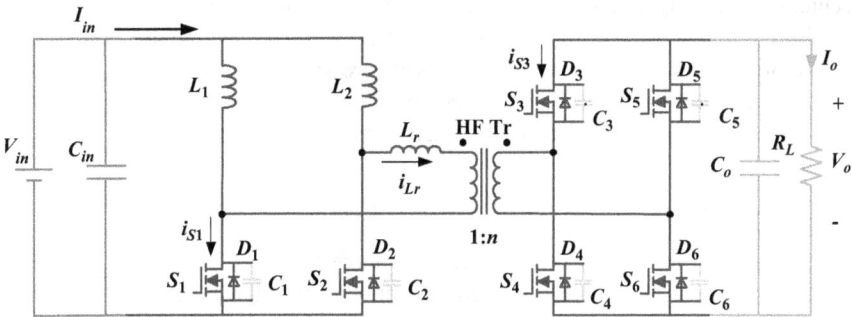

FIGURE 5.31 Topology proposed in [32].

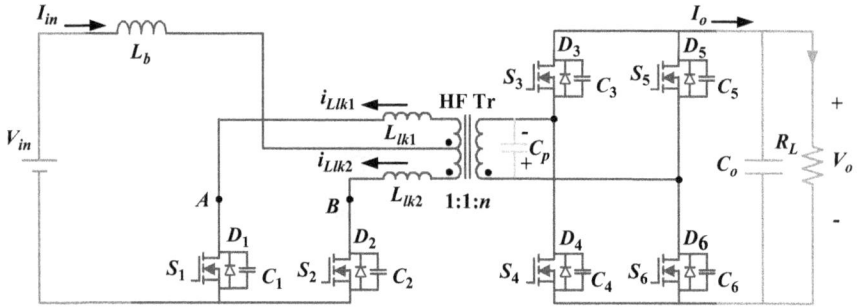

FIGURE 5.32 Topology proposed in [34].

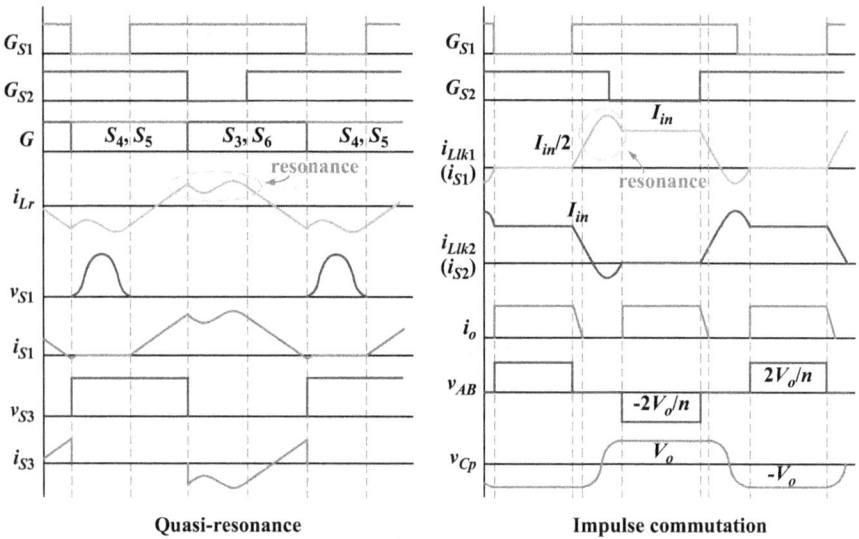

Quasi-resonance

Impulse commutation

FIGURE 5.33 Waveforms of quasi-resonant converter [32] and impulse commutated converter [34].

technique can be applied to three basic double-ended topologies push-pull, L-L type half-bridge and L-L type full-bridge. Figure 5.32 shows an impulse commutated push-pull topology (the operating waveforms are shown in Figure 5.33). Reduced peak and circulating currents compared to resonant converters with identical LC tank are achieved for impulse commutated current-fed converter.

5.3.3 NATURALLY CLAMPED CURRENT-FED CONVERTER

For current-fed bidirectional converter, the voltage-fed (high voltage) side switches provide the flexibility to preset the current flowing through leakage inductance to the boost inductor current before the commutation of current-fed side switches thus reducing or eliminating the need of snubber circuits. This is referred as active commutation technique or natural clamping technique [35,36]. For the soft-commutation method proposed by [35], the energy consumption of the passive snubber can be reduced but can't be completely eliminated. In [36], the commutation time has to be precisely controlled which limits the practical applications.

Akshay K. Rathore *et al.* have contributed a lot to the development of a new family of current-fed converters: active commutated or naturally clamped current-fed converters [21,37–39]. The main principle of the secondary-modulation-based naturally clamping technique is to utilize the reflected output voltage across the primary winding of the HF transformer. The reflected output voltage diverts the input boost inductor current from one switch (or switch pair) to the other switch (or switch pair) through the HF transformer. A minor circulating current allows body diode conduction to ensure ZCS and natural turn-off. Low voltage side switches are naturally voltage clamped by reflected output voltage and achieve ZCS turn-off and nearly ZVS turn-on, while the high voltage side switches realize ZVS turn-on. Soft switching and voltage clamping are inherent and maintained independent of the load and voltage variations. Owing to the low clamped voltage of low-voltage side devices, low-voltage-rating devices with low on-state resistance can be used, resulting in low conduction losses and higher efficiency.

Two types of secondary-modulation have been implemented over current-fed L-L type half bridge topology and full bridge topology as shown in Figure 5.34 [37] and Figure 5.35 [38]. The key operating waveforms of these two secondary-modulation techniques have been displayed in Figure 5.36(a) and (b). Low voltage side switches are controlled by identical gating signals (greater than 0.5) phase-shifted by 180° with an overlap. Both of them can achieve natural voltage clamping while for [37] the current flowing through HF transformer is discontinuous unlike [38]. The continuity of leakage inductance current is determined by the gating signal of high-voltage side switches. The turn-off moment of high-voltage switches is synchronous with the turn-off of corresponding low-voltage switches. While for [37], the duty cycle is necessarily less than $\varphi/2$ at full load and for [38], the duty cycle should be large enough even up to 0.5. Thus the control method proposed in [37] suffers from oscillation due to the resonance between the leakage inductance of HF transformer and device capacitances when the current in the transformer reaches zero.

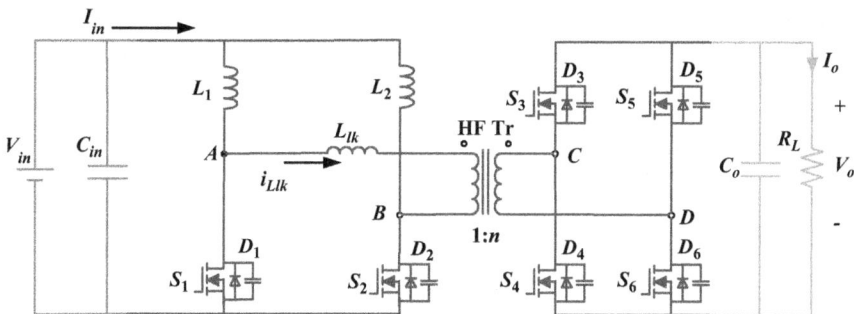

FIGURE 5.34   Naturally clamped current-fed L-L type half bridge topology.

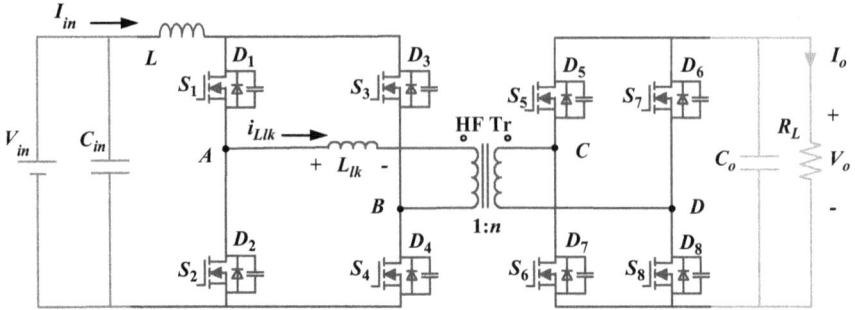

FIGURE 5.35 Naturally clamped current-fed full bridge topology.

For both of these two techniques [37,38], at light loads conditions, the peak current through the leakage inductance of HF transformer and the low voltage side switches is much higher than the input current, i.e., the excess current or circulating current is relatively larger, compared to that at rated load. The value of this peak current is constant and fixed for a given operating condition, irrespective of any power level or load change. Such peculiarity leads to a higher current available for soft-switching than that required, a high circulating current, and high conduction and reverse recovery losses of primary switches' body diode at light loads. As a result, the performance of the converter is reduced at light loads.

The modulation methods implemented in [21,37,38] can be seen as a single phase-shift (SPS) modulation as discussed in [39]. The control variable φ is the phase shift between the gating signals of the low-voltage switches and the corresponding high-voltage switches as shown in Figure 5.36(b) and (c). The deficiency of the single variable control is the unregulated peak current especially at light load conditions as discussed above. [39] proposes dual phase shift (DPS) modulation implemented over the same topology in Figure 5.35. In addition to φ, the phase difference α_s between the diagonal switches on the high-voltage side is also controlled as shown in Figure 5.36(c). With this modified modulation strategy, the value of the aforementioned peak current is greatly reduced and thus the performance of the converter is improved at light load conditions. For DPS modulation, the voltage across the secondary winding of the transformer is three level which is beneficial to reduce the circulating current and the power transfer is controlled by φ and α_s simultaneously.

In summary, the merits of naturally clamping are no need for auxiliary circuits, ZVZCS for low voltage side devices and ZVS for high voltage devices for wide operating range, inherent natural voltage clamping, and higher efficiency and power density of the converter.

5.4 COMPARISON AND EVALUATION

5.4.1 COMPARISON WITH VOLTAGE-FED CONVERTERS

The voltage-fed converters have low switch voltage ratings enabling the use of switches with low on-state resistance. This can significantly reduce the conduction loss of primary side switches. However, voltage-fed converters suffer from several limitations, i.e., high pulsating current at the input, limited soft-switching range, rectifier diode ringing, duty cycle loss (inductive output filter), high circulating current through devices and magnetics, and relatively low efficiency for high voltage amplification and high input current specifications. Voltage-fed topologies employ considerably large electrolytic capacitor to suppress the large input current ripple resulting in large size, high cost, and shortened lifetime. Compared to conventional voltage-fed isolated DC/DC converters, current-fed converters are meritorious owing to lower input current ripples, inherent short circuit protection, lower high-frequency (HF) transformer turns ratio, high step-up ratio, no duty cycle loss, and easier current controllability.

5.4.2 COMPREHENSIVE COMPARATIVE EVALUATION OF CURRENT-FED BIDIRECTIONAL DC/DC CONVERTER

For the performance evaluation of a power electronic converter, five significant aspects of converters are considered: cost, volume, weight, losses, and power density. The cost models and loss models are presented in the references [40]. The component cost models for MOSFET, transformer, inductor, capacitor, heatsink, and gate driver ICs of the converters are included. Different loss models, including switching and conduction losses of MOSFET, inductor losses, and transformer losses, are presented as well.

A quantitative and comprehensive performance comparison is conducted for the seven typical converters (single-phase active clamped circuit in Figure 5.13 [11], dual half-bridge circuit in Figure 5.17 [18], L-L type dual active bridge circuit in Figure 5.25 [26], resonant type circuit in Figure 5.30 [31], L-L type naturally clamped circuit in Figure 5.34 [37], naturally clamped full-bridge circuit in Figure 5.35 [39], and three-phase active clamped circuit in [41]), which are selected from each type of all reviewed current-fed topologies. In order to have a fair comparison of these topologies, a virtual prototype of each topology has been designed for the same specification: 20–40 V low voltage input, 400 V high voltage output, 2 kW power rating, 100 kHz switching frequency, 10 A input current ripple, and 4V output voltage ripple. To verify the design parameters of each virtual prototype and assure that they coincide closely with the original papers, simulations have been performed on PSIM 9.1 as well [40].

Current-fed converter would be a smart choice for applications with low voltage high current. Typical applications include auxiliary DC/DC converter to interface low voltage energy storage (12/24/48 V) and high voltage dc bus in the range of 200–900 V for transportation electrification, renewable energy generation such as solar photovoltaic and fuel cells with low voltage energy storage, and hybrid dc microgrid with multiple voltage level, etc. Different application scenario has different electrical specifications and performance requirements in terms of cost, efficiency, weight, volume, and reliability, which results in different selections of reviewed current-fed converter.

As illustrated by Figure 5.34, naturally clamped and L-L type dual active bridge current-fed converters provide low-cost solutions. However, if efficiency and lifetime can be compromised, dual half-bridge current-fed converter has the potential of lowest cost. For the selected dual half-bridge current-fed converter, capacitor, inductor, and transformer contribute to a large proportion of the total cost. If electrolytic capacitor is allowed to be employed to mitigate input current ripple and achieve voltage clamping, the cost of capacitor and inductor can be reduced enormously. Compared with dual half-bridge current-fed converter in Figure 5.17, the cost of devices and transformer can be much reduced. With the considerable reduction of capacitors, inductors, transformers, and devices, the dual half-bridge current-fed converter is capable to achieve the lowest cost. Dual half-bridge current-fed converter is suitable for applications like small urban city driving or residential driving (driving range of 100 km) where cost is valued more than efficiency and lifetime.

The naturally clamped current-fed converter represents the optimal topologies amongst the selected options and for the given system specifications and design constraints as shown in Figures 5.37, 5.38, 5.39, 5.40, 5.41. They can achieve high efficiency and high power density while saving a large part of costs without the need of external snubbers. Since voltage clamping is achieved completely through software control, its reliability under different operating scenarios like extreme load, wide transient, and failure conditions needs to be researched and evaluated further. Another issue is that control schemes are different for bidirectional power transfer, and the seamlessly transfer between boost mode and buck mode has not yet been reported. Until these issues are solved, it is recommended to apply current-fed converter to applications like hybrid dc microgrid, where cost and efficiency are weighted more than reliability.

As the energy demand from low voltage dc source on EVs/More Electric Aircraft (MEA)/ All Electric Aircraft (AEA) etc. keeps increasing, the need for bidirectional DC/DC converters capable of handling large currents with high

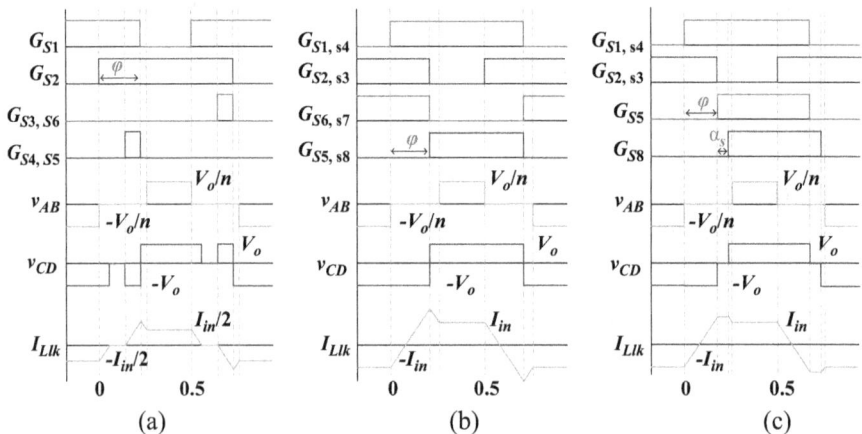

FIGURE 5.36 Key operating waveforms of three types of secondary-modulation-based naturally clamping techniques [37–39].

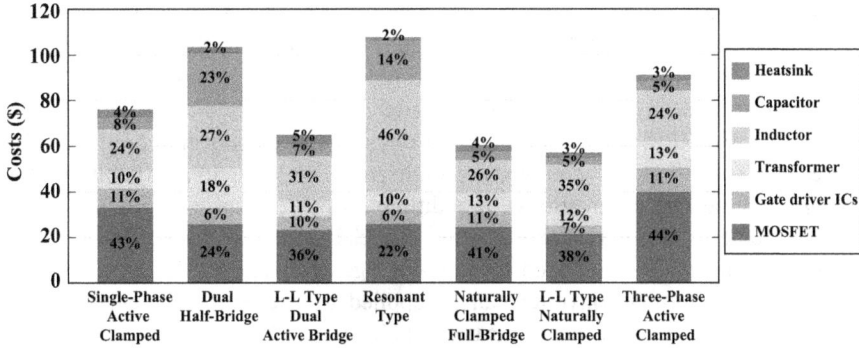

FIGURE 5.37 Costs distribution and comparison of different types of current-fed converter.

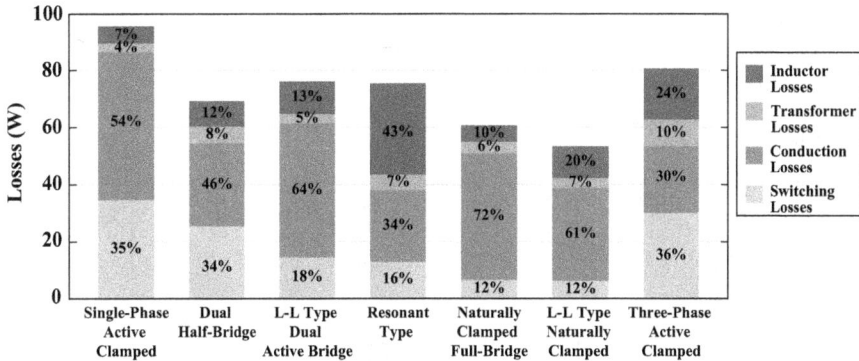

FIGURE 5.38 Losses distribution and comparison of different types of current-fed converter.

reliability, high efficiency, high power density, and low cost is growing. Compared with naturally clamped current-fed converter, control of bidirectional power transferring of L-L type dual active bridge is symmetrical and easy to be implemented. L-L type dual active bridge can achieve high efficiency over a wide range of input voltage/output voltage/loads variations, and hardware voltage clamping is more reliable. This is quite critical for a lot of applications. For example, the voltage across the energy storages varies under different operating conditions, and in many cases, the high voltage dc bus is not fixed. In electrical vehicles, high voltage dc bus varies in a wide range to achieve the highest "fuel economy." For renewable energy generation systems, if the high-voltage bus is directly connected to solar photovoltaic and fuel cells, its output voltage varies. Therefore, L-L type dual active bridge is quite advantageous for these applications. With the exception of naturally clamped topology, L-L type dual active bridge Current-fed converter provides optimized overall performances as shown in Figure 5.41, which makes it the best candidate for the transportation electrification application and renewable energy generation system.

L-L type dual active bridge is quite flexible and can be easily modified into multiport structure as well. This feature makes it suitable for the applications like hybrid energy storage system or the hybrid renewable energy generation system.

For the resonant type current-fed converter, owing to circulating current, peak (above 2x) and RMS current through the devices and components are high. This makes the ratings of the components high and their selection difficult for low voltage high current applications. In addition, its power density is limited by the passive elements of the resonating bank, especially for low voltage high current condition. This can be alleviated for high-voltage applications. Therefore, the resonant type current-fed converter can be applied to interface high-voltage power battery in transportation electrification application. Or like wireless charging application, the heavy and bulky devices are placed off-the-board.

For low voltage high current application, voltage stress of primary switches is quite critical to achieve high efficiency. As given by the table, voltage stress of half-bridge derived topology doubles the full-bridge derived topology and push-pull

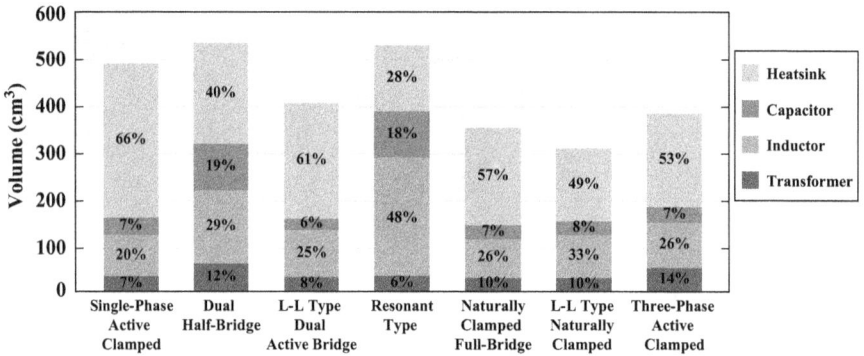

FIGURE 5.39 Volume distribution and comparison of different types of current-fed converters.

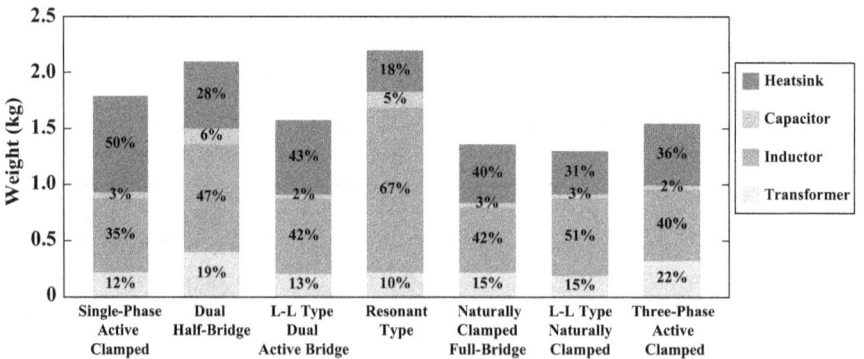

FIGURE 5.40 Weight distribution and comparison of different types of current-fed converters.

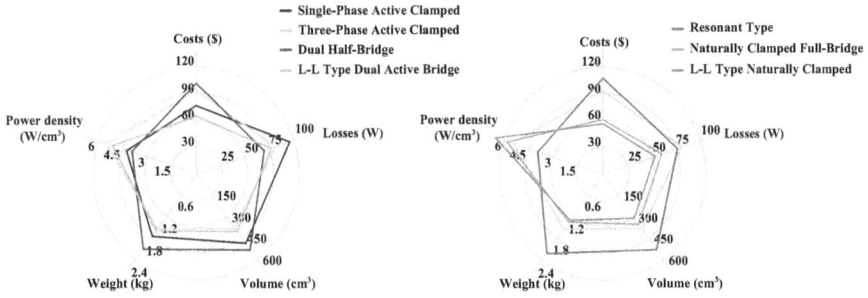

FIGURE 5.41 Comparison of different types of current-fed converters regarding costs, losses, volume, weight, and power density.

topology. Thus, for the low voltage application less than 30 V (like 12/24 V storage), half-bridge derived topology is competitive. Once the voltage specification goes higher than 30 V, the choice between half-bridge-derived topology and full-bridge-derived topology needs to be evaluated comprehensively. Generally, L-L type dual active bridge is one special type of active clamped topology. Other types of active clamped current-fed converter have the potential to be applied to transportation electrification application and renewable energy generation system as well with different design specifications.

As discussed above, three-phase topology is promising for high-power applications with the advantages of reduced current rating in active components, and smaller size of reactive components compared with the corresponding single-phase converter. The three-phase converter is capable of working with partial power even if one of the phases fails, thus providing higher reliability. Three-phase current-fed converter plays an important role in the high power application that values more about reliability, efficiency, and power density more than cost. Three-phase converters can be applied to transportation electrification applications with the need for heavy power and high reliability.

In practical design, it depends on the engineering design and markets constraints whether particular topologies are favorable for the specific design specifications.

5.4.3 FUTURE TREND

In view of the discussion and evaluation of state-of-the-art current-fed converters in previous sections, many researchers have focused on the topology, modulation strategy, soft-switching solution, and hardware design optimization. In the future, the improvement of the converter efficiency and power density while lowering the costs and weight will be the trend. In addition, reliability and modularity are the other two significant indicators, which are concerned with the development of smart grid and energy internet. Furthermore, as the mediation module, converter should achieve wide input voltage variation and output voltage variation range on the premise of high efficiency. New topologies and corresponding modulation and soft-switching techniques are expected to be further researched for wide input voltage and wide output voltage applications. The utilization of SiC/GaN power devices will allow the design of higher switching frequency and lower switching loss of converters, thus contributing to higher power density. But it is worth noting that presently SiC/GaN-based power semiconductors can't compete with Si devices in terms of the current rating, while the conduction loss plays a dominant role in low voltage high current applications instead of the switching loss. Therefore, a tradeoff between device selection and usage should be made in practice.

REFERENCES

[1] Han, S. K., Yoon, H. K., Moon, G. W., Youn, M. J., Kim, Y. H. & Lee, K. H., "A new active clamping zero-voltage switching PWM current-fed half-bridge converter," *IEEE Trans. Power Electron.*, vol. 20, no. 6, pp. 1271–1279, 2005.

[2] Wu, T. F., Chen, Y. C., Yang, J. G. and Kuo, C. L., "Isolated bidirectional full-bridge DC–DC converter with a flyback snubber," *IEEE Trans. Power Electron.*, vol. 25, no. 7, 2010, pp. 1915–1922.

[3] Qiao, C. and Smedley, K. M., "An isolated full bridge boost converter with active soft switching," in *Proc. 32nd IEEE Annual Power Electronics Specialists Conference (PESC)*, vol. 2, pp. 896–903, 2001.

[4] Jang, S. et al, "Fuel cell generation system with a new active clamping current-fed half-bridge converter," *IEEE Trans. Energy Conversion.*, vol. 22, no. 2, pp. 332, 2007.

[5] Averberg, A., Meyer, K. R. and Mertens, A., "Current-fed full bridge converter for fuel cell systems," in *Proc. IEEE Power Electronics Specialists Conference (PESC)*, pp. 866–872, 2008.

[6] Leung, A. S. W., Chung, H. S. H. and Chan, T., "A ZCS isolated full-bridge boost converter with multiple inputs," in *Proc. IEEE Power Electronics Specialists Conference (PESC)*, pp. 2542–2548, 2007.

[7] Wang, H. et al, "A ZCS current-fed full-bridge PWM converter with self-adaptable soft-switching snubber energy," *IEEE Trans. Power Electron.*, vol. 24, no. 8, pp. 1977–1991, 2009.

[8] Mousavi, A., Das, P. and Moschopoulos, G., "A comparative study of a new ZCS DC–DC full-bridge boost converter with a ZVS active-clamp converter," *IEEE Trans. Power Electron.*, vol. 27, no. 3, pp. 1347–1358, 2012.

[9] Wu, T. F., Chen, Y. C., Yang, J. G. and Kuo, C. L., "Isolated bidirectional full-bridge DC-DC converter with a flyback snubber," *IEEE Trans. on Power Electronics*, vol. 25, no. 7, pp. 1915–1922, July 2010.

[10] Averberg, A., Meyer, K. R. and Mertens, A., "Current-fed full bridge converter for fuel cell systems," in *Proc. IEEE Power Electronics Specialists Conference (PESC)*, pp. 866–872, 2008.

[11] U R, P. and Rathore, A. K., "Extended range ZVS active-clamped current-fed full-bridge isolated DC/DC converter for fuel cell applications: Analysis, design, and experimental results," *IEEE Trans. on Industrial Electronics*, vol. 60, no. 7, pp. 2661–2672, July 2013.

[12] Rathore, A. K., Bhat, A. K. S. and Oruganti, R., "Analysis, design and experimental results of wide range ZVS active-clamped L-L type current-fed DC/DC converter for fuel cells to utility interface," *IEEE Trans. on Industrial Electronics*, vol. 59, no. 1, pp. 473–485, January 2012.

[13] Shi, Y. Y., Li, R., Xue, Y. and Li, H., "Optimized operation of current-fed dual active bridge DC-DC converter for PV applications," *IEEE Trans. on Industrial Electronics*, vol. 62, no. 11, pp. 6986–6995, November 2015.

[14] Jang, S. et al, "Fuel cell generation system with a new active clamping current-fed half-bridge converter," *IEEE Trans. Energy Conversion.*, vol. 22, no. 2, pp. 332, 2007.

[15] Shimada, T., Shoji, H. and Taniguchi, K., "Two novel control methods expanding input-output operating range for a bi-directional isolated DC-DC converter with active clamp circuit," in *2012 IEEE Energy Conversion Congress and Exposition (ECCE)*, Raleigh, NC, pp. 2537–2543, 2012.

[16] Shimada, T., Shoji, H. and Taniguchi, K., "A novel scheme for a bi-directional isolated DC-DC converter with a DC-link diode using reverse recovery current," in *2014 16th European Conference on Power Electronics and Applications*, Lappeenranta, pp. 1–7, 2014.

[17] Al-Atbee, O. Y. K. and Bleijs, J. A. M., "Improved modified active clamp circuit for a current fed DC/DC power converter," in *2015 IEEE 6th International Symposium on Power Electronics for Distributed Generation Systems (PEDG)*, Aachen, pp. 1–7, 2015.

[18] Peng, F. Z., Li, H., Su, G.-J. and Lawler, J. S., "A new ZVS bidirectional DC-DC converter for fuel cell and battery application," *IEEE Trans. on Power Electronics*, vol. 19, no. 1, pp. 54–65, January 2004.

[19] Ma, G., Qu, W., Yu, G., Liu, Y., Liang, N. and Li, W., "A zero-voltage-switching bidirectional DC–DC converter with state analysis and soft-switching-oriented design consideration," *IEEE Trans. on Industrial Electronics*, vol. 56, no. 6, pp. 2174–2184, June 2009.

[20] Sun, X., Wu, X., Shen, Y., Li, X. and Lu, Z., "A current-fed isolated bidirectional DC–DC converter," *IEEE Trans. on Power Electronics*, vol. 32, no. 9, pp. 6882–6895, September 2017.

[21] Rathore, A. K. and Prasanna, U. R., "Analysis, design, and experimental results of novel snubberless bidirectional naturally clamped ZCS/ZVS current-fed half-bridge DC/DC converter for fuel cell vehicles," *IEEE Trans. on Industrial Electronics*, vol. 60, no. 10, pp. 4482–4491, October 2013.

[22] Xiao, H. and Xie, S., "A ZVS bidirectional DC-DC converter with phase-shift plus PWM control scheme," *IEEE Trans. on Power Electronics*, vol. 23, no. 2, pp. 813–823, March 2008.

[23] Sha, D., Wang, X. and Chen, D., "High efficiency current-fed dual active bridge DC–DC converter with ZVS achievement throughout full range of load using optimized switching patterns," *IEEE Trans. on Power Electronics*, vol. 33, no. 2, pp. 1347–1357, February 2018.

[24] Guo, Z., Sun, K., Wu, T. and Li, C., "An improved modulation scheme of current-fed bidirectional DC–DC converters for loss reduction," *IEEE Trans. on Power Electronics*, vol. 33, no. 5, pp. 4441–4457, May 2018.

[25] Sha, D., Wang, X., Liu, K. and Chen, C., "A current-fed dual-active-bridge DC–DC converter using extended duty cycle control and magnetic-integrated inductors with optimized voltage mismatching control," *IEEE Trans. on Power Electronics*, vol. 34, no. 1, pp. 462–473, January 2019.

[26] Zhang, J. and Sha, D., "A current-fed dual active bridge DC-DC converter using dual PWM plus double phase shifted control with equal duty cycles," in *2016 Asian Conference on Energy, Power and Transportation Electrification (ACEPT)*, Singapore, pp. 1–6, 2016.

[27] Sha, D., Xu, Y., Zhang, J. and Yan, Y., "Current-fed hybrid dual active bridge DC-DC converter for a fuel cell power conditioning system with reduced input current ripple," *IEEE Trans. on Industrial Electronics*, vol. 64, no. 8, pp. 6628–6638, August 2017.

[28] Ding, Z., Yang, C., Zhang, Z., Wang, C. and Xie, S., "A novel soft-switching multiport bidirectional DC–DC converter for hybrid energy storage system," *IEEE Trans. on Power Electronics*, vol. 29, no. 4, pp. 1595–1609, April 2014.

[29] Samanta, S., Rathore, A. K. and Thrimawithana, D. J., "Bidirectional current-fed half-bridge (C)(LC)-(LC) configuration for inductive wireless power transfer system," *IEEE Trans. on Industry Applications*, vol. 53, no. 4, pp. 4053–4062, July–August 2017.

[30] Wu, H., Sun, K., Li, Y. and Xing, Y., "Fixed-frequency PWM-controlled bidirectional current-fed soft-switching series-resonant converter for energy storage applications," *IEEE Trans. on Industrial Electronics*, vol. 64, no. 8, pp. 6190–6201, August 2017.

[31] Li, Y., Xing, Y., Lu, Y., Wu, H. and Xu, P., "Performance analysis of a current-fed bidirectional LLC resonant converter," in *IECON 2016 – 42nd Annual Conference of the IEEE Industrial Electronics Society*, Florence, pp. 2486–2491, 2016.

[32] Noh, Y. S., Won, C. Y., Oh, M. S., Jeon, J. Y. and Jung, Y. C., "Design and analysis of isolated bi-directional DC/DC converter using quasi-resonant ZVS," in *2014 International Power Electronics Conference (IPEC-Hiroshima 2014 –ECCE ASIA)*, Hiroshima, pp. 166–171, 2014.

[33] Jegal, J.-H., Noh, Y.-S. and Won, C.-Y., "Control method for increasing efficiency of current-fed isolated bi-directional dual half-bridge converter using resonant switch," in *2016 IEEE Transportation Electrification Conference and Expo, Asia-Pacific (ITEC Asia-Pacific)*, Busan, pp. 302–307, 2016.

[34] Sree, K. R. and Rathore, A. K., "Impulse commutated zero-current switching current-fed push–pull converter: Analysis, design, and experimental results," *IEEE Trans. on Industrial Electronics*, vol. 62, no. 1, pp. 363–370, January 2015.

[35] Zhu, L., "A novel soft-commutating isolated boost full-bridge ZVS-PWM DC–DC converter for bidirectional high power applications," *IEEE Trans. on Power Electronics*, vol. 21, no. 2, pp. 422–429, March 2006.

[36] Reimann, T., Szeponik, S., Berger, G. and Petzoldt, J., "A novel control principle of bi-directional DC-DC power conversion," in *Proc. IEEE PESC'97 Conf.*, pp. 978–984, 1997.

[37] Prasanna, U. R., Rathore, A. K. and Mazumder, S. K., "Novel zero-current-switching current-fed half-bridge isolated DC/DC converter for fuel-cell-based applications," *IEEE Trans. on Industry Applications*, vol. 49, no. 4, pp. 1658–1668, July–August 2013.

[38] Xuewei, P. and Rathore, A. K., "Novel bidirectional snubberless naturally com-mutated soft-switching current-fed full-bridge isolated DC/DC converter for fuel cell vehicles," *IEEE Trans. on Industrial Electronics*, vol. 61, no. 5, pp. 2307–2315, May 2014.

[39] Bal, S., Yelaverthi, D. B., Rathore, A. K. and Srinivasan, D., "Improved modulation strategy using dual phase shift modulation for active commutated current-fed dual active bridge," *IEEE Trans. on Power Electronics*, vol. 33, no. 9, pp. 7359–7375, September 2018.

[40] Pan, X., Li, H., Liu, Y., Zhao, T., Ju, C. and Rathore, A. K., "An overview and comprehensive comparative evaluation of current-fed-isolated-bidirectional DC/DC converter," in *IEEE Trans.s on Power Electronics*, vol. 35, no. 3, pp. 2737–2763, March 2020.

[41] Song, Y. et al, "A current-fed three-phase half-bridge dc-dc converter with active clamping," in *2009 IEEE Energy Conversion Congress and Exposition*, San Jose, CA, pp. 1362–1366, 2009.

6 Wireless Inductive Power Transfer

Suvendu Samanta and Akshay Kumar Rathore

CONTENTS

DOI: 10.1201/9781003330134-6

6.1　INTRODUCTION

Research on wireless power transfer (WPT) technology has gained significant popularity in recent years due to the availability of high-power and high-switching frequency semiconductor devices. The advantages of this technology are as follows [1,2]:

1. very convenient, safe, and reliable due to the elimination of direct electrical contact,
2. the power transfer is unaffected in hostile environments such as snow, water, dirt, wind, and chemicals, and
3. it provides galvanic isolation.

Besides these general advantages, the WPT technology has merits specific to particular applications. In biomedical implants, e.g., in heart pump battery recharging, WPT technology is the most practical and convenient [2,3]. Similarly, WPT has found wide acceptance for recharging batteries of electronic gadgets, lighting, chemical plants, underwater vehicles, etc., due to its flexible usage and the ability to prevent damage to the charging port.

Implementation of wireless charging in EV applications provides remarkable outcomes. Along with the merits mentioned above, it can reduce the battery storage requirement to 20% through opportunistic charging techniques [1,4]. For EVs, opportunistic charging is possible by placing wireless chargers in parking areas, e.g., homes, offices, services, shopping complexes, and other general parking areas. Also, these chargers can be installed in the traffic signal areas for quick recharging. WPT chargers can be installed in bus terminals, bus stops, and traffic signals for recharging electric buses. These types of chargers are called static WPT chargers. Intense research is being carried out on dynamic WPT technology, where the EV battery can recharge while the vehicle is in motion [5,6]. However, there are several challenges at the technical and infrastructural or initial investment levels.

6.2 TYPES OF WPT TECHNOLOGIES

This section presents a brief overview and qualitative comparison of several possible WPT technologies. Based on this study, the most effective technology is selected for medium power (fraction of kW to several kW) and mid-range airgap (about 100–350 mm) applications, which are especially suitable for EV battery charging.

6.2.1 Inductive WPT

Inductive charging systems generally consist of a primary side power converter, an inductive interface transformer, and a secondary converter [7]. The interface transformer is separable along the magnetic circuit so that one of the windings can be physically removed. This eliminates the need for ohmic, i.e., metal-to-metal contact of electric wires. In such a transformer, the shape and location of the magnetic core material and windings are crucial design choices. This technology is already implemented in EV charging systems such as the GM EV1 [4], as shown in Figure 6.1. The charging paddle (the primary coil) of the inductively coupled charger is sealed in epoxy as it is done in the secondary. The paddle inserted into the center of the secondary coil permitted charging of the EV1 without any contacts or connectors at either 6.6 kW or 50 kW. As depicted in Figure 6.1, this system is connector-less but not wireless.

6.2.2 Capacitive Power Transfer

Wireless capacitive power transfer (CPT) technology is also an alternate WPT solution [8]. Figure 6.2 shows a typical CPT system fed from a half-bridge voltage

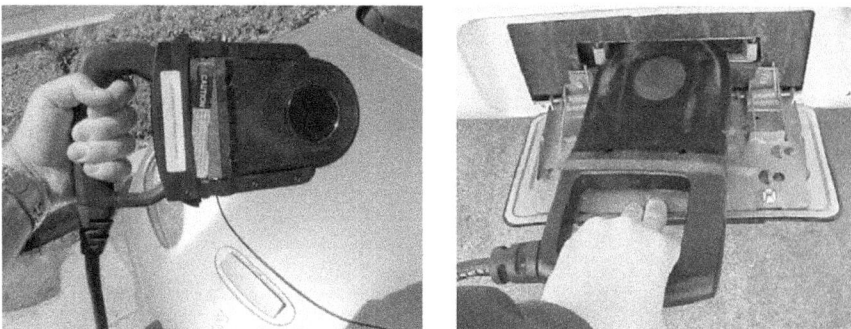

FIGURE 6.1 Inductive charger paddle (left) and electric vehicle charging through the paddle (right) [4].

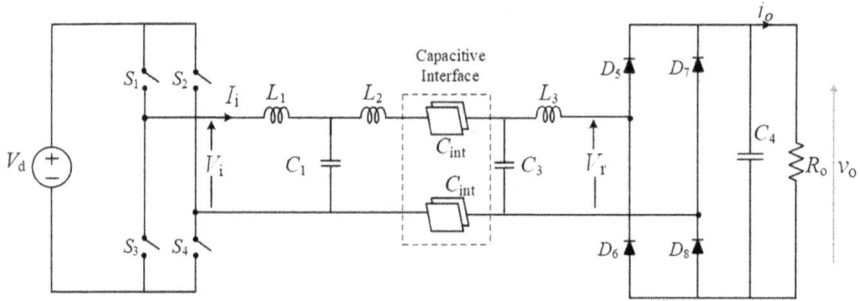

FIGURE 6.2 A typical CPT system with LCL compensation on the primary side and LC compensation on the secondary [8].

source inverter, and the primary and secondary side compensation networks are *LCL* and *LC* types, respectively. As shown in Figure 6.2, the CPT interface is constructed around a pair of coupling capacitors. The operating principle is the same as usual parallel capacitors, where the dielectric medium is only air. CPT may compete with inductive coupling at a higher operating frequency. The former can offer equally good galvanic isolation and does not require a costly, high-frequency-rated magnetic core. This technology finds suitable applications in low power levels such as biomedical implants or charging of space-confined systems such as robots or mobile devices [1,8,9], etc. Its design flexibility and low cost make it ideal for power delivery in reconfigurable and moving systems, such as robot arms, latches, and in-track-moving systems [8]. However, owing to lower power density, CPT technology is not preferred for higher power applications such as EV charging [4].

6.2.3 RESONANT INDUCTIVE POWER TRANSFER

In inductive coupling technology, if a significant air gap is introduced, the power transfer between the primary and secondary coil becomes ineffective. This fact is illustrated using a simple two-winding coupled inductor model shown in Figure 6.3. The mutual inductance, M reduces very fast with an increase in the air gap between the coils. Therefore, to have a significant amount of induced voltage $(=\omega M i_1)$ in the secondary side, the operating frequency (ω) and primary coil current (i_1) must be high. Owing to the wide availability of high voltage, high current, and high switching frequency semiconductor devices, ω can range from kHz to hundreds of kHz. However, the high leakage impedance of the primary coil introduces another

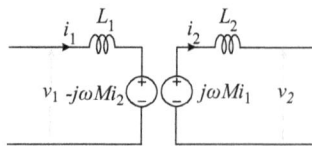

FIGURE 6.3 Two winding coupled inductor model.

difficulty in driving a significant amount of current (i_1) through the coil. A reactive power compensation circuit can be placed on the primary side to drive a higher coil current. Therefore, the primary side converter does not need to feed high voltage to the coil. Although with these arrangements, it is possible to induce a significant amount of voltage at the secondary coil, most of it is dropped across the high secondary leakage impedance. Therefore, another reactive power compensation is also required on the secondary side to transfer power effectively. Usually, these compensation elements on both sides resonate with the coil inductances; therefore, this technology is called resonant inductive power transfer (RIPT). This technology was pioneered and patented by Nikola Tesla [10]. This technology is most suitable for transferring power with an airgap from a few mm to hundreds of mm, where the power level varies from a fraction of a watt to hundreds of kW. Owing to high power density, high efficiency, wide power range, and acceptable airgap length, RIPT technology is the most demanding among all other WPT technologies. It finds applications in EV charging, electronic gadgets, implants, chemical factory, underwater vehicles, lighting, etc. [1–3]. In the literature, this technology is often called inductive power transfer (IPT), and throughout this chapter, it is named IPT.

6.2.4 Resonant Antennae Power Transfer

Resonant Antennae Power Transfer (RAPT) was also pioneered and patented by Nikola Tesla and has recently been studied by MIT [10] and Intel. The fundamental operating principle of this technology is similar to IPT. RAPT uses two or more resonant antennae tuned to the same frequency. The resonant capacitances and inductances are integrated into the antennae. These systems often have large WPT coils (antennae), often helical with controlled separation between the turns to obtain a distributed and integrated, resonant capacitance. The airgap length can be much longer than the IPT system due to the use of high-quality factor coils and the high frequency of operation. Acceptably, efficient power transfer is possible at distances up to approximately 10 meters, and the operating frequency is in the MHz range [11]. However, for several kW power transfers with airgap suitable for EV applications, this RAPT technology is essentially the same as IPT technology.

6.3 COUPLED INDUCTOR MODEL OF IPT

Considering the advantages of IPT technology, a detailed theory is included. The basic operating principle is like a two-winding transformer (or a coupled inductor), where the magnetizing inductance (or coil coupling factor) is much lower than a conventional iron core transformer (or inductor) [4]. From Figure 6.3 coupled inductor model, the voltage across primary and secondary terminals is given as

$$v_1 = L_1 \frac{di_1}{dt} + M \frac{di_2}{dt} \tag{6.1}$$

$$v_2 = L_2 \frac{di_2}{dt} + M \frac{di_1}{dt} \tag{6.2}$$

where $M = k\sqrt{L_1 L_2}$; k is the coefficient of coupling between TC (primary) and RC (secondary); L_1 and L_2 are self-inductances of primary and secondary coils, respectively. For a given primary coil current, I_1, the open-circuit induced voltage, and short circuit current through L_2 are given as

$$V_{oc} = \omega M I_1, \quad I_{sc} = \frac{V_{oc}}{\omega L_1} = I_1 \frac{M}{L_2}. \tag{6.3}$$

When the system is tuned at resonance frequency with compensation capacitors, the available power is derived as [4]

$$P = \frac{V_{oc}^2}{R_e} = \omega \frac{M^2}{L_2} I_1^2 Q = \omega L_1 I_1 I_2 \frac{M^2}{L_1 L_2} Q \tag{6.4}$$

where $Q = \omega L_2 / R_e$ and R_e is the load resistance.

Several IPT topologies are reported in the literature based on compensation topologies, power converter topologies, IPT coils, etc. Generally, all these IPT systems follow a similar power conversion stage, as shown in Figure 6.4. Generally, the input is line frequency ac, which is rectified by an active rectifier to draw power at the unity power factor (UPF). Next, the dc-ac inverter injects high-frequency ac into the primary compensation network. The power is extracted effectively using another compensation network on the secondary side. Finally, this ac power is rectified either by active or passive rectifiers. A detailed study of existing IPT systems is included, and descriptions are reported based on different classifications.

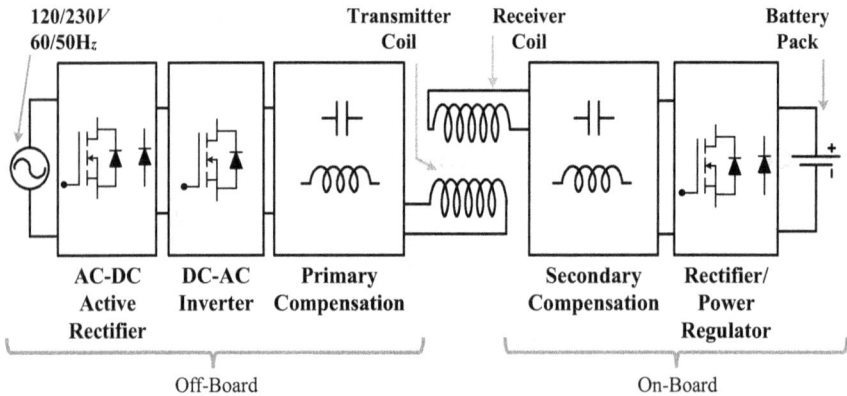

FIGURE 6.4 A typical electric vehicle wireless charging system.

6.4 IPT COMPENSATION TOPOLOGIES

The existing IPT topologies can be broadly classified based on basic series and parallel compensations or combinations of these basic compensations.

6.4.1 BASIC COMPENSATION TOPOLOGIES

With an airgap ranging from 150 mm to 300 mm for EV battery charging applications, the TC to RC coupling is generally very low (typically below 0.3), and compensation on the TC and RC side is mandatory. By adding one capacitor to each side of the coils, four basic types of compensation networks are formed, as shown in Figure 6.5. Several research papers provide a detailed study of these networks, especially for the series-series (S/S) and series-parallel (S/P). From Figure 6.5, the selection of inverter type is dependent on the TC tank network. When the TC side compensation is series type, the inverter is VSI, whereas the inverter is CSI when TC side compensation is parallel type.

The S/S topology is the simplest design, where compensation is load and coupling independent. However, the series RC tank network provides almost zero reflected impedance to the TC coil during no load or light load. Therefore, the VSI output current surges when RC is uncoupled from TC, and the system becomes unstable [12]. This issue is generally taken care of with a fast closed-loop control. The parallel LC tank at the RC side in S/P compensation always reflects some impedance to the TC coil; hence, the current surge does not arise. However, surge current will also appear at VSI output for this case if the RC coil is moved out. As per the study, the S/S topology provides higher efficiency than the S/P topology in a wide range of load resistance [13–15]. Moreover, a parallel-compensated system has a large reactive current in the receiver coil, and the reactive power is reflected on the primary side [16]. Considering all these issues, most IPT topologies use S/S compensation for EV charging applications.

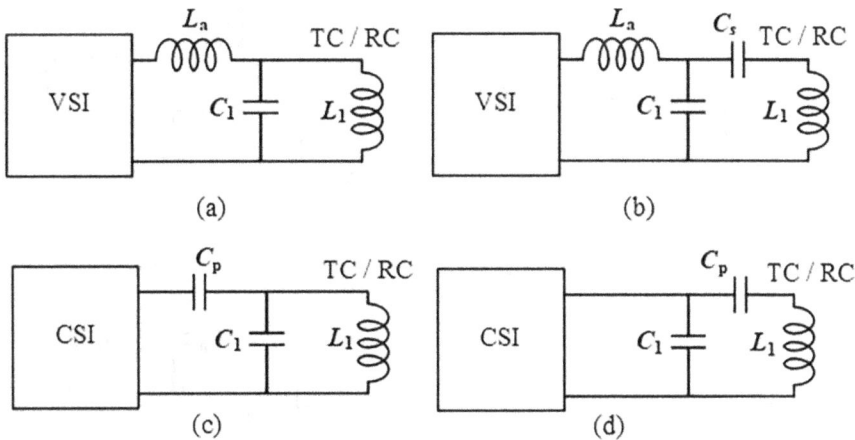

FIGURE 6.5 Basic compensation topologies for IPT circuit (a) Series-Series (S/S); (b) Series- Parallel (S/P); (c) Parallel- Series (P/S); (d) Parallel- Parallel (P/P).

The P/S and P/P compensation networks are realized with the current source inverter (CSI). Generally, a dc link inductor is required to function CSI [2,17–20]. Because the previous stage PFC output is generally stiff voltage; hence, this dc link inductor is an extra bulky component. Research on P/S and P/P compensation networks was quite limited to only lower power. However, in applications where the stiff current is readily available, P/S and P/P topologies are comparatively more suitable than the S/S and S/P topologies. The major advantages of these topologies are as follows:

1. High magnitude TC current circulates through the parallel capacitor without flowing through the devices [2]. In parallel compensation, the effective magnetizing impedance becomes very high, and therefore, the current stress on inverter devices is lower;
2. The coil current corresponding to a parallel tank is very close to sinusoidal because higher-order harmonic currents predominantly flow through the lower impedance offered by the capacitor;
3. Unlike S/S and S/P compensations, in P/S and P/P compensation, the inverter output never gets shorted during coil uncoupled conditions;
4. The CSI dc link inductor provides natural short-circuit protection.

6.4.2 COMBINATIONS OF BASIC TOPOLOGIES

Considering the merits and demerits of the basic compensation techniques, several complex combinations of these four topologies are reported to improve performance. Figure 6.6 shows the most frequently reported compensation networks for their added advantages. Figure 6.6 shows three general compensation network diagrams. Any of them could be selected to form either TC or RC compensation networks. Often, one side tank network is selected from one of these three topologies, and the other side is selected as a basic series or parallel compensation network. It is clear from here that

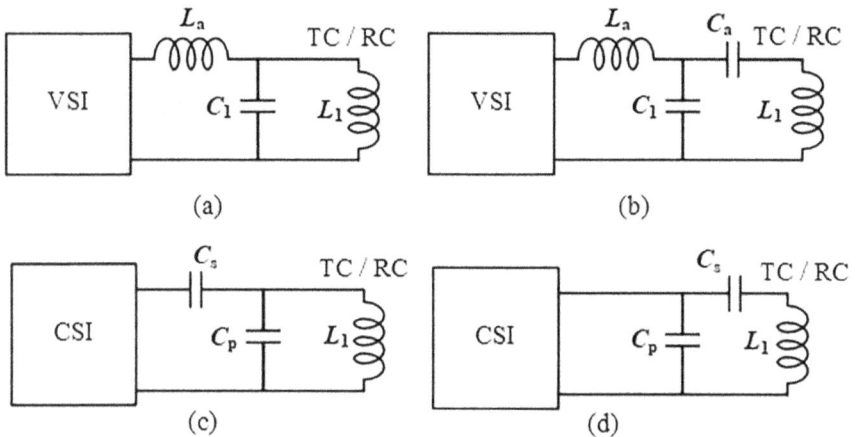

FIGURE 6.6 Combination of basic compensation topologies. (a) *LCL* tank; (b) *LCLC*; (c) Series; (d) parallel (SP) tank.

several IPT topologies are possible based on the combinations of these compensation networks. In fact, as per the existing research on IPT topologies is concerned, almost all these possible networks are already studied or are being studied.

Figure 6.6a *LCL* tank network is very common for its simple yet comparatively improved performances over simple parallel or series tank [12,21–25]. It includes the advantages of parallel *LC* tank networks while eliminating the limitations of the series tank. Unlike parallel tanks, this topology is fed from VSI; hence, the bulky dc link inductor is eliminated. However, a lightweight inductor, L_f, in the high-frequency tank network is required. Unlike the dc inductor used in CSI-fed parallel tank, L_f carries AC and is highly sensitive to effective power transfer; therefore, L_f requires high precision, which incurs additional cost. When the LCL tank is used on the TC side, it does not create instability issues like the series tank during no-load conditions [12]. This tank is well used for bi-directional IPT applications [21,22].

Y. Yao *et al.* [23] report that the VSI output current with T type compensation topology contains a significant amount of lower order harmonics (3rd, 5th, and 7th, etc.), and they have 90° phase differences from their respective harmonic voltages for e.g., 3rd harmonic current and voltage has 90° phase difference. Therefore, although the fundamental component voltage and current are in the same phase, significant harmonic content in the current deviates significantly from the UPF operation of the inverter. Although this fact is very well visible in several reported works, [23] reports the details of impedance analysis to justify this fact. Therefore, UPF operation with an *LCL* tank is not as significant as the parallel compensated primary topology, where the inverter output voltage is very close to sinusoidal. From [2,17,18], it is evident that the parallel tanks are mostly attempted to operate at the ZPA point.

In *LCL* topology, adding one extra capacitor in series with the coil makes the *LCCL* tank, as shown in Figure 6.6b [26,27]. Even though the analysis of this topology is like an *LCL* tank, this is an improved version. This extra capacitor directly reduces the coil leakage impedance on the transmitter side. This improves power transfer capability [28]. Ts tank is used to achieve a unity power factor pickup [16].

Figure 6.6c compensation network is suitable on the TC side when the inverter is a current-fed type. This compensation network is termed a series-parallel–series (SPS) topology in [29]. However, this converter is tested with a voltage source inverter where an extra inductor is added in series with the series capacitor, C_p to make it compatible with VSI. This compensation on the TC side and series *LC* compensation on the RC side have the capacity to deliver rated power with wider coil misalignments [29].

6.5 IPT POWER CONVERTER TOPOLOGIES

The input power of the IPT inverter is usually supplied from a dc source, such as PV modules or a rectified ac grid, as shown in Figure 6.4. This converter aims to feed high-frequency ac to the primary resonant tank and maintain load power to the desired level. Several IPT power converter topologies are reported in the literature, and these are classified as follows.

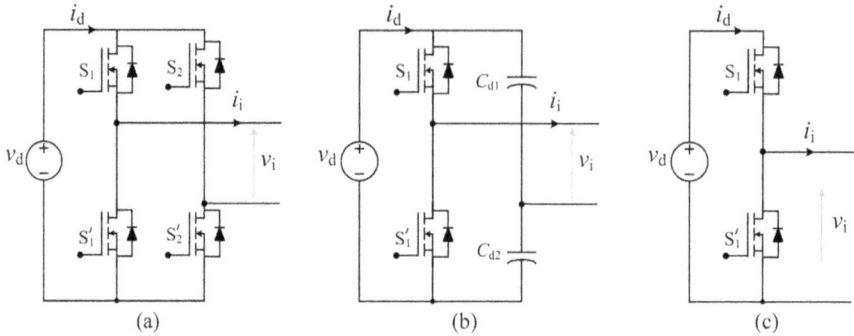

FIGURE 6.7 (a) Full-bridge, (b) half-bridge, and (c) semi-bridgeless VSI topologies for IPT systems.

6.5.1 VSI TOPOLOGIES

The voltage-fed H-bridge and half-bridge power converters are shown in Figure 6.7a and b [30], respectively. The half-bridge converter employs two capacitors in the second leg to provide the neutral point of input dc voltage. Compared with an H-bridge converter, a half-bridge converter is simpler to control and reduces the number of switching devices, which reduces switching losses. However, the output voltage of the neutral point between two capacitors is normally unbalanced during the switching process of two switches. Another limitation of the half-bridge topology is that the converter can only generate ac output with the amplitude of $\pm v_d/2$, which limits its application only in lower power. Therefore, in practice, H-bridge converters are preferable in most applications.

In IPT systems VSI is compatible with series, *LCL*, or *LCCL* compensation networks. Another power converter can also be used to control the output power on the secondary side. If another H-bridge converter is employed on the secondary side, it becomes a bidirectional IPT system. The power level, power flow direction, and input power factor can be controlled by adjusting the phase shift angle between two converters [21,22]. Figure 6.7c shows a semi-bridgeless VSI, which is reported to have very high efficiency (94.4% at1 *k*W) with phase shift control [31].

6.5.2 DIRECT AC-AC CONVERTER TOPOLOGIES

To improve overall efficiency and reduce the component count of the complete converter system, direct ac-ac converter is reported to generate switching frequency ac directly from line-frequency ac [26–37]. Therefore, compared with a two-stage power conversion, it is single-stage power conversion, as shown in Figure 6.8. Also, a short-lived bulky dc-link capacitor is removed from the circuit. Usually, the conductive charging of EV batteries is more efficient and less expensive compared with the wireless inductive method, mainly due to higher copper loss and expensive materials used in IPT coils. Successful implementation of the direct ac-ac converter can compensate for those two demerits of the existing multi-stage IPT power

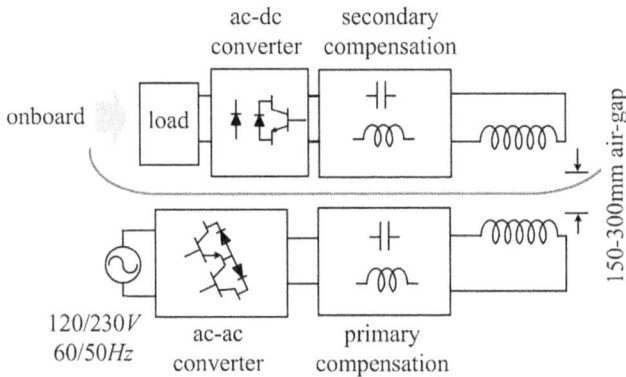

FIGURE 6.8 IPT topology with single-stage direct ac-ac converters.

supplies. However, there are some obvious limitations of IPT topology with direct ac-ac converters, such as

1. The components are required to be rated for peak power, where the rating of the components in the tank network, including IPT pads, is quite high due to poor coil coupling, and
2. Since the grid current is not directly controlled; therefore, high-quality source current is not ensured.

These topologies could be more suitable for low-power applications, electronic gadgets, body implants, and other low-power industrial applications. This is because the IPT circuit components can easily handle the peak power of such low power while keeping the converter cost low due to fewer components.

6.5.3 Current Source Inverter Topologies

Occasionally, current source inverter (CSI) is also used in IPT systems [2,17–19,38]. Figure 6.9 shows the existing IPT systems fed from a current-fed push-pull inverter, where the transmitter coil tank network is parallel *LC* type. The following merits of these systems are reported in the literature:

1. Suitably designed parallel capacitor supplies the reactive power consumed by TC without flowing through the inverter devices. Therefore, the inverter device current is lower;
2. Coil current quality with the presence of a parallel capacitor is almost sinusoidal because the higher-order harmonics primarily pass through this capacitor;
3. Achieves soft-switching of all the inverter devices;
4. In CSI topology, the inductor in dc-link limits short circuit current during fault.

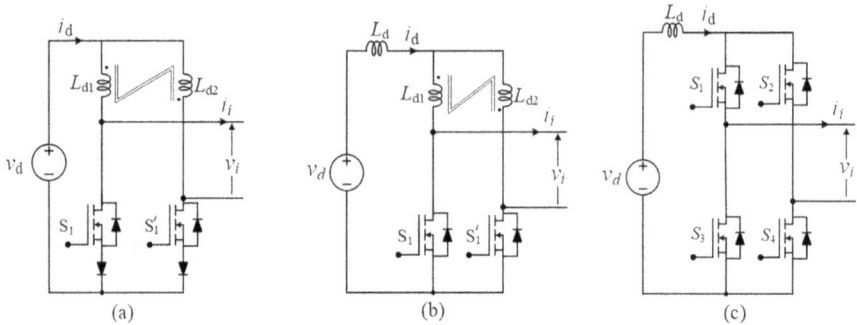

FIGURE 6.9 Existing IPT topology using current-fed: (a) push-pull inverters without extra input inductor and with reverse voltage blocking devices, (b) push-pull inverters with extra input inductor and without reverse voltage blocking devices, and (c) H-bridge.

However, there are some demerits of these systems, which are listed as follows:

1. Bulky dc link inductor is needed to get stiff DC at the inverter input. However, in an application where the stiff DC input is readily available such as solar cell output, there this topology will find a suitable application.
2. Conventional design of parallel resonant tanks is load-dependent. Therefore, complex control is required to tune the inverter switching frequency to the tank resonance frequency. This dynamic tuning is usually carried out either by adopting variable frequency control of the inverter or by varying tank capacitances dynamically.
3. Due to this inverter control, the dynamic load demand is generally met by an additional dc-dc chopper connected before the load, increasing power conversion stages.
4. Variable frequency control experiences converter start-up problems and frequency bifurcation (i.e., multiple operating frequencies) issues.

6.6 IPT COILS

Magnetic coupling is an important factor in designing a wireless IPT system. The regular 2-winding transformer or coupled inductors cannot be used here. The IPT coil geometry widely varies from conventional magnetic structures to achieve high coupling inductance with a large airgap. Also, additional design challenges include coil misalignments, especially when one or both coils are in motion. Generally, IPT pads are made both with and without a ferromagnetic core. Though the IPT pad with ferrite core is comparatively expensive, it has some advantages. The major advantage of having a ferrite core is minimizing magnetic field emissions around the coils by reducing the stray or fringe fields because ferrites keep the magnetic field between the coils. Various types of flux-guided IPT pads using ferrite blocks or bar are discussed in [6,13,39–46]. Figure 6.10a and 6.10b show circular IPT pads, and Figure 8c shows a UU type of IPT pad. Figure 6.10a circular pad is

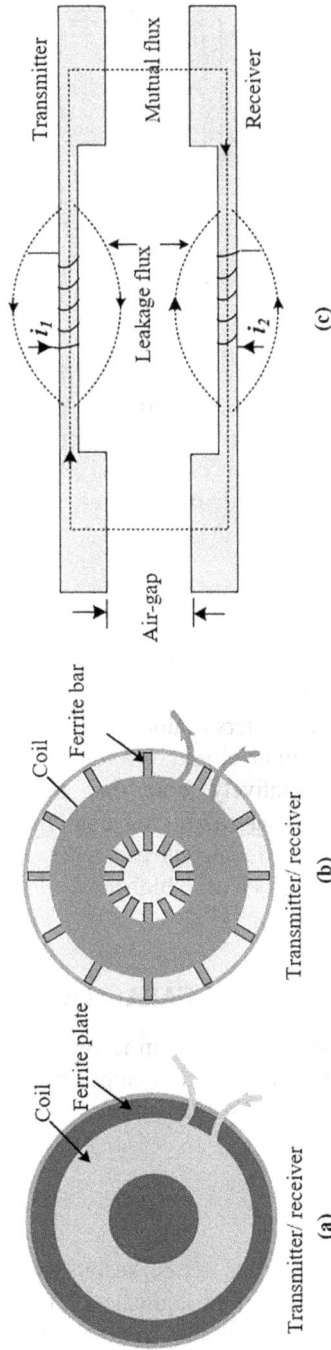

FIGURE 6.10 2-D structure and different unipolar IPT pads. (a) Circular with solid ferrite core, (b) circular with ferrite bars, and (c) UU-type IPT pad.

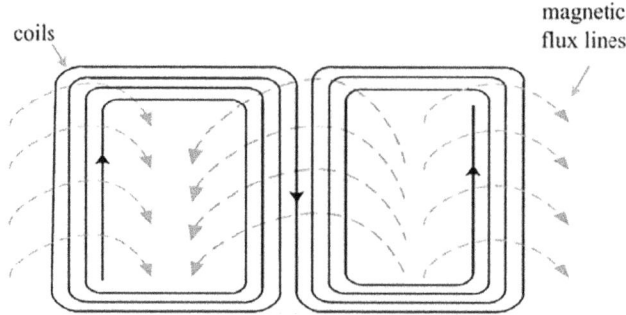

FIGURE 6.11 A bipolar IPT coil [49].

developed with solid ferrite plate where Figure 6.10b is pad is an improved structure for EV applications. Figure 6.10b structure is extensively used by the University of Auckland and Oak Ridge National Laboratory (ORNL) researchers [13,40,47], whereas Figure 6.10c structure is extensively used both commercially and in research by Korea Advanced Institute of Science and Technology (KAIST) [39].

6.6.1 BIPOLAR IPT COILS

Figure 6.10b shows a multi-coil coupler, which is commonly known as a bipolar coil [41]. Its operation may be visualized by considering two coils lying on the striated ferrite such that the line of centers is along the direction of the ferrite. Since this structure has a comparatively more closed path for the magnetic flux; therefore, the coefficient of coupling is relatively more than the unipolar arrangement. Although both the loops of the bipolar coil in this diagram carry the same current, they can be controlled to carry different currents with the help of two power converters [48]. The bipolar structures have a higher tolerance to coil misalignment than unipolar coils (Figure 6.11).

6.7 DESIGN OF COMPENSATION CAPACITANCES

The coupling coefficient of a typical IPT system is very low ($k<0.3$). This system requires dual-side compensation. The values of compensation circuit parameters for different IPT topologies are different. The derivation of some of the important IPT circuits is discussed here.

6.7.1 S/S COMPENSATION

The secondary side compensation for series secondary is $L_2 = 1/\omega_o^2 C_2$. The reflected impedance is the impedance of the equivalent impedance of the dependent voltage, and it is derived by dividing the current from the induced voltage (Figure 6.12).

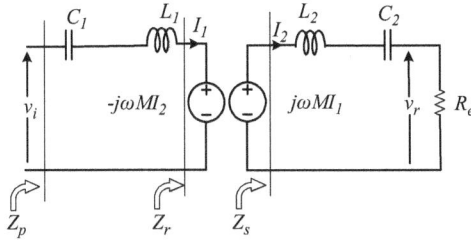

FIGURE 6.12 S/S IPT Topology.

$$Z_r = \frac{-j\omega_o M I_2}{I_1} = \frac{\omega_o^2 M^2}{R_e}. \tag{6.5}$$

In the above expression, I_1/I_2 part is replaced with $j\omega_o M/R_e$ which is derived by applying KVL in the secondary side circuit. The overall input impedance of the circuit is

$$Z_p = j\omega_o L_1 + \frac{1}{j\omega_o C_1} + \frac{\omega_o^2 M^2}{R_e}. \tag{6.6}$$

With the resonance condition, i.e., by making the complex part of the above expression equal to 0, the primary compensation is derived as

$$C_1 = \frac{1}{\omega_o^2 L_1} = \frac{L_2 C_2}{L_1}. \tag{6.7}$$

6.7.2 P/S COMPENSATION

For the P/S tank, the secondary side compensation, Z_s and Z_r, is the same as the S/S tank. Because the primary side is parallel, the admittance analysis makes finding the resonance condition easier. The input admittance is (Figure 6.13).

$$\frac{1}{Z_p} = j\omega_o C_1 + \frac{1}{j\omega_o L_1 + \frac{\omega_o^2 M^2}{R_e}}. \tag{6.8}$$

From this admittance, the primary compensation and the input impedance at resonance are derived as

$$C_1 = \frac{L_1}{\left(\frac{\omega_o^2 M^2}{R_e}\right)^2 + (\omega_o L_1)^2}; \quad Z_p = \frac{(\omega_o L_1)^2 + \omega_o^4 M^4/R_e^2}{\frac{\omega_o^2 M^2}{R_e}}. \tag{6.9}$$

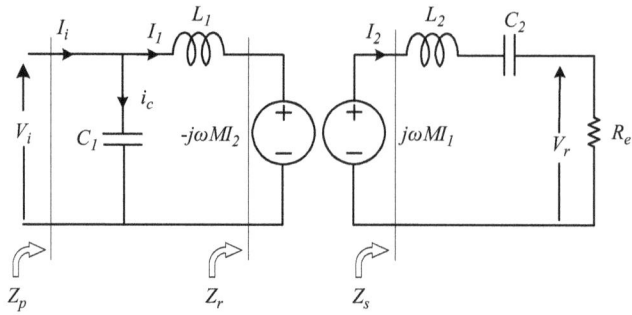

FIGURE 6.13 Parallel series tank.

6.7.3 SERIES-PARALLEL

For the series-parallel tank, the secondary side compensation is chosen as $C_2 = \frac{1}{\omega_o^2 L_2}$. Therefore, secondary impedance and reflected impedance are

$$Z_s = \frac{x^2}{R_e - jx}, \quad Z_r = \frac{\omega_o^2 M^2 (R_e - jx)}{x^2}. \tag{6.10}$$

By analyzing the input impedance of this circuit and deriving the resonant condition, the primary capacitance is derived as (Figure 6.14).

$$C_1 = \frac{L_2^2 C_2}{L_1 L_2 - M^2}. \tag{6.11}$$

6.7.4 LCL COMPENSATION

IPT topologies with LCL compensation networks can be subdivided into several topologies. A few of them are S/LCL, P/LCL, LCL/S, LCL/P, LCL/LCL, etc. The

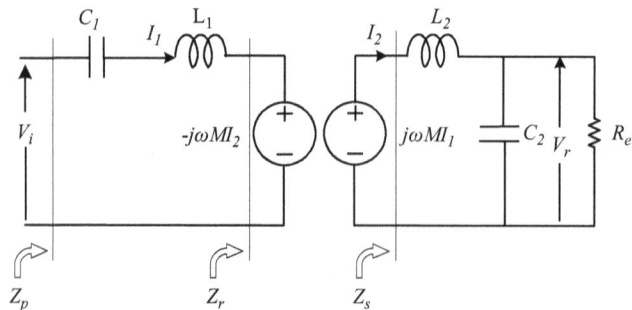

FIGURE 6.14 Series-parallel tank.

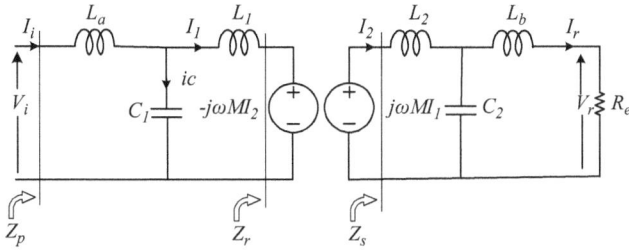

FIGURE 6.15 IPT Circuit with a dual-side LCL compensation.

determination of the compensation circuit parameters is quite similar to earlier, except that there is an extra inductor to design in the LCL tank. Figure 6.15 is an LCL/LCL WPT circuit with two extra inductors, L_a and L_b. Generally, an IPT circuit is preferred to have an operating point where the resonance/ compensation is load-independent. However, not all IPT topologies have this feature. The S/S, LCL/S, S/LCL, and LCL/LCL topologies can be designed to have a load-independent resonance. The secondary side impedance of the LCL tank is

$$Z_s = j\omega_o L_2 + \left(\frac{1}{j\omega_o C_2}\right) \| (j\omega_o L_b + R_e) \tag{6.12}$$

This expression can be arranged and written in the following format to determine the conditions for a load-independent resonant point.

$$Z_s = \frac{[R + (1 - \omega_o^2 L_b C_2)\{ j\omega_o L_2 (1 - \omega_o^2 L_b C_2) + j\omega_o L_b\} + j\omega_o C_2 R_e^2 (-1 + \omega_o^2 L_2 C_2)]}{(1 - \omega_o^2 L_b C_2)^2 + \omega_o^2 C_2^2 R_e^2} \tag{6.13}$$

The first condition is,

$$1 - \omega_o^2 L_b C_2 = 0 \text{ i. e. } \omega_o L_b = \frac{1}{\omega_o C_2}. \tag{6.14}$$

This condition doesn't have a load impedance term. Similarly, the second condition is

$$-1 + \omega_o^2 L_2 C_2 = 0 \text{ i. e. } \omega_o L_2 = \omega_o L_b = \frac{1}{\omega_o C_2} \tag{6.15}$$

These conditions for resonance provide the required values of the compensation capacitor and additional inductor values. The secondary impedance and the reflected impedance for the secondary LCL tank are derived as (Table 6.1).

TABLE 6.1

Circuit Parameters for Various Compensation Networks

Type	C_2	Z_r	C_1	Z_p	$\left\|\dfrac{v_r}{v_i}\right\|$
SS	$\dfrac{1}{\omega_0^2 L_2}$	$\dfrac{\omega_0^2 M^2}{R_e}$	$\dfrac{1}{\omega_0^2 L_1}$	$\dfrac{\omega_0^2 M^2}{R_e}$	$\dfrac{R_e}{\omega_0 M}$
SP	$\dfrac{1}{\omega_0^2 L_2}$	$\dfrac{M^2 R}{L_2^2} - \dfrac{M^2 \omega_0}{L_2}$	$\dfrac{L_2^2 C_s}{L_1 L_2 - M^2}$	$\dfrac{M^2 R_e}{L_2^2}$	$\dfrac{L_2}{M}$
PS	$\dfrac{1}{\omega_0^2 L_2}$	$\dfrac{\omega_0^2 M^2}{R_e}$	$\dfrac{L_1}{\left(\dfrac{\omega_0^2 M^2}{R_e}\right)^2 + (\omega_0 L_1)^2}$	$\dfrac{\omega_0^2 M^2}{R_e} + R_e \left(\dfrac{L_1}{M}\right)^2$	$\dfrac{1}{\sqrt{\left(\dfrac{L_1}{M}\right)^2 + \left(\dfrac{\omega_0 M}{R_e}\right)^2}}$
PP	$\dfrac{1}{\omega_0^2 L_2}$	$\dfrac{M^2 R}{L_2^2} - \dfrac{M^2 \omega_0}{L_2}$	$\dfrac{(L_1 L_2 - M^2)L_1^2 C_2}{(L_1 L_2 - M^2)^2 + \dfrac{M^4 R_e^2 C_2}{L_2}}$	$\dfrac{\dfrac{M^2 R}{L_2^2}}{\left(\dfrac{M^2 R}{L_2^2}\right)^2 + \left(\omega_0 L_1 - \dfrac{\omega_0 M^2}{L_2}\right)^2}$	$\dfrac{M R_e C_1}{L_2} \times \sqrt{\left(\dfrac{M^2 R_e}{(L_1 L_2 - M^2)}\right)^2 + \omega_0^2}$
LCL/LCL	$\dfrac{1}{\omega_0^2 L_b}$	$\dfrac{M^2 R_e}{L_2^2}$	$\dfrac{1}{\omega_0^2 L_1} = \dfrac{1}{\omega_0^2 L_a}$	$\dfrac{(\omega_0 L_1)^2}{\left(M^2/L_2^2\right) R_e}$	$\dfrac{M R_e}{\omega_0 L_1 L_2}$
LCC/LCC	$\dfrac{1}{\omega_0^2 L_b}$	$\dfrac{M^2 R_e}{L_b^2}$	$\omega_0 L_a = \omega_0 L_1 - \dfrac{1}{\omega_0 C_p}$	$\dfrac{(\omega_0 L_a)^2}{\left(M^2/L_b^2\right) R_e}$	$\dfrac{M R_e}{\omega_0 L_a L_b}$

$$Z_s = \frac{1}{\omega_o^2 C_2^2 R_e} = \frac{(\omega_o L_2)^2}{R_e}; \; Z_r = \frac{(\omega_o M)^2}{Z_s} = \frac{M^2}{L_2^2} R_e \tag{6.16}$$

The secondary side impedance is resistance, which gets reelected as a resistive impedance in the primary circuit. Therefore, the primary LCL tank will also have a similar impedance, and the condition for load-independent resonance will provide the information for compensation circuit parameter values. The final expressions are given as

$$\omega_o L_a = \frac{1}{\omega_o C_1} = \omega_o L_1; \; \text{and } Z_p = \frac{(\omega_o L_1)^2}{Z_r} = \frac{(\omega_o L_1)^2}{\frac{M^2}{L_2^2} R_e} \tag{6.17}$$

The LCCL (or LCC) resonant tank is also another popular IPT compensation network. Here another extra compensation capacitor is used in series with the coil. Essentially it is an LCL resonant tank only but with improved performance because this series capacitor directly compensates for part of the reactive coil power. For the secondary LCCL IPT topology, the condition for load-independent resonance will provide the design values and these are

$$\omega_o L_b = \frac{1}{\omega_o C_2} = \omega_o C_2 - \frac{1}{\omega_o C_s}. \tag{6.18}$$

Here, the Cp and Cs are the additional capacitors that are connected in series with the primary and secondary coils, respectively. Similarly, for the primary side, the design parameters are

$$\omega_o L_a = \frac{1}{\omega_o C_1} = \omega_o L_1 - \frac{1}{\omega_o C_p}. \tag{6.19}$$

6.7.5   LOAD INDEPENDENT TANK DESIGN FOR P/S AND P-S/S

The relative phase angle of the fundamental component of voltage and current at inverter output is zero at the tuned resonant frequency. This is commonly known as zero phase angle (ZPA) operation. ZPA operation with VSI and CSI are shown in Figure 6.16. The merits of ZPA operations are as follows.

1. Unity displacement PF operation at inverter output ensures the least volt-amp (VA) loading on the inverter.
2. It ensures the lowest voltage and current ratings and lower power loss of inverter devices.
3. For CSI topology, ZPA operation facilitates zero voltage switching (ZVS) of inverter devices, whereas the devices of VSI topology experience zero current switchings (ZCS).

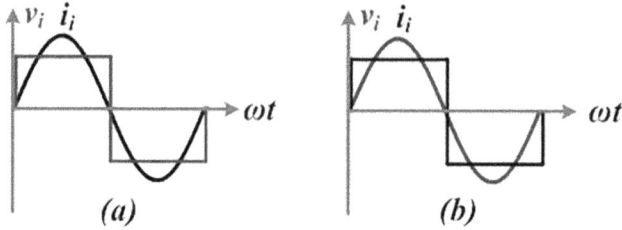

FIGURE 6.16 Ideal inverter output voltage and current profiles at ZPA operating point for (a) voltage source inverter (VSI) (b) current source inverter (CSI).

Due to this load-dependent compensation for a P/S or PS/S, the inverter switching frequency has to vary dynamically, or C_1 has to change dynamically to operate the converter at a resonant point. Figure 6.18a shows the magnitude and phase of the input impedance of a typical PS/S or P/S converter with the conventional design. Clearly, the resonant point shifts with the change of load power. Therefore, fixed switching frequency operation fails to operate the system efficiently. This is especially prominent at light load due to the large power factor angle at inverter output, as shown in Figure 6.18a. However, Figure 6.18a shows that there exists a load-independent resonance frequency, which can enable a simple fixed-frequency operation of the inverter.

6.7.5.1 PS/S IPT Circuit

A rigorous mathematical analysis of Figure 6.17 circuit is performed to determine this particular operation. Since the selection of RC side compensation using (6.9) leads to load-dependent resonance; therefore, in this new design method, $\omega_o L_2 = 1/\omega_o C_2$ simplification is not done. The input impedance of the Figure 6.17 circuit—i.e., the output impedance of the inverter is derived as

$$Z_p = \frac{1}{Y_{re} + jY_{im}} = \left[j\omega L_m // \left\{ R'_{eq} + j\left(\omega L'_{2k} - \frac{1}{\omega C'_2}\right) \right\} + j\left(\omega L_{1k} - \frac{1}{\omega C_s}\right) \right]$$

$$// \frac{1}{j\omega C_p},$$

$$(6.20)$$

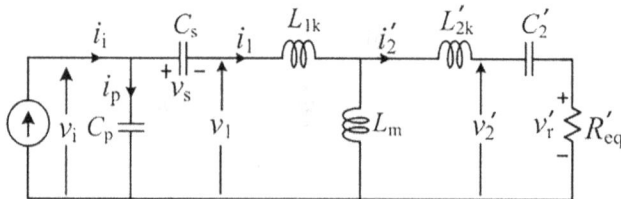

FIGURE 6.17 Equivalent circuit with respect to the transmitter side.

where the equivalent magnetizing and leakage impedances of the coupled coils are derived as

$$
\begin{aligned}
L_{1k} &= L_1 - nM, \\
L_m &= nM, \\
L_{2k} &= L_2 - M/n,
\end{aligned}
\tag{6.21}
$$

and TC to RC turns ratio is n. From (6.20) the complex (imaginary) part of input impedance is derived as

$$
\begin{aligned}
Y_{im} &= \frac{1}{t^2 + R_{eq}'^2 \{X_m + (X_{1k} - X_s)\}^2} \left[R_{eq}'^2 \{X_m + (X_{1k} - X_s)\} \left[\{X_m + (X_{1k} - X_s)\}/X_p - 1 \right] \right. \\
&\quad \left. - t \left[X_m - (X_{2k}' - X_2') + t/X_p \right] \right]
\end{aligned}
\tag{6.22}
$$

where all 'X' represents the impedance of a particular element, e.g., $X_m = \omega L_m$, $X_2' = 1/\omega C_2'$, etc. and

$$
t = X_m(X_{2k}' - X_2') + X_m(X_{1k} - X_s) + (X_{1k} - X_s)(X_{2k}' - X_2')
\tag{6.23}
$$

From (6.22), it is clear that to make Y_{im} zero, the two parts of the equation have to be zero individually, i.e.,

$$
R_{eq}'^2 \{X_m + (X_{1k} - X_s)\} \left[\{X_m + (X_{1k} - X_s)\}/X_p - 1 \right] = 0,
\tag{6.24}
$$

$$
t \left[X_m - (X_{2k}' - X_2') + t/X_p \right] = 0.
\tag{6.25}
$$

From (6.24) the first condition for resonance is derived as

$$
\left[\{X_m + (X_{1k} - X_s)\}/X_p - 1 \right]
\tag{6.26}
$$

This expression directly provides the value of TC side compensation capacitors as

$$
C_p//C_s = 1/\omega_s^2 L_1
\tag{6.27}
$$

The second condition of resonance in (6.25) has two parts. Since making the second part of (6.25) zero leads to $C_2 = 0$; therefore, the first part of it has to be zero, i.e., $t = 0$. Considering $t = 0$ and using (6.23) and (6.27), the second condition leads to the appropriate value of the RC side compensation capacitor as

$$X_2' = X_{2k}' - \frac{X_m(X_{1k} - X_s)}{X_m + (X_{1k} - X_s)}. \tag{6.28}$$

Clearly, both the conditions derived in (6.27) and (6.28) do not contain load impedance. Therefore, this passive selection of components leads to load-independent resonance. To verify the performance of this newly designed tank, the input impedance and phase are plotted in Figure 6.18b. The power factor at inverter output always remains unity, irrespective of wide load variations. This operation directly reduces the control effort of the inverter compared with existing dynamic tuning methods. Therefore, inverter control is fully focused on meeting load demand, whereas the existing parallel compensated typologies require an extra dc-dc chopper at the output side to achieve it.

6.7.5.2 P/S IPT Circuit

Figure 6.19 shows the transformer equivalent circuit of the $(L)(C)$ transmitter and (LC) receiver tank network, referred to as the transmitter side. Following the same design steps, the load-independent compensation circuit parameters are derived. The final expressions are given as

$$C_t = 1/\omega_s^2 L_1 \tag{6.29}$$

$$C_2' = \left[\omega_s \left(X_{2k}' - \frac{X_m X_{1k}}{X_m + X_{1k}} \right) \right]^{-1} \tag{6.30}$$

6.8 DESIGN OF IPT CIRCUIT PARAMETERS

This section derives the design of the magnetic circuit parameters such as L_1, L_2, and M. Using the expressions derived in the previous section, the compensation circuit parameters C_1 and C_2 can be calculated. To find magnetic circuit parameters for a specific IPT topology, the following parameters have to be given

- Battery (output) voltage = V_o
- Battery (output) current = I_o
- Input Voltage = V_d

For a practical EV charger, the battery voltage (V_o), battery current (I_o), and input voltage (V_d) is always given. Furthermore, any IPT topology can be chosen and designed for any type of battery charging application. However, their circuit

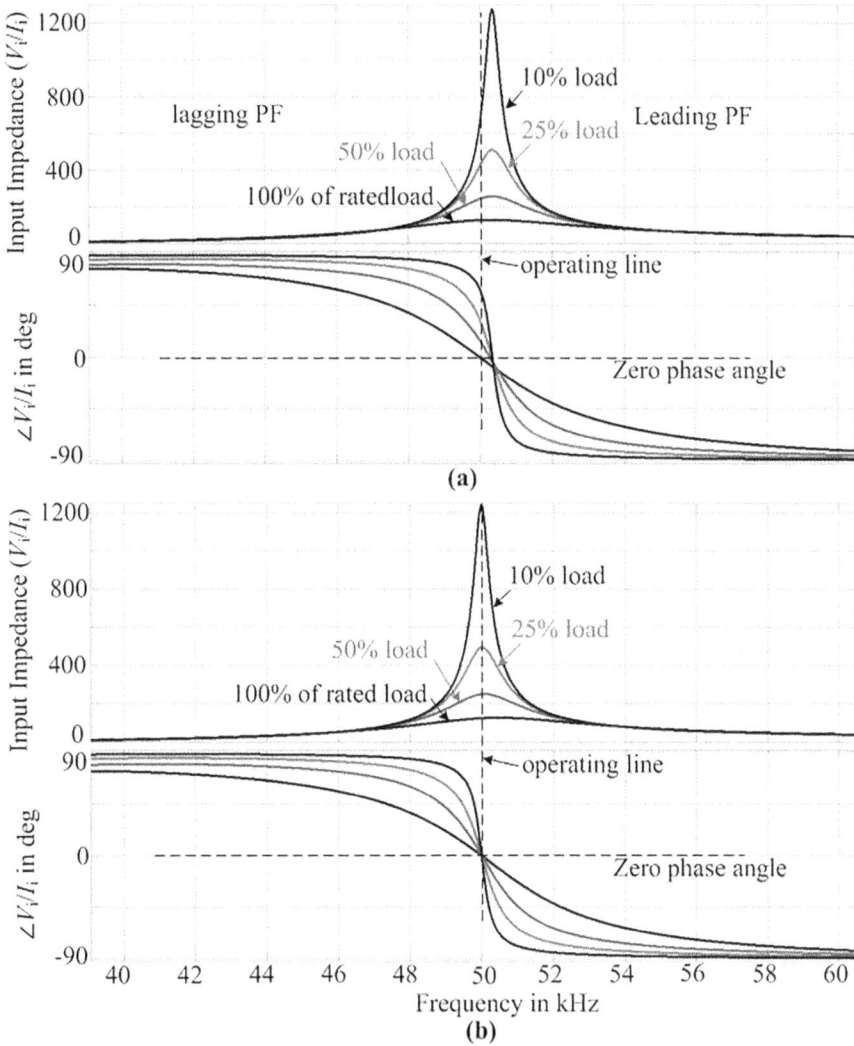

FIGURE 6.18 ZPA operation with (a) conventional design and (b) new design.

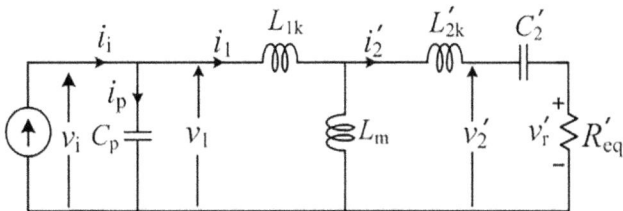

FIGURE 6.19 Equivalent circuit with respect to the transmitter side.

parameters will be completely different. This section discusses an S/S and a P/S IPT circuit design.

6.8.1   SS IPT Circuit Design

For a perfectly tuned secondary series IPT circuit, assuming the IPT circuit to be ideal. Such that the transferred power is completely available at the output then, the expression for the power transferred from TC to RC can be given as

$$P = \omega_o M I_1 I_2 = \omega_o M I_1 \times \frac{\pi}{2\sqrt{2}} I_o = V_o I_o \qquad (6.31)$$

The relation between I_2 and I_o is derived assuming a full-bridge diode rectifier. Considering an H-bridge inverter, the fundamental component of V_i will be

$$V_{i1} = \frac{4}{\pi} V_d \sin \omega_o t \qquad (6.32)$$

The input power will be the same as the output power for the ideal circuit. So, input power expression will be

$$P = V_{i1} I_1 = \frac{2\sqrt{2}}{\pi} (V_d) I_1 \qquad (6.33)$$

Therefore, from (6.31) and (6.33), the value of mutual inductance is derived as

$$M = \frac{8}{\pi^2} \frac{V_d}{\omega_o I_o} \qquad (6.34)$$

To find the value of L_1 and L_2, the coupling (k) and the coil-to-coil turn ratio (n) should be given. If it is not given, then a suitable value can be chosen. The value of L_1 and L_2 is derived from the following two expressions.

$$\frac{L_1}{L_2} = n^2; \quad M = k\sqrt{L_1 L_2} \qquad (6.35)$$

6.8.2   PS IPT Circuit Design

Since the secondary side has series compensation; therefore, the transferred power expression (6.31) is valid for P/S topology too. Applying KVL in the primary side loop

$$V_1 = (j\omega_o L_1 + \omega_o^2 M^2 / R_e) I_1 \qquad (6.36)$$

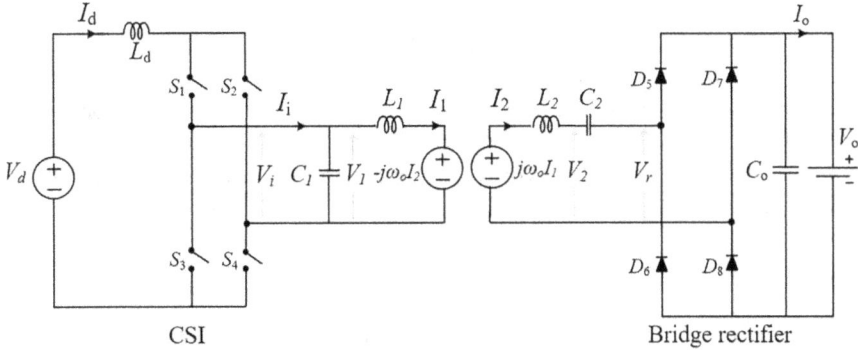

FIGURE 6.20 A typical P/S IPT circuit fed from a full-bridge CSI.

where, $R_e = \frac{8}{\pi^2} \frac{V_o}{I_o}$. The next step is to find the relation between V_1 and V_d. This relation can be found by applying flux (volt-sec) balance across the inductor, L_d. Considering the profile of V_1 is perfectly sinusoidal and the IPT circuit is perfectly tuned, the H-bridge side pole voltage of the inductor, L_d, will be the rectified V_1. Therefore, V_1 can be defined as

$$\langle |V_1| \rangle = V_d \text{ so, } V_1 = \frac{\pi}{2\sqrt{2}} V_d \tag{6.37}$$

By deriving values of I_1 from (6.31), i.e., $\left[I_1 = \frac{2\sqrt{2}}{\pi} \frac{V_o}{\omega_o M} \right]$ and L_1 from (6.35) i.e., $[L_1 = nM/k]$ and substituting these in (6.36), the magnitude of V_1 is calculated as

$$|V_1| = \left[\sqrt{(n\omega_o M/k)^2 + (\omega_o^2 M^2 / R_e)^2} \right] \frac{2\sqrt{2}}{\pi} \frac{V_o}{\omega_o M} \tag{6.38}$$

Hence, the value of M can be derived from (6.38) as (Figure 6.20).

$$M = \frac{R_e}{\omega_o^2} \sqrt{\left(\frac{\pi}{2\sqrt{2}} |V_1| \omega_o M / V_o \right)^2 - (n\omega_o/k)^2} \tag{6.39}$$

6.9 IPT CIRCUIT MODELING AND CONTROL

Along with the derivation of innovative power converter topologies, dynamic modeling and closed-loop control are also an integral part of the practical implementation of the converters. The resonant converts can either be controlled through the variable switching frequency fixed duty cycle method or the fixed frequency variable duty cycle method. The frequency modulation technique ensures the ZVS at turn-on of all the inverter devices. However, the inverter output power factor deteriorates when the operating frequency drifts from the tuned resonance frequency. In variable duty cycle modulation, half of the inverter switches lose ZVS at turn-on, and the inverter output power factor is maintained near to unity power factor. This section includes small-signal modeling and the closed-loop control of a S/S and P/S IPT topology. The control input is the inverter duty cycle, and the output is the output current (in constant current mode) or the voltage (for constant voltage mode).

6.9.1 S/S IPT Topology

The converter circuit shown above is controlled through a single-loop control method. The controller controls the duty ratio of the switches. The plant transfer function is given as (Figure 6.21).

$$G(s) = \frac{\tilde{V}_o(s)}{\tilde{d}(s)}, \tag{6.40}$$

$$G(s) = \frac{\tilde{\widetilde{V}}_i(s)}{\tilde{d}(s)} \cdot \frac{\tilde{\widetilde{V}}_r(s)}{\tilde{\widetilde{V}}_r(s)} \cdot \frac{\tilde{V}_o(s)}{\tilde{\widetilde{V}}_r(s)} = G_{inv} G_t G_{rec}. \tag{6.41}$$

6.9.1.1 Inverter Transfer Function

The input to the loop is the duty cycle of the inverter, and the output is the inverter voltage. Figure 6.22 shows a typical voltage, and current waveform at the voltage-source inverter output where the current profile is sinusoidal and voltage is quasi-square. 'd' is the equivalent duty cycle that is created due to phase-shift.

The peak of the fundamental component at the inverter output is given as

$$\widehat{V}_i = \frac{4}{\pi} V_d \sin \pi d. \tag{6.42}$$

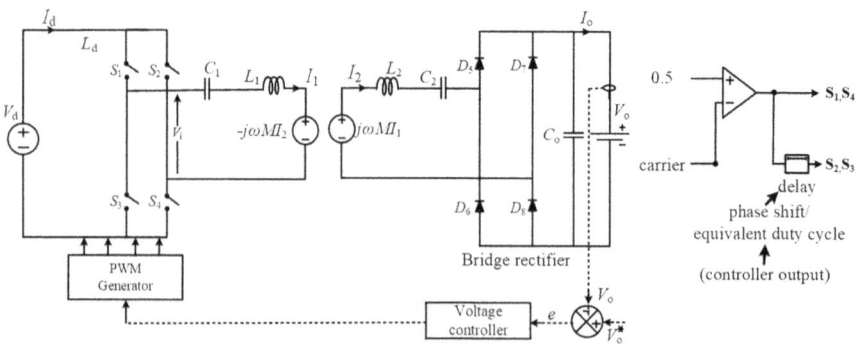

FIGURE 6.21 S/S topology with a full bridge rectifier.

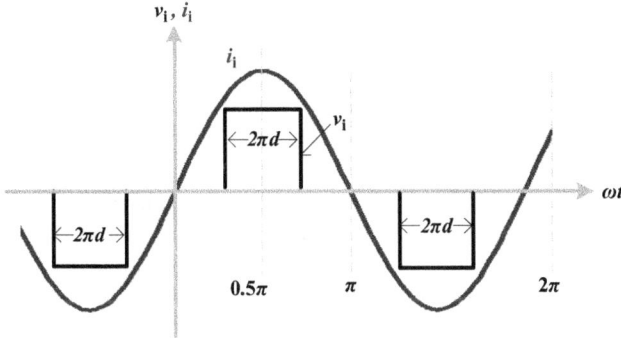

FIGURE 6.22 Inverter output voltage and current waveform.

Introducing small perturbations ($\tilde{i}_d$, $\tilde{v}_d$, $\tilde{d}$), around an equilibrium point (I_d, V_d, d) and neglecting second-order terms, we have inverter transfer function as,

$$G_{\text{inv}} = \frac{\widetilde{V}_i(s)}{\tilde{d}(s)} = 4V_d \cos \pi d. \tag{6.43}$$

Tank gain $\dfrac{\widetilde{V}_r(s)}{\widetilde{V}_i(s)}$ at the tuned resonance frequency

$$G_t = A = \left| \frac{R_e}{j\omega_o M} \right|. \tag{6.44}$$

6.9.1.2 Modeling of H-Bridge Rectifier

The dynamic expressions of the output capacitors for a given inverter switching cycle are given as

$$C_o \frac{dV_o}{dt} = \langle i_2 \rangle - i_o, \tag{6.45}$$

where C_o is the output capacitor, V_o is the output voltage, and R is the load resistance. All these state variables are considered to be average values over a switching cycle. Since, i_2 is switching frequency ac quantity, and its average value over a switching cycle is zero; therefore, the half-cycle average of i_2 i.e.,

$$C_o \frac{d\tilde{v}_o}{dt} = \frac{\pi}{4} \frac{\widehat{V}_r}{R} - \frac{\tilde{v}_o}{R} \tag{6.46}$$

$\langle i_2 \rangle$ is considered

$$G_{rec} = \frac{\tilde{V}_o(s)}{\widetilde{V}_r(s)} = \frac{\pi}{4} \times \frac{1}{sC_oR + 1} \tag{6.47}$$

Introducing small perturbations and applying Laplace transformation, the above expression is derived as

Therefore, the overall plant transfer function is derived as

$$G(s) = \frac{V_d \sin \pi D}{sC_oR + 1} A \tag{6.48}$$

6.10 MODELING AND CONTROL OF PS/P AND P/S

6.10.1 STEADY-STATE OPERATION WITH UNIPOLAR PWM

Unlike voltage source inverter, in this boost-derived inverter, complementary switching signals are given between top devices (S_1 and S_2)and between bottom devices (S_3 and S_4) (Figure 6.23). This arrangement enables us to get characteristics exactly like the boost chopper. A slight overlap between the complementary switching signals is always maintained to provide a continuous path for the input inductor current. Figure 6.24 shows equivalent circuits during different switching intervals, and Figure 6.25 shows important voltage and current waveforms with a typical unipolar PWM. Because the secondary side voltage doubler converter simply provides passive rectification; therefore, this part is not elaborated on. Simply, based on the polarity of RC current, i_2, each rectifier diode conducts accordingly and feeds load.

Interval 1 (t_0–t_1): During this interval, devices S_1 and S_4 are ON and S_3 and S_2 are OFF. In this interval, the source inductor, L_d is directly connected to the TC side tank network, as shown in Figure 6.24a. This is similar to the turn-off interval of the conventional boost converter. The inverter output voltage is very close to sinusoidal, but the current, i_i is quasi-square. Clearly, the voltage and current profiles of

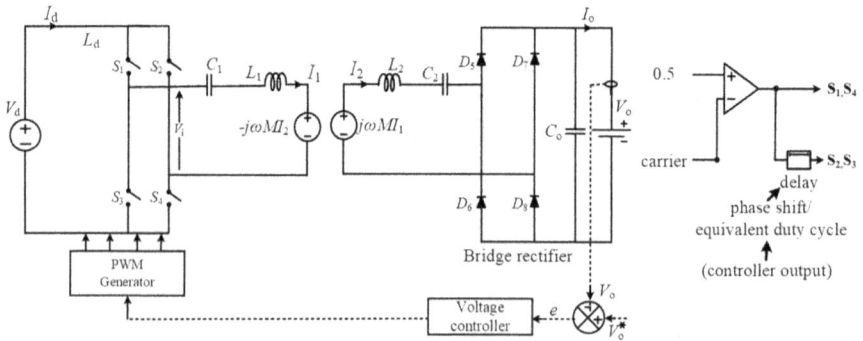

FIGURE 6.23 P/S topology with a full bridge CSI.

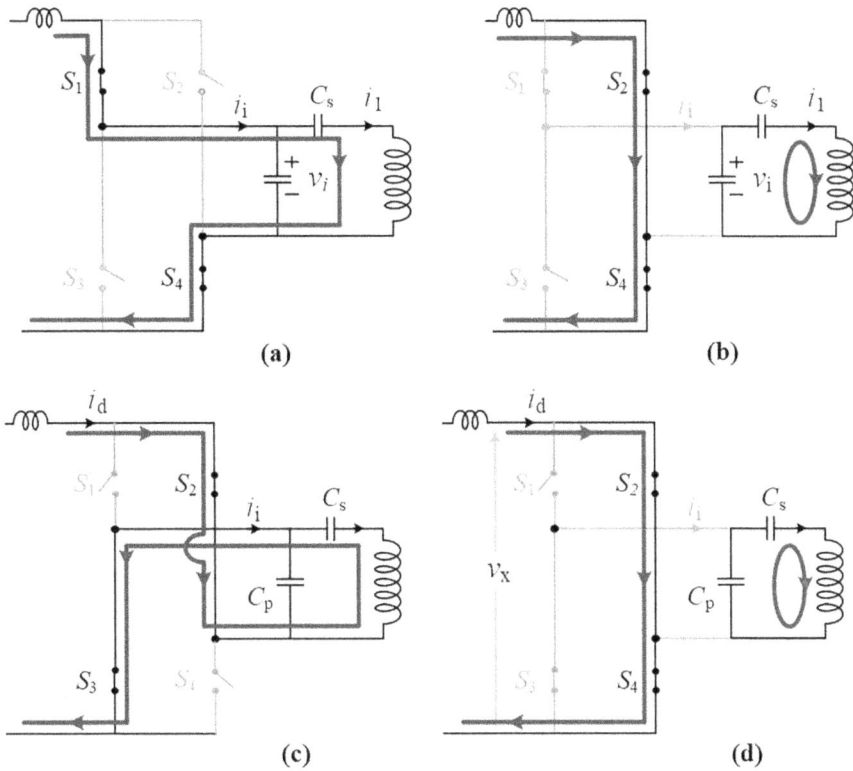

FIGURE 6.24 Equivalent circuits during steady-state operation.

the remaining tank elements are also sinusoidal, as shown in Figure 6.25. The voltage and current expressions for this duration are given as

$$L_d \frac{di_d}{dt} = v_d - v_i, \tag{6.49}$$

$$i_i = i_d, \quad v_x = v_i. \tag{6.50}$$

Interval 2 (t_1–t_4): The first part of this interval has a slight overlap (t_1–t_2) between S_1 and S_2 as shown in Figure 6.25. At instant t_1, the voltage across the incoming device S_2 is positive; therefore, it immediately takes inverter current, commutating device S_1. This transfer of inverter current at t_1 leads to a hard turn-on of S_2. Gate-pulse of S_1 is withdrawn at instant t_2, but the current through S_1 is already zero. Therefore, zero-current-switching (ZCS) turn-off of S_1 is achieved. The duration of overlap (t_1–t_2) is solely dependent on the turn-on and turn-off delay of the devices. It should be sufficient enough to allow the turn-on and turn-off of incoming and outgoing devices but small enough to have an insignificant impact on the overall performance of the converter.

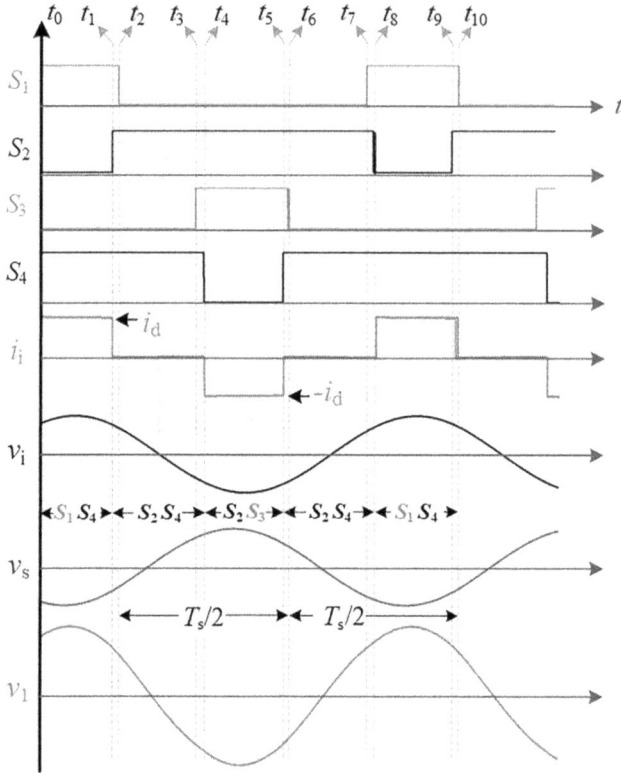

FIGURE 6.25   Steady state operating waveforms.

Referring to Figure 6.25, till time instant t_3, devices S_2 and S_4 carry complete inverter current and the corresponding equivalent circuit is shown in Figure 6.24b. At the instant t_3, switching overlap of bottom devices (S_3 and S_4) starts. Owing to the presence of negative voltage, the incoming device S_3 does not take the inverter current immediately at instant t_3. Therefore, during overlap time (t_3–t_4), S_2 and S_4 keep on conducting and this enables zero voltage switching (ZVS) turn-on of device S_3.

Throughout the interval t_1–t_4, the input inductor is directly connected across the source, which is equivalent to the turn-on period of the regular boost chopper. The voltage and current expressions during this interval are given as

$$L_d \frac{di_d}{dt} = v_d, \tag{6.51}$$

$$i_i = 0, \quad v_x = 0. \tag{6.52}$$

Interval 3 (t_4–t_5): At the instant t_4, overlap period ends when the gate pulse of S_4 withdrawn. Device S_4 experiences hard turn-off at instant t_4, and from this time onward, S_2 and S_3 take the inverter current as shown in the equivalent circuit Figure 6.24c. Similar to interval 1, the source is directly connected to the inverter output side network, which is similar to the turn-off interval of a conventional boost converter. Voltage and current expressions for this duration are given as

$$L_d \frac{di_d}{dt} = v_d + v_i, \tag{6.53}$$

$$i_i = -i_d, \quad v_x = -v_i. \tag{6.54}$$

Interval 3 (t_5–t_8): At instant t_5 S_4 is given pulse and it immediately starts conducting commutating device S_3, owing to the presence of positive voltage across S_4. This leads to a hard turn-on of S_4. But, due to the transfer of the current from S_3 before withdrawing its gate pulse leads to ZCS at turn-off. Although at an instant t_7, S_1 is triggered, due to the presence of negative voltage, it does not take the inverter current and the devices S_2 and S_3 keep on conducting as shown in the equivalent circuit Figure 6.24d. Therefore, S_1 experiences ZVS turn-on at instant t_7. At instant t_8 gate pulse of S_2 is withdrawn and S_1 is forced to take inverter current, and the corresponding equivalent circuit is the same as Figure 6.24a. The voltage and current expressions during interval t_5–t_8 are the same as (6.51) and (6.52).

This completes one complete inverter switching cycle. Steady-state operation of the IPT topology with $(L)(C)$ transmitter and (LC) receiver tank is exactly similar; therefore, it is not repeated.

6.10.2 Two-Loop Control for PS/S and P/S IPT Topologies

The converter circuit is controlled through a two-loop control method, where the outer output current loop meets the load demand and the inner input current loop controls the input inductor current. Figure 2.1 shows a complete control loop diagram, and this control scheme is the same for both IPT topologies. This type of two-loop control for current-fed full-bridge topology in fuel-cell applications has been reported in [50–52]. The advantage of this two-loop control is that the stability margin of the system is very high. Also, the power transmission can be reduced abruptly from full load to very light load using the faster inner current control loop. This feature is very useful for practical implementation. Furthermore, this control of source current with a boost-derived inverter can provide a single-stage solution, where direct source current control using a boost-chopper is required to extract energy from sources, e.g., solar PV, effectively. The derivation of the open-loop transfer function and controller design for both topologies are given.

6.10.3 (LC) Transmitter and (LC) Receiver Tank

6.10.3.1 Inner Input Current Loop

The input to this loop is the duty cycle of the inverter, and the output is the inductor current. Figure 6.26 shows a typical voltage and current waveform at the

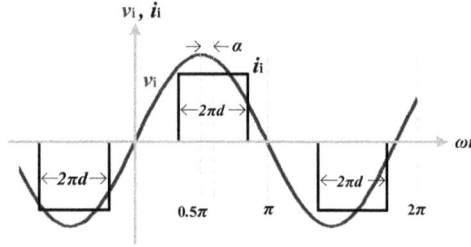

FIGURE 6.26   Inverter output voltage and current waveform.

current-source inverter output. The voltage profile is sinusoidal, and the current is quasi-square. '*d*' is the duty cycle of the device S_1 or S_3, '*a*' represents the phase-lag of current with respect to voltage. Although the system is designed to achieve unity power factor at inverter output, parameter variation of tank elements may lead to slight drift. This practical aspect is considered in the modeling of the converter. The dynamic expression of the input inductor voltage is given as

$$L_d \frac{di_d}{dt} = v_d - v_x \tag{6.55}$$

Considering the duty cycle of inverter device S_1 as $0 \le d \le 0.5$, the instantaneous input voltage of the inverter in terms of output RMS voltage is derived as

$$v_x = \frac{2\sqrt{2}}{\pi} V_i \times \cos\alpha \times \sin\pi d \tag{6.56}$$

The storage elements $(L_1, L_2, C_s, C_p, C_2)$ in the tank network predominantly carry a fundamental switching frequency component. Therefore, their dynamics are much faster compared with the dc side parameters. Considering negligible power losses in tank network, the power balance expression is given as

$$V_i I_i \times \cos\alpha = V_r I_2 \tag{6.57}$$

Since the inverter output voltage is very close to sinusoidal, the active power transfer occurs only with a fundamental component of I_i. Therefore, (6.57) is modified as

$$V_i \cos\alpha = \frac{V_r^2}{I_i R_{eq}} = \left(\frac{V_r}{I_i}\right)^2 \times \frac{I_i}{R_{eq}} = \frac{A^2}{R_{eq}}\left[\frac{2\sqrt{2}}{\pi}i_d \times \sin\pi d\right], \tag{6.58}$$

where, $A = V_r/I_i$ = gain of the tank network at the operating frequency. Feeding (6.56) and (6.58) in (6.55), the dynamic expression is modified as

$$L_d \frac{di_d}{dt} = v_d - \frac{4A^2}{\pi^2 R_{eq}} \times i_d (1 - \cos 2\pi d) \tag{6.59}$$

Introducing small perturbations ($\tilde{i}_d$, $\tilde{v}_d$, $\tilde{d}$), around an equilibrium point (I_d, V_d, d) and neglecting second-order terms, the small-signal dynamic expression is derived from (6.59) as

$$L_d \frac{d\tilde{i}_d}{dt} = \tilde{v}_d - \frac{4A^2}{\pi^2 R_{eq}} [\tilde{i}_d (1 - \cos 2\pi d) + 2\pi I_d \tilde{d} \sin 2\pi d], \tag{6.60}$$

Applying Laplace transformation, the control to the output transfer function of the inner loop is derived as

$$G_i(s) = \frac{\tilde{i}_d(s)}{\tilde{d}(s)} = -\frac{\frac{8A^2}{\pi R_{eq}} I_d \sin 2\pi d}{sL_d + \frac{4A^2}{\pi^2 R_{eq}}(1 - \cos 2\pi d)}, \tag{6.61}$$

Tank gain ($\frac{\tilde{V}_r(s)}{\tilde{I}_i(s)}$) at the tuned resonance frequency— Applying KVL in transmitter coil (TC),

$$V_i = \left[\left(\frac{L_1}{M} R_{eq} \right) - j\omega_o M \right] \tag{6.62}$$

Primary admittance at the resonance frequency,

$$Y_P = \frac{\left(\frac{\omega_o^2 M^2}{R_{eq}} \right)}{\omega_o^2 L_1^2 + \frac{\omega_o^4 M^4}{R_{eq}^2}} \tag{6.63}$$

Applying KCL in transmitter coil (TC),

$$i_i = \left[\left(\frac{L_1}{M} R_{eq} \right) - j\omega_o M \right] i_r. \; Y_P \tag{6.64}$$

Gain ($\frac{\tilde{V}_r(s)}{\tilde{I}_i(s)}$) at resonance frequency is,

$$G_t = A = \left| \frac{\left(\frac{L_1}{M}^2 R_{eq}^2 \right) + (\omega_o^2 M^2)}{\frac{L_1}{M} R_{eq} - j\omega_o M} \right| \tag{6.65}$$

6.10.3.2 Outer Current Loop

The input to the outer loop is i_d, and the output is load current, i_o. The outer loop control to output transfer function can be split as

$$G_o(s) = \frac{\tilde{i}_o(s)}{\tilde{i}_d(s)} = \frac{\tilde{\tilde{I}}_i(s)}{\tilde{i}_d(s)} \cdot \frac{\widetilde{V}_r(s)}{\tilde{\tilde{I}}_i(s)} \cdot \frac{\tilde{i}_o(s)}{\widetilde{V}_r(s)} = G_{inv} G_t G_{rec} \qquad (6.66)$$

These three transfer functions are the inverter, resonant tank, and rectifier. In (6.66), the dc side parameters are considered as the average value over an inverter switching cycle, where the ac side parameters are the amplitude of that switching cycle. Extracting the fundamental component of the quasi-square shaped current in Figure 2.2 and introducing small perturbation, the inverter transfer function is derived as

$$G_{inv} = \frac{\tilde{\tilde{I}}_i(s)}{\tilde{i}_d(s)} = \frac{4}{\pi} \sin \pi d \qquad (6.67)$$

Modeling of H-bridge rectifier—the dynamic expressions of the output capacitors for a given inverter switching cycle are given as

$$C_o \frac{dV_o}{dt} = \langle i_2 \rangle - i_o. \qquad (6.68)$$

All these state variables are considered to be average values over a switching cycle. Since, i_2 is switching frequency ac quantity, and its average value over a switching cycle is zero; therefore, the half-cycle average of i_2 i.e., i_2 is considered,

$$C_o \frac{d(E + i_o r_b)}{dt} = \frac{2}{\pi} \frac{\widehat{V}_r}{R_{eq}} - i_o, \qquad (6.69)$$

where, r_b and E are battery internal resistance and e.m.f., respectively. Introducing small perturbations and applying Laplace transformation, the gain expression is derived as

$$G_{rec} = \frac{\tilde{i}_o(s)}{\widetilde{V}_r(s)} = \frac{2}{\pi} \times \frac{1/R_{eq}}{sC_o r_b + 1} \qquad (6.70)$$

Therefore, the overall plant transfer function for the outer loop is derived as

$$G_o(s) = \frac{8}{\pi^2} \frac{A}{R_{eq}} \sin \pi d \times \frac{1}{sC_o r_b + 1} \qquad (6.71)$$

6.11 SINGLE-STAGE IPT

Generally, in the IPT system, the power is processed through multiple power transfer stages, leading to lower efficiency and higher cost of the system. Recent research shows that using a direct ac-ac converter in the IPT system compensates for these limitations significantly. However, one of the major challenges of an IPT circuit with a direct ac-ac converter is to achieve multiple control goals through a single converter. These include load power requirement, maintaining a high-quality source current, and achieving soft switching of inverter switches.

6.11.1 CHALLENGES WITH VSI-DERIVED SINGLE STAGE IPT TO ACHIEVE UNITY P.F.

In conventional multi-stage IPT topologies shown in Figure 6.27, the converter before the resonant tank is generally a voltage source inverter (VSI). VSI imposes to select primary tank network as series LC, LCL, or LCCL, etc. Generally, these same tanks are used to realize the IPT topology with a direct ac-ac converter because their properties are well established in the literature [26,32–34]. However, when these tanks are used in a single-stage direct ac-ac converter, the input to the converter is required to be stiff voltage.

Figure 6.27a shows a general direct matrix converter structure with buck-derived configuration when viewed from the source side. Most of the studies have been carried out considering the load is resistive, and it is connected either directly with the RC side tank or at the output of the rectifier [26,33,34]. Therefore, the matrix converter perceives the load as linear, as shown in Figure 6.27c. These topologies will fail to control input current when a stiff dc voltage load such as the battery is connected to the rectifier output. This is because the equivalent load impedance varies significantly with this type of load, as shown in Figure 6.27c. The practical interpretation is that the buck-derived ac-ac converter topology does not get sufficient input voltage around zero crossing to feed power to high voltage output dc bus. This challenge is similar to regular buck-derived PFCs in that near-zero crossing of source voltage fails to boost the input voltage to the required output voltage. Like regular buck-derived PFCs, the current source quality with the existing ac-ac converters is highly compromised. The boost-derived topology will be suitable for this application to achieve power factor correction, as shown in Figure 6.27b.

6.11.2 IPT FOR UNITY P.F. OPERATION

Figure 6.28 shows the complete circuit of the selected IPT topology where the TC side tank is parallel-series (CCL), and the RC side is series LC type. The basic operation of this converter is like a boost-derived PFC. When the current is passed through the source inductor, L_s needs to be raised, then the ac-ac converter is switched such that L_s directly gets connected to source voltage, v_{ac} through converter switches. The L_s is connected to the transmitter side tank network through ac-ac converter switches. Because the input to the ac-ac converter is stiff current;

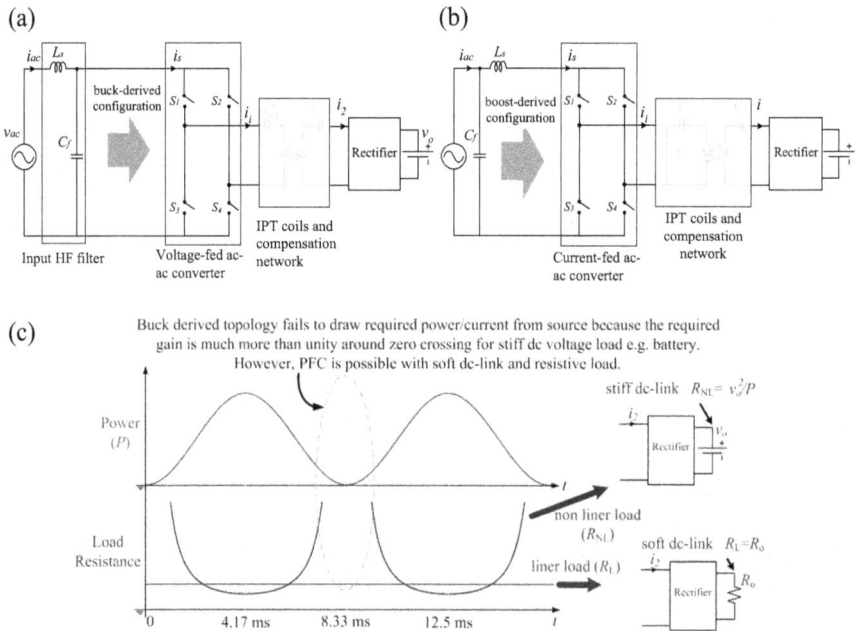

FIGURE 6.27 (a) General powertrain of IPT topology with direct ac-ac converter, (b) Required structure of single-stage IPT topology.

therefore, the transmitter side tank network is required to be parallel. The parallel tank has several advantages in IPT systems. Compared with the conventional parallel LC tank, the selected converter has an extra series capacitor connected with TC to reduce the switch voltage stress and improve TC current quality.

On the receiver side, a capacitor, C_2, is connected in series with the receiver coil (RC) to achieve the required compensation, and this ensures the least number of components on the RC side. The rectifier on the RC side is selected to be a voltage doubler to achieve higher voltage gain while reducing the number of rectifier diodes.

6.11.2.1 Comparisons of Single-Stage IPT Topologies

Table 6.2 lists several important aspects of different IPT topologies with direct ac-ac converter reported in the literature. This gives a clear picture of the contributions of this research work compared with other reported IPT topologies with matrix converters. With this understanding, the preferred selection of an ac-ac converter will be a current-source topology.

6.11.3 Steady-State Operation

To explain the operation, consider that S_1–S_4 are matrix converter switches, i.e., two switches connected in the reverse direction to achieve bi-directional voltage and current controlling facility. Only four switches are given switching frequency

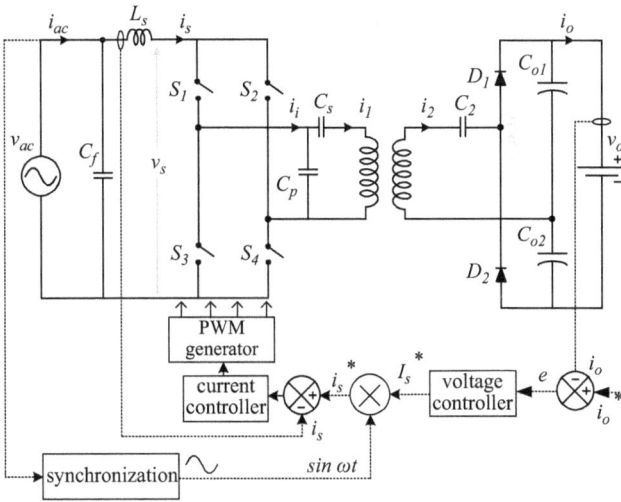

FIGURE 6.28 current-source inverter-based single-stage based IPT topology with two-loop control.

pulses during the positive half of the input voltage, vac. These switches are named S_{1P} to S_{4P}, respectively. Similarly, during the negative half of input voltage, the remaining 4 switches of ac-ac converter are given switching frequency pulses and these are named S_{1N} to S_{4N}, respectively. This section presents only one switching cycle of the ac-ac converter during the positive half of source voltage in detail. Operation of the converter during the negative half of source voltage is exactly similar where S_{1N}–S_{4N} takes the position of S_{1P}–S_{4P}, respectively. During the positive half of source voltages, S_{1N}–S_{4N} are kept permanently off, whereas the negative half cycle of source voltages, S_{1P}–S_{4P}, are kept off permanently. It is confirmed that operating the inverter at the lagging power factor region ensures ZVS turn-on of the switches [54,55]. Therefore, in the steady-state operation, ac-ac converter output voltage, v_i is considered to be leading with respect to current, i_i. To achieve the control goals, the unipolar modulation scheme is adopted, i.e., the turn-on time of S_1 and S_3 are the same, but the phase-shifted by 180°. Similarly, the turn-on time of S_2 and S_4 are the same but phase shifted by 180°. Also, a slight overlap between complementary switch pairs S_1–S_2 and S_3–S_4 is always maintained to provide a continuous current path to the input inductor, L_s. Compared with a voltage-fed converter, one difference is that the complementary switching signals are not given within the leg switches; instead, it is given to two top switches and two bottom switches. This technique is usually applied to current-fed inverters.

Interval 1 (t_0–t_1): During this interval, the ac-ac converter switches S_{1P} and S_{4P} are on. Therefore, the input inductor current i_s flows through the transmitter coil (TC) side tank network. Figure 6.29 shows voltage and current waveforms of different circuit components and Figure 6.30a shows an equivalent circuit of this switching interval. The voltages and currents of TC side components are given as

TABLE 6.2

Qualitative Comparison of Different IPT Topologies with Direct ac-ac Converter Reported in the Literature

Topology	TC and RC tank configuration	Control technique	f_s (kHz)	Load demand fulfilled	Dynamic model Reported	Achieved high-quality source current	Soft switching	Power level
Single-stage 3phase [33]	TC: LC series RC: LC parallel	Variable f_s;	12.3; ZCD required	Yes, using Power Regulation Control Mode	No	THD 14.3%, P.F. 0.59–0.67	Yes, ZCS of all devices	267 W
SiC-based matrix converter [34]	TC: (LC) RC: (LC)	Fixed f_s phase shift modulation	50	Not reported	No	No	Not achieved	300 W
Direct ac-ac Converter for IPT [35]	TC: (LC) RC: Not reported	Variable f_s; energy-injection and free oscillation	30; ZCD required	Not reported	No	Not reported	Yes, ZCS of all devices	10 W
Matrix Converter with LCLC tank [26]	TC: LCCL RC: LCCL	Fixed f_s and variable duty cycle modulation	20; ZCD required	Yes, with phase shift control	No	Yes, but only for R load	Not reported	1 kW
Boost derived ac-ac converter [53]	TC: parallel-series (CCL) RC: LC series	Fixed f_s and variable duty cycle	50	Yes, using output current loop	Yes	Yes, THD <5% PF: UPF	Yes, ZVS & ZCS of two switches	1.2 kW

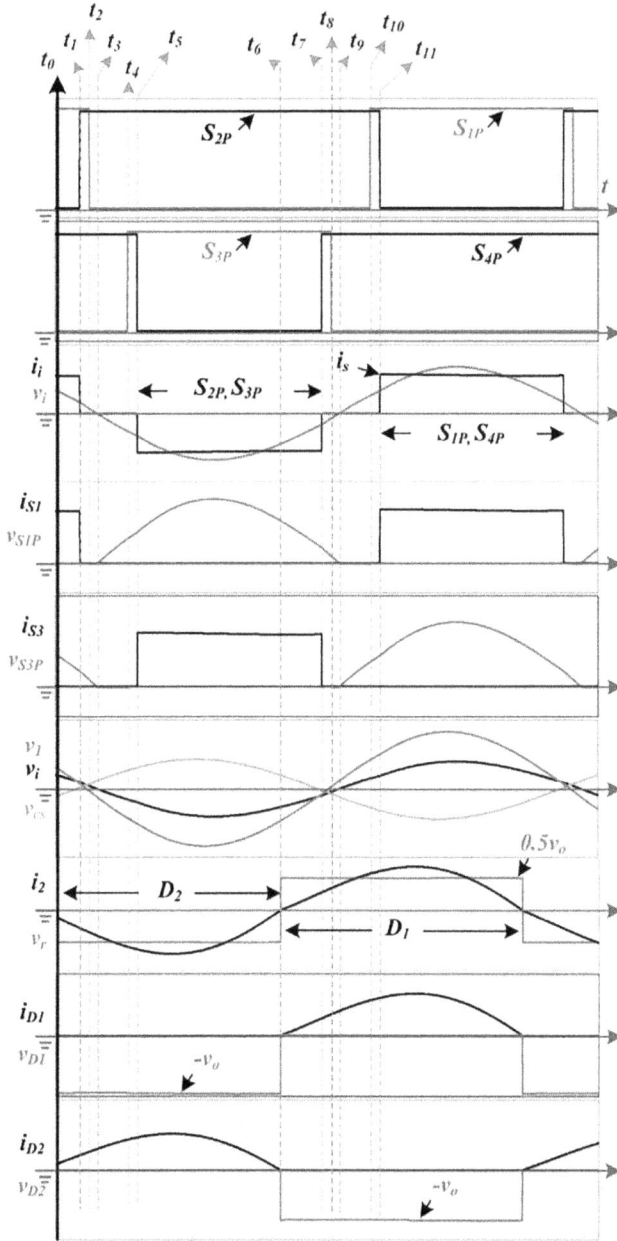

FIGURE 6.29 Steady-state operating waveforms of one switching cycle during the positive half of source voltage.

$$L_s \frac{di_s}{dt} = v_{ac} - v_i, \quad i_{S1} = i_{s4} = i_s. \tag{6.72}$$

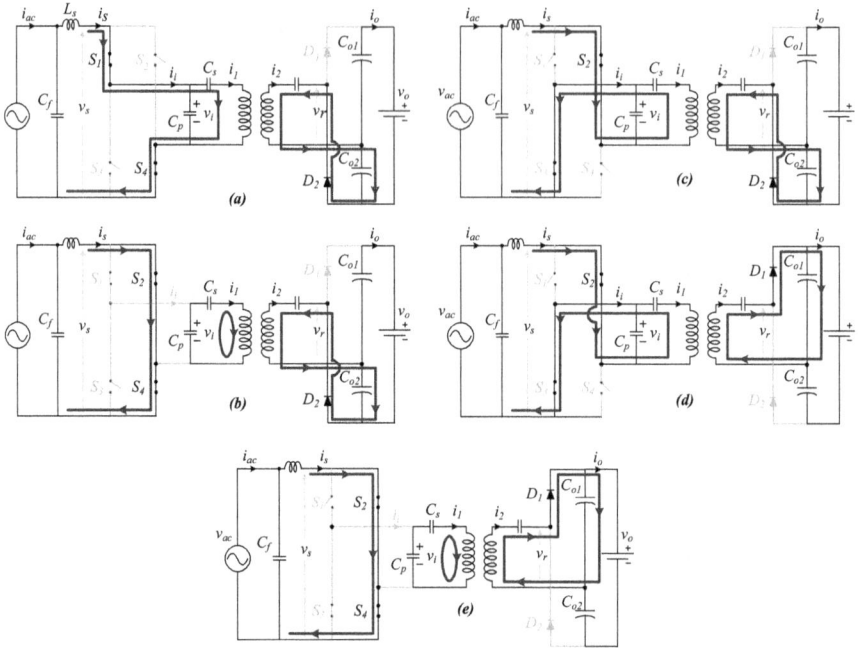

FIGURE 6.30 Equivalent circuit during the different intervals of operations.

During this interval, on the receiver side, the RC current is rectified by diode D_2. The voltages and currents of RC side elements are given as

$$v_{D1} = v_{o1} + v_{o2}, \quad i_{D2} = i_2, \tag{6.73}$$

$$C_{o1} \frac{dv_{o1}}{dt} = -i_o, \quad C_{o2} \frac{dv_{o2}}{dt} = i_2 - i_o ., \tag{6.74}$$

Interval 2 (t_1–t_5): interval t_1–t_2 is the switching overlap period between switches S_{1P} and S_{2P}. The overlap period's duration is almost negligible compared to one complete switching cycle but sufficient to transfer the input current from an incoming device to an outgoing device. Since the voltage across S_{2P} is positive at time instant t_2; therefore, S_{2P} immediately starts conducting, and current through S_{1P} is transferred even before its gate pulse is removed. Thus, at the instant t_1 switch, S_{2P} experiences hard turns-on, whereas, at instant t_2, S_{1P} experiences zero-current turn-off. The equivalent circuit of this interval is shown in Figure 6.30b. The voltages and currents of TC side components are given as

$$L_s \frac{di_s}{dt} = v_{ac}, \quad i_{S3} = i_{s2} = i_s. \tag{6.75}$$

At instant t_2, the gate pulse of S_{1P} is removed. However, the equivalent circuit of the converter remains the same as in Figure 6.30b because the current commutation from switch S_{1P} to switch S_{2P} has already occurred. At instant t_4 the overlap period of S_{3P} and S_{4P} starts. However, at this instant, the converter output voltage, v_i is negative, and the voltage across S_3 is negative. Although S_{3P} is triggered, S_{3N} blocks this negative voltage present across S_3. Thus, S_{4P} keeps on conducting without transferring current i_i to S_{3P}. Therefore, S_{3P} is switched on at zero voltage. The overlap period of S_{3P} and S_{4P} gets over at instant t_5, and the equivalent circuit of the converter from t_1 to t_5 remains the same as in Figure 6.30b. This interval (t_1-t_5) is similar to the conventional boost converter turn-on period.

Interval 3 (t_5–t_6): At the instant t_5 gating signal of S_{4P} is removed, and it experiences a hard turn-off. Although the voltage across S_{3P} is negative, the source inductor current, i_s forces S_{3P} to conduct because i_s has no alternate path. In this interval t_5–t_6, the source current flows through the TC tank network. The equivalent circuit of the converter circuit is shown in Figure 6.30c. It is clear that this interval is similar to the turn-off time period of a conventional boost converter, where the input inductor is directly connected to the output. The current and voltage expressions of this interval are given as

$$L_s \frac{di_s}{dt} = v_{ac} + v_i, \quad i_{S2} = i_{S3} = i_s. \tag{6.76}$$

Interval 4 (t_6–t_7): Although at time instant t_6, there is no switching transition on the TC side, the current through the RC changes its polarity at this instant. Thus, the rectifier diode D_1 commutates D_2 and rectifies the RC coil current. The equivalent circuit of this interval is shown in Figure 6.30d. the voltage and current expressions of receiver side components during this interval are given as

$$v_{D2} = v_{o1} + v_{o2}, \quad i_{D1} = i_2, \tag{6.77}$$

$$C_{o1} \frac{dv_{o1}}{dt} = i_2 - i_o, \quad C_{o2} \frac{dv_{o2}}{dt} = -i_o, \tag{6.78}$$

Interval 5 (t_7–t_{11}): At the instant t_7, S_{4P} is triggered, and the interval t_7–t_{11} is switching overlap period. Similar to interval-2 the switch S_{4P} immediately commutates S_{3P} because of the positive voltage present across S_{4P}. Therefore, S_{4P} experiences a hard turn-on at instant t_7, and S_{3P} experiences ZCS turn-off at instant t_8. It is clear that S_{1P} and S_{3P} experience both turn-on and turn-off soft-switching whereas both S_{2P} and S_{4P} experience hard turn-on and hard turn-off. The equivalent circuit of the converter during intervals t_7 to t_{11} is shown in Figure 6.30e. The steady-state operation of the converter circuit repeats in this order.

6.11.4 VOLTAGE AND CURRENT RATINGS

Applying KCL on the RC side of Figure 6.28b converter equivalent circuit, the rectifier input current in terms of output current is derived as

$$I_2 = \frac{\pi}{\sqrt{2}} \cdot I_o, \tag{6.79}$$

Applying power balance between the output and input of the rectifier and using (6.79), the rectifier input ac voltage RMS is derived as

$$V_r = \frac{2\sqrt{2}}{\pi} \times \frac{V_o}{2} = \frac{\sqrt{2}}{\pi} \times V_o. \tag{6.80}$$

Since this RC side rectifier is a passive rectifier; therefore, the voltage V_r and current I_2 are in the same phase, and these phasors are considered reference phasors. Referring to the coupled inductor equivalent circuit of the coupled IPT coils as shown in Figure 6.28b and applying KVL at the RC side loop, the TC current is derived as

$$I_1 = -j\frac{V_r}{\omega_o M}. \tag{6.81}$$

Using this current expression and adding the induced voltage in TC due to RC current, the TC voltage is derived as

$$V_1 = \frac{L_1}{M}V_r - j\frac{\omega_o M}{R_{oeq}}V_r. \tag{6.82}$$

Using (6.81) and (6.82) and applying KVL and KCL at the TC tank network, the ac-ac converter output voltage and current are derived as

$$V_i = \left(\frac{L_1}{M} - \frac{1}{\omega_o^2 M C_s}\right)V_r - j\frac{\omega_o M}{R_{oeq}}V_r, \tag{6.83}$$

$$I_i = \frac{\omega_o^2 M C_p}{R_{oeq}}V_r + j\frac{V_r}{\omega_o M}\left(\omega_o^2 C_p L_1 + \frac{C_p}{C_s} - 1\right). \tag{6.84}$$

6.11.5 SOFT SWITCHING

Referring to steady-state operation, to achieve ZVS turn-on and ZCS turn-off of converter switches S_1 and S_3, the operating power factor of ac-ac converter output is required to be lagging. Therefore, it is important to know the operating power factor of the converter. From Figure 6.28b equivalent circuit, the impedance at the input of the tank network is derived as

$$Z_i = \frac{1}{j\omega C_p} // \left[\left\{ j\omega (L_1 - M) + \frac{1}{j\omega C_p} \right\} \right.$$

$$\left. + \left\{ j\omega M // \left(j\omega (L_2 - M) + \frac{1}{j\omega C_p} + R_{oeq} \right) \right\} \right]. \tag{6.85}$$

Figure 6.31 shows the variation of tank network input impedance and phase angle with a change in operating frequency. The figure shows that below the resonance frequency, the operating power factor is lagging, and on the other side, it is the leading power factor. This is an advantage of this tank network that the lagging power factor operation is achieved below the resonance frequency, whereas for the series LC tank, it occurs above the resonance frequency.

6.11.6   TWO LOOP CONTROL

The selected IPT topology is controlled through a two-loop control method. From steady-state operation, it is clear that the basic operation of this ac-ac converter is similar to a boost converter. When switches in the same leg are O_N simultaneously, i.e., either S_1–S_3 or S_2–S_4 are O_N it is similar to boost converter switch turn-on interval. Similarly, when either diagonal (S_1–S_4) or off-diagonal (S_2–S_3) switch

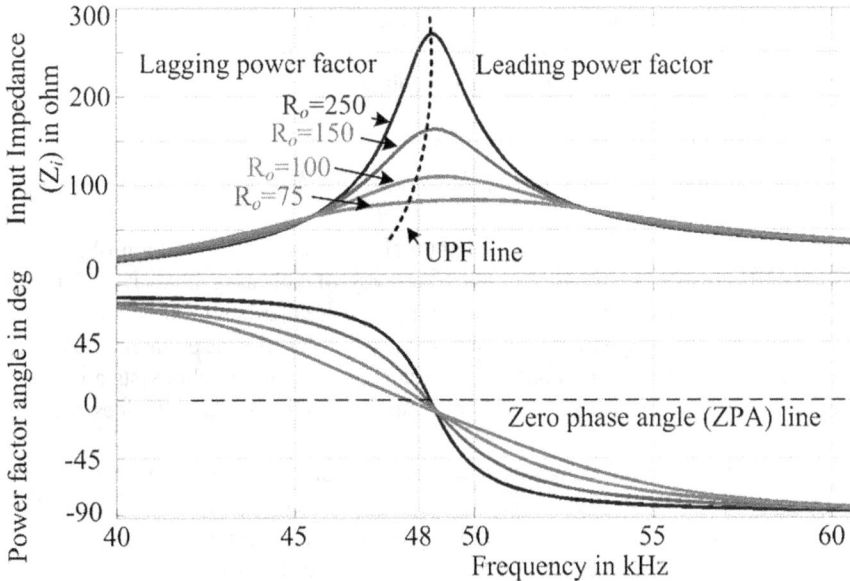

FIGURE 6.31   Variation of tank network input impedance with switching frequency.

pairs are O_N, the converter input gets directly connected to the output capacitor. It is equivalent to a boost converter turn-off time interval.

The outer output current loop is used to regulate the converter output current, whereas the inner input inductor current loop is used to achieve unity power factor at ac-ac converter input. The detailed control loop diagram is presented in Figure 6.28. The input voltage polarity determines which switch set will be triggered. When input voltage polarity is positive, switches S_{1P}-S_{4P} are triggered, and similarly, S_{2N}–S_{4N} is triggered when the input voltage is negative.

6.11.7 EXPERIMENTAL RESULTS

To verify the operation and control of the selected IPT circuit experimental validation is required. In this section, a set of experimental results is presented.

6.11.7.1 Experimental Set-Up

The ac-ac converter switches are MOSFETs with manufacturer part number C2M0080120D. The tank capacitors are all Epcos make, 700V RMS film capacitors. The IPT coils are circular type and the airgap between the TC and RC is around 25 cm. Since, the major focus of this section is to verify the performances of this converter and not the IPT coil design; therefore, the details of circular coil design are not included here and it can be found in [40,43]. It is already proven that the leakage magnetic flux can be kept well within the specified limit with the use of proper aluminum shielding [40]. However, when this IPT pad is used for direct ac-ac converter, there might be some low frequency (60Hz, 120Hz) flux present around the coil surroundings. Nonetheless, from the fundamental point of view, this pulsating low-frequency leakage flux cannot impact any object when the switching frequency leakage flux i.e., the flux responsible for effective power transfer is kept within a regulated limit. This is because the RMS values of the induced voltage in this foreign object due to TC and RC are given as

$$V_{1F} = 2\pi f M_{1F} I_1 \text{ and } V_{2F} = 2\pi f M_{2F} I_2 \tag{6.86}$$

respectively, where, M_{1F} and M_{2F} are mutual couplings of the foreign object with TC and RC, respectively and f is the frequency of pulsating flux. Therefore, the impact of this low-frequency pulsating magnetic field is about 1000 times lower than the switching frequency field. This is very similar to a case when a conductor carries line frequency current and a low-frequency magnetic field exists around the conductor. This does not have a significant impact on regular devices working nearby.

The rectifier side diodes are 400V fast recovery diodes and the manufacturer part number is SCS215KGC. The ac-ac converter MOSFETs are driven with Semikron make SKHI 61(R) gate driver with a rated switching frequency 50 kHz. Figure 6.32 shows the experimental set-up and 60 cm diameter circular coil.

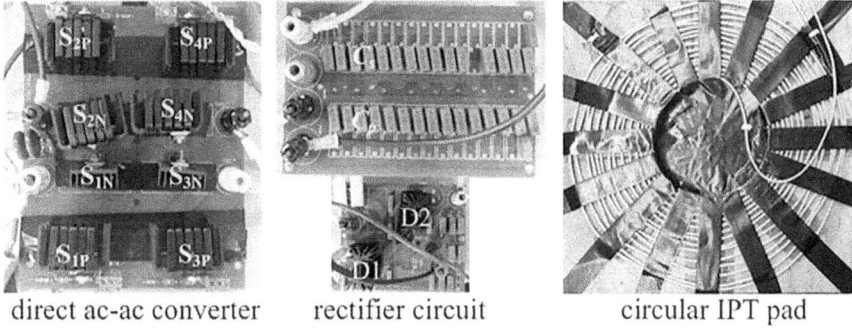

direct ac-ac converter   rectifier circuit   circular IPT pad

FIGURE 6.32 Experimental set-up.

6.11.7.2 Balancing Uneven Power Loss of Inverter Devices

From a practical implementation point of view, non-uniform power loss distribution leads to an uneven structure of the converter in terms of heat sink design. Therefore, a simple logic circuit can be used such that during a positive half cycle of line frequency, the switches S_1 and S_3 experience soft switching, and during a negative half cycle S_2 and S_4 experience soft switching. Therefore, this will make the converter size regular and suitable for practical implementation. Figure 6.33 shows this logic structure where the usual switching sequence is shown in red color and it leads to soft switching of S_1 and S_3 and hard switching of S_2 and S_4, repeatedly. However, during the negative half of input voltage, the switching sequence can be interchanged as shown in in Figure 6.33, leading to uniform switching loss distribution among all four sets of devices.

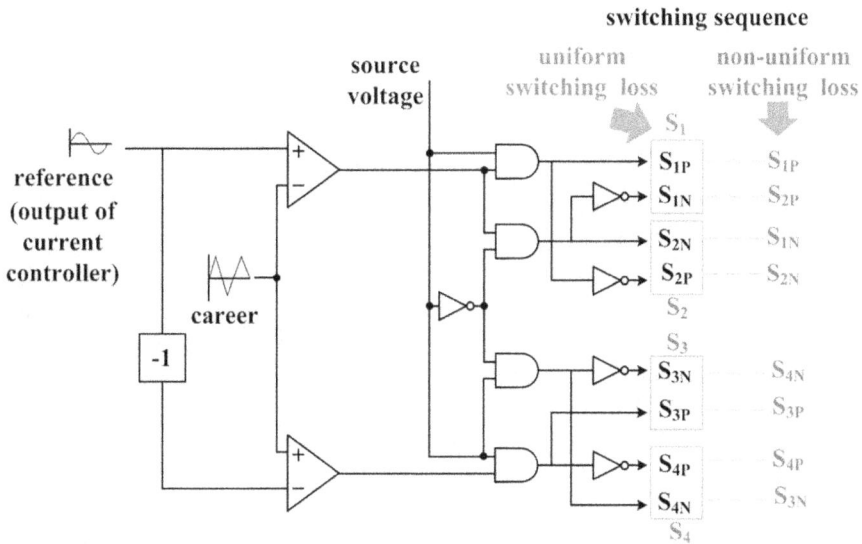

FIGURE 6.33 Switching sequence generation circuit for uniform switching loss distribution.

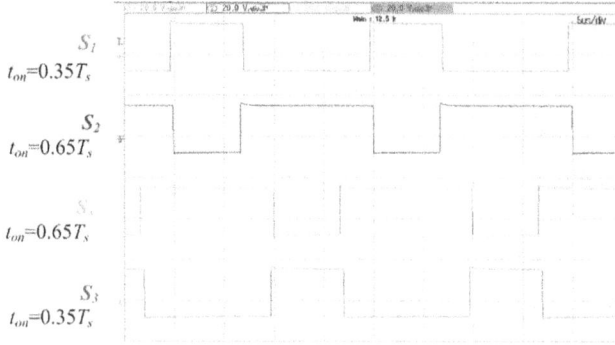

FIGURE 6.34   Gate pulse sequence.

6.11.7.3   Gate Pulse Sequence

Figure 6.34 shows typical gating signals of the ac-ac converter switches when the duty cycle of S_1 (or S_3) is D = 0.35. The overlap duration between the complementary gating signals S_1 and S_2 is around 250 ns, and this is enough to turn on and turn off the MOSFETs. The overlap time is mainly dependent on the turn-on time of incoming and turn-off time of outgoing devices and vice versa. This overlapping time should be small enough such that converter duty cycle utilization is close to 100% but long enough to successfully turn-on and turn-off the incoming and outgoing devices. In the experimental setup, the devices are MOSFETs; therefore, this duration is significantly less. However, with an IGBT-based converter circuit, this overlapping time will be slightly longer due to comparatively larger device turn-on and turn-off time. S_1 and S_3 have the same length turn-on time, but 180° phase shifted. Similarly, S_2 and S_4 have the same duty cycle, but 180° phase shifted. From Figure 6.34, it is clear that only when S_1 and S_3 are on, the ac-ac converter input is connected to TC side tank network. The rest of the time, the input inductor, L_s is directly connected across the source through the second leg of the ac-ac converter switches. Input voltage polarity is used to determine whether to provide switching frequency pulses to positive switch sets (S_{1P}–S_{4P}) or negative switch sets (S_{1N}–S_{4N}).

6.11.7.4   Results with Resistive Load

Figure 6.35 through Figure 6.40 shows experimental results of the converter when the load is resistive, whereas Figure 6.41 through Figure 6.44 shows results when the load is stiff dc voltage. Since all the results have a line frequency and switching frequency components, line frequency views are shown in the middle figures, whereas switching frequency views of the waveforms are shown on two sides. The zoomed view on the left side of every figure shows zoomed view at the line frequency peak, whereas the right-side figure shows zoomed view at the off-peak of the line frequency.

Figure 6.35 shows experimental results of source voltage, current, ac-ac converter output voltage and current for 1200 W power output, and the corresponding input voltage is 200 V, 60 Hz ac and output voltage is 300 V dc. Since the ac-ac converter input current predominantly contains line frequency and switching frequency

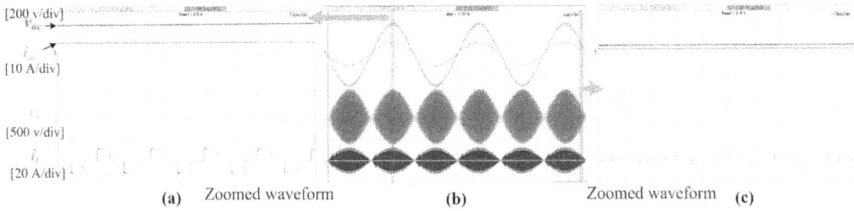

FIGURE 6.35 Experimental results of grid voltage, current and ac-ac converter output voltage and current waveforms when V_{ac} = 200V ac, P_o = 1.2 kW, V_o = 300 V. (a) zoomed view at line frequency peak, (b) line frequency view, and (c) zoomed view at off-peak of line frequency.

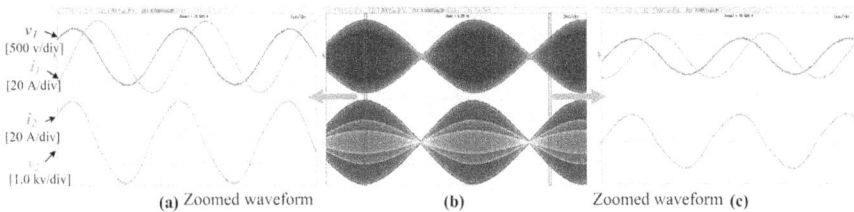

FIGURE 6.36 Experimental results: Transmitter and receiver coil voltages and currents at when V_{ac} = 200V ac, P_o = 1.2 kW, V_o = 300 V.

components; therefore, the small input filter is sufficient to filter out the switching frequency components. The total harmonic distortion (THD) of the source current is around 4.0% and it is well within IEEE 519-1992 specified standard (5%). From Figures 6.35a and 6.35c zoomed waveform, the ac-ac converter out current waveform is a quasi-square, but the voltage is very close to sinusoidal.

Figure 6.36 shows the voltage and current waveforms of both the transmitter and receiver coils. Figures 6.36a and 6.36c zoomed waveforms show switching frequency view at around the source voltage peak (90°) and 45° from zero-crossing, respectively. Due to the presence of the parallel capacitor, the TC coil voltage and current profiles are very close to sinusoidal. Compared with the parallel capacitor in a conventional parallel LC tank, the parallel capacitor, C_p in the CCL tank, offers much lower impedance to higher-order harmonics. This is because the capacitance value of C_p in the CCL tank is approximately twice that of a simple parallel LC tank.

Figure 6.37 shows the gating signal of S_{1P} and voltages across TC side tank elements. These results show the advantage of the CCL tank network over the simple parallel LC tank. Without the presence of the series capacitor, C_s, the converter switches would get directly TC coil voltage which is quite high. However, since the series capacitor, C_s partially compensate for the TC coil leakage impedance; therefore, the parallel capacitor has to provide only the remaining amount of reactive power to the TC coil. Thus, the converter switches get only $v_{i,}$ and it is a fraction of TC coil voltage, v_1, as shown in Figure 6.37.

Figure 6.38 shows input and output voltages and currents of the CCL primary and LC series secondary tank network. The operating power factor at the

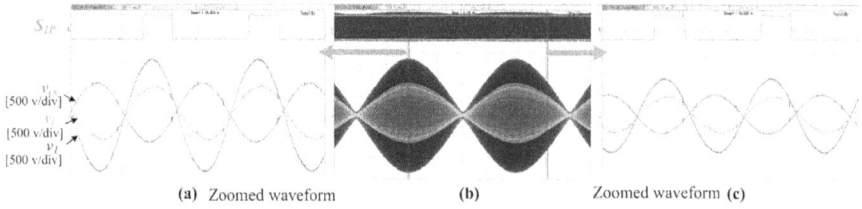

(a) Zoomed waveform   (b)   Zoomed waveform (c)

FIGURE 6.37 Experimental results: Gate pulse of switch S_{1P} and voltages across different elements (C_s, C_p, TC) in the TC side tank network.

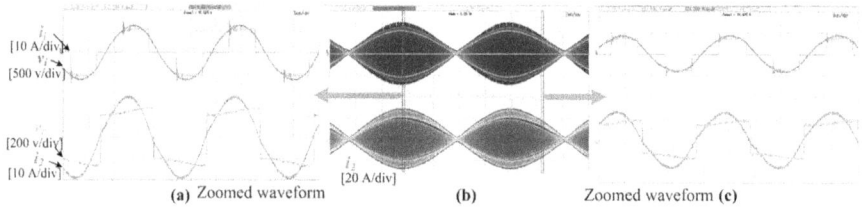

(a) Zoomed waveform   (b)   Zoomed waveform (c)

FIGURE 6.38 Experimental results: Input and output voltages and currents of the resonant tank network.

ac-ac converter output is lagging, and it is suitable for soft switching operation. Also, on the receiver side, the rectifier diodes turns-on and turns-off at zero current. This ensures zero reverse recoveries of these rectifier diodes. There is some surge current present in the ac-ac converter output current. In Figure 6.38, experimental results show this current profile with the full bandwidth of digital storage oscilloscope (DSO). However, in Figure 6.35, the bandwidth of DSO is kept at 20 MHz to show the exact line frequency envelope profile of the ac-ac converter output current. However, without this setting, this current profile is not very clear.

Figure 6.39 shows the soft-switching performance of converter switches S_{1P} and S_{3P}. In the lagging power factor, the switch S_{1P} is triggered when the voltage across switch S_1 is negative, and S_{1N} blocks this negative voltage. Therefore, switch S_{1P} does not start conducting immediately, resulting in ZVS of S_{1P}. However, after the overlap period, the complementary switch of S_1, i.e., S_2 is turned off, and S_1 is forced to take the input current, i_s. Also, before the turn-off of S_1, the complementary switch, S_2, is triggered to maintain the required overlap. Since the voltage

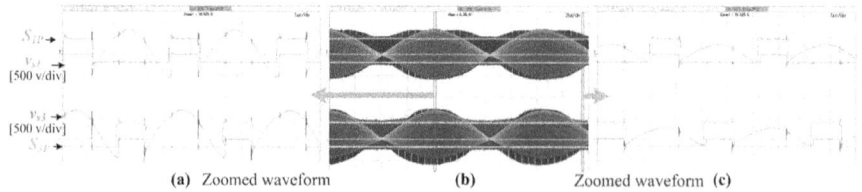

(a) Zoomed waveform   (b)   Zoomed waveform (c)

FIGURE 6.39 Experimental results: Soft switching of ac-ac converter switches S_{1P} and S_{3P}.

across S_2 is positive; therefore, $S2$ immediately commutates S_1. Thus, switch S_1 turns off at zero current. During this operation, the switch S_2 experiences hard switching both during turn-on and turn-off. Similar soft-switching turn-on and turn-off characteristics are obtained for switch S3, as shown in Figure 6.39.

Figure 6.40 shows the dynamic response of the converter when a load step commend from 70% of rated load (0.84 kW) to rated load (1.2 kW) is applied. From Figure 6.40, it is clear that the outer loop is capable of meeting the load requirements while the inner input current loop ensures high-quality source current. The load current settles at around 0.75 s, which verifies the converter's dynamic model. Since there is no bulky electrolytic capacitor to filter out the second harmonic; therefore, the load current contains second harmonics. These results are significant when the load is resistive such as the lighting load. In case of a fault or other emergency, such as the living object detection (LOD) function detects an intrusion into the active region, then the load power has to be reduced abruptly from full load to light load. In this situation, the control command can be directly applied to the inner input current control loop for faster response and safety.

Literature study shows that this rectified sinusoidal current is acceptable for several battery charging applications. However, if the load does not accept this current profile, then a reasonable size dc capacitor can be connected to the converter's output. The analysis and experimental results show that the converter can deliver power in either case. The first part of the experimental results shows the load current as rectified sinusoidal, and the later part of the experimental results shows the ripple-free dc load current.

6.11.7.5   Results with Stiff Voltage Load

To verify the converter's performance for battery charging applications, a stiff dc voltage load is connected to the converter output. Figure 6.41 shows experimental results of grid voltage, current, and ac-ac converter output voltage and current waveforms when $V_{ac} = 200$ V ac, $P_o = 1260$ W, $V_o = 270$ V fixed dc. It is clear that the inner input current loop is capable of maintaining a high-quality grid current. The THD of this current is calculated to be around 4.5%, and it is well within IEEE 519-1992 specified standards. From Figure 6.41a, it is seen that, unlike resistive load, here, the ac-ac converter output voltage does not follow the line frequency sinusoidal envelope. This voltage does not reduce significantly toward the zero crossing of the line frequency. This is because the load is stiff dc voltage, and to pump the charge to this high

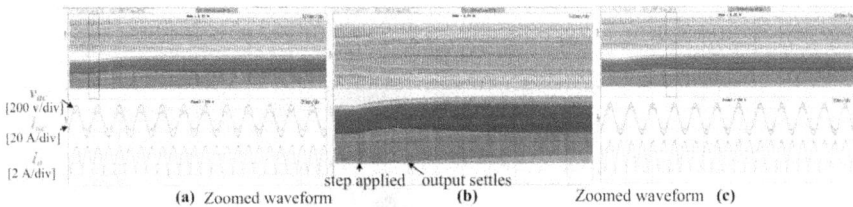

FIGURE 6.40   Dynamic response of the converter for a load step from 70% of rated load to rated load.

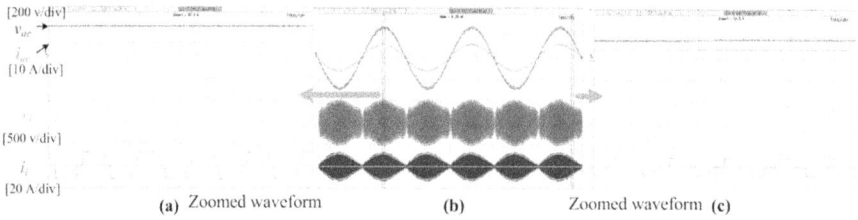

FIGURE 6.41 Experimental results of grid voltage, current and ac-ac converter output voltage and current waveforms when $V_{ac} = 200$V ac, $P_o = 1260$ W, $V_o = 270$ V stiff dc (a) zoomed view at line frequency peak, (b) line frequency view, (c) zoomed view at off-peak of line frequency.

dc voltage toward line frequency zero crossing, the TC side has to maintain a significant amount of voltage. This fact is evident from Figures 19a and 19c zoomed waveform of i_i that toward zero crossing the boosting feature of ac-ac converter becomes high. This phenomenon of the converter is very similar to boost-derived PFCs.

Figure 6.42 shows TC and RC voltages and current profiles. The TC coil receives a high-quality sinusoidal current, similar to earlier results. As discussed, the TC side coil voltage and current envelope do not follow the line frequency sinusoidal trend because of the stiff dc output voltage. Figure 6.43 shows the gating signal of switch S_{2P} and voltages across TC tank network elements. Similar to earlier, the magnitude of the voltage across the parallel capacitor, V_i, is a fraction of TC voltage V_1 because of the presence of a series capacitor, C_s. Since V_i directly determines ac-ac converter switch voltage stress; therefore, the CCL tank is superior in terms of inverter switch voltage rating.

From Figures 6.41 and 6.43, it is clear that, unlike resistive load, the duty cycle of the ac-ac converter devices varies significantly throughout the line frequency.

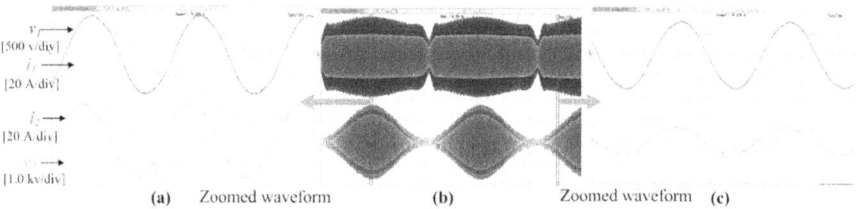

FIGURE 6.42 Transmitter and receiver coil voltages and currents at when $V_{ac} = 200$ V ac, $P_o = 1.2$ kW, $V_o = 270$ V stiff dc.

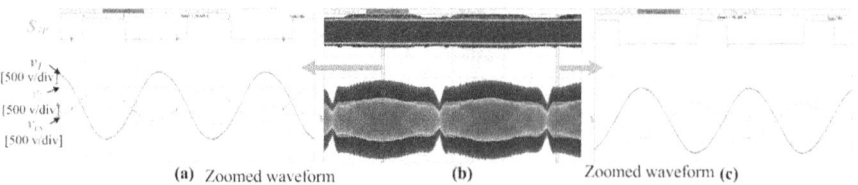

FIGURE 6.43 Gate pulse of switch S_{3p} and voltages across different elements (C_s, C_p, TC) in the TC side tank network.

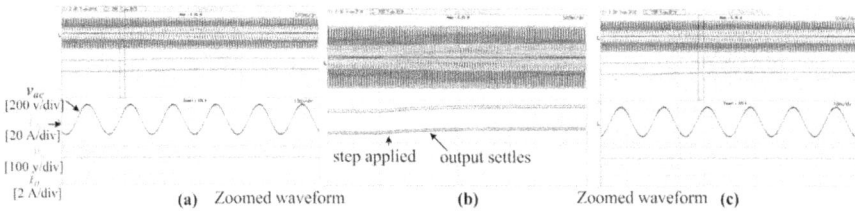

		step applied output settles	
(a) Zoomed waveform	(b)	Zoomed waveform (c)	

V_{ac}
[200 v/div]
[20 A/div]
[100 y/div]
I_p
[2 A/div]

FIGURE 6.44 Dynamic performance of the converter for a step-change in output current reference from 4.4 A to 5.5 A.

The duty cycle of switches S_1 and S_3 slowly reduces when the input voltage moves from peak to zero. From steady-state operation interval II, it is clear that current through the S_1 is transferred to S_2 before the gate pulse of S_1 is withdrawn. Therefore, S_1 experiences a soft turn-off, and S_2 experiences a hard turn-on. Again, in interval II of steady-state operation, the device S4 keeps on conducting when S_3 is given gating pulse. This leads to a soft turn-on of S_3 and a hard turn on S_4.

Figure 6.44 shows the converter's dynamic performance when a step-change in load current reference is given from 4.4A to 5.5A. The outer output current loop can meet load demand within the designed settling time, i.e., around 0.75 s, while the inner loop maintains the high-quality source current.

6.12   PRACTICE PROBLEMS

Q1. Using the conventional method, find the values of compensation capacitors such that the input voltage and current of the following WPT circuit are in the same phase.

Draw a possible inverter and rectifier circuit for this IPT circuit.

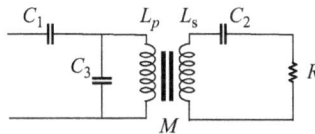

Q2. What will happen to S/S, S/P, P/S, and P/P compensated WPTs if a) the load is removed instantly and b) the secondary coil and associated circuitry are removed suddenly.

Q3. Design a Parallel-Series compensated WPT Charger for Okinawa i-praise battery.

Given: Okinawa i-praise battery: 72 V Li-ion, charging power 720 W, input 400 V DC, $k = 0.25$ and choose any suitable value of L_1/L_2.

Q4. Design a series-series compensated WPT Charger for a Tesla make EV Battery rated for 400 V, 25 A and drawing power from a 400 V DC bus. Assume coil-to-coil coupling $k = 0.2$, $f_s = 85$ kHz. Choose any suitable value of L_1/L_2.
  • If the load impedance (R_e or R_o) is variable, derive the output power and efficiency expression as a function of R_e for the above WPT charger. Assume coil resistances: $r_p = r_s = 100$ mΩ
  • Plot P_o-verses- R_e, and efficiency-verses-R_e.
  • Find the value/s of R_e for which maximum efficiency and maximum P_o are obtainable. What are the values of maximum efficiency and maximum P_o?

- Find voltage and current ratings of each element as necessary for manufacturing/ or selecting components.

Q5. In the following WPT circuit, using the conventional method, find C_2 expression.
- Considering this value of C_2, is it possible to design the C_1 such that its value is not dependent on the load (load-independent resonant tank design)?
- Derive the control (duty cycle) to output (output voltage) transfer function (TF) for the following circuit.

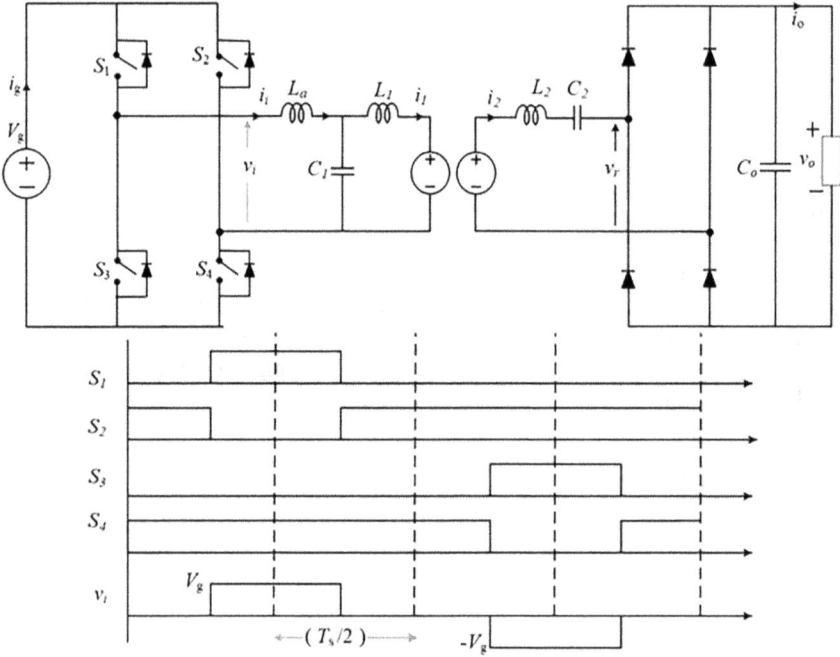

Q6. Assuming the L_1, L_2, and C_s impedance are the same at the resonance frequency, find the primary compensation capacitor value for the following circuit.
- Find the inverter $(\tilde{I}_i/\tilde{i}_g)$ and rectifier transfer functions $(\tilde{v}_o/\tilde{V}_s)$.

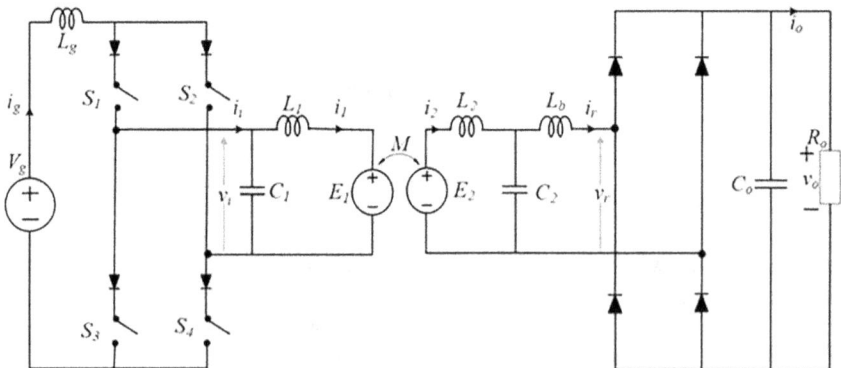

REFERENCES

[1] S. Li and C. Mi, "Wireless power transfer for electric vehicle applications," *IEEE J. Emerging Sel. Topics Power Electron.*, vol. 3, no. 1, pp. 4–17, Mar. 2015.

[2] G. A. Covic and J. T. Boys, "Inductive power transfer," *Proceedings of the IEEE*, vol. 101, no. 6, pp. 1276–1289, Jun. 2013.

[3] P. Si, A. P. Hu, S. Malpas and D. Budgett, "A frequency control method for regulating wireless power to implantable devices," *IEEE Trans. Biomedical Circuits and Systems*, vol. 2, no. 1, pp. 22–29, Mar. 2008.

[4] F. Musavi and W. Eberle, "Overview of wireless power transfer technologies for electric vehicle battery charging," *Power Electronics, IET*, vol. 7, no. 1, pp. 60–66, Jan. 2014.

[5] C. C. Mi, G. Buja, S. Y. Choi and C. T. Rim, "Modern advances in wireless power transfer systems for roadway powered electric vehicles," *IEEE Trans. Ind. Electron.*, vol. 63, no. 10, p. 6533–6545, Oct. 2016.

[6] S. Y. Choi, B. W. Gu, S. Y. Jeong and C. T. Rim, "Advances in wireless power transfer systems for roadway-powered electric vehicles," *IEEE J. Emerging Sel. Topics Power Electron.*, vol. 3, no. 1, pp. 18–36, Mar. 2015.

[7] K. W. Klontz, A. Esser, R. R. Bacon, D. M. Divan, D. W. Novotny and R. D. Lorenz, "An electric vehicle charging system with 'universal' inductive interface," Conference Record of the Power Conversion Conference – Yokohama 1993, Yokohama, Japan, 1993, pp.," *IEEE Power Conversion Conf. Yokohama, Japan*, pp. 227–232, 1993.

[8] M. P. Theodoridis, "Effective capacitive power transfer," *IEEE Trans. Power Electron.*, vol. 27, no. 12, pp. 4906–4913, Dec. 2012.

[9] S. K. Mishra, R. Adda, S. Sekhar, A. Joshi and A. K. Rathore, "Power transfer using portable surfaces in capacitively coupled power transfer technology," *IET Power Electronics*, vol. 9, no. 5, pp. 997–1008, April 2016.

[10] A. Kurs, A. Karalis, R. Moffatt, J. D. Joannopoulos, P. Fisher and M. Soljacic, "Wireless power transfer via strongly coupled magnetic resonances," *Science*, vol. 317, pp. 83–86, 2007.

[11] F. Musavi, D. S. Gautam, W. Eberle and W. G. Dunford, "A simplified power loss calculation method for PFC boost topologies," *Transportation Electrification Conference and Expo (ITEC), 2013 IEEE*, pp. 1–5, June 2013.

[12] Y. Liao and X. Yuan, "Compensation topology for flat spiral coil inductive power transfer systems," *Power Electronics, IET*, vol. 8, no. 10, pp. 1893–1901, October 2015.

[13] J. M. Miller, O. C. Onar and M. Chinthavali, "Primary-side power flow control of wireless power transfer for electric vehicle charging," *IEEE J. Emerging Sel. Topics Power Electron.*, vol. 3, no. 1, pp. 147–162, March 2015.

[14] P. Ning, J. M. Miller, O. C. Onar, C. P. White and L. D. Marlino, "A compact wireless charging system development," *IEEE Applied Power Electronics Conference and Exposition (APEC)*, pp. 3045–3050, 2013.

[15] G. Buja, M. Bertoluzzo and H. K. Dashora, "Lumped track layout design for dynamic wireless charging of electric vehicles," *IEEE Trans. Ind. Electron.*, vol. 63, no. 10, p. 6631–6640, Oct. 2016.

[16] N. Keeling, G. Covic and J. Boys, "A unity-power-factor IPT pickup for high-power applications," *IEEE Trans. Ind Electron.*, vol. 57, no. 2, pp. 744–751, Feb. 2010.

[17] A. Kamineni, G. A. Covic and J. T. Boys, "Self-tuning power supply for inductive charging," *IEEE Trans. Power Electron.*, vol. 32, no. 5, pp. 3467–3479, May 2017.

[18] A. P. Hu, G. A. Covic and J. T. Boys, "Direct ZVS start-up of a current-fed resonant inverter," *IEEE Trans. Power Electron.*, vol. 21, no. 3, pp. 809–812, May 2006.

[19] A. P. Hu, J. T. Boys and G. Covic, "Frequency analysis and computation of a current-fed resonant converter for ICPT power supplies," *Power System Technology, 2000. Proceedings. PowerCon 2000, International Conference*, vol. 1, pp. 327–332, 2000.

[20] C. Tang, X. Dai, Y. S. Z. Wang and A. P. Hu, "Frequency bifurcation phenomenon study of a soft switched push-pull contactless power transfer system," *6th IEEE Conference on Industrial Electronics and Applications*, pp. 1981–1986, 2011.

[21] U. K. Madawala and D. J. Thrimawithana, "A bidirectional inductive power interface for electric vehicles in V2G systems," *IEEE Trans. Ind. Electron.*, vol. 58, no. 10, pp. 4789–4796, Oct. 2011.

[22] U. K. Madawala, M. Neath and D. J. Thrimawithana, "A power–frequency controller for bidirectional inductive power transfer systems," *IEEE Trans. Ind. Electron.*, vol. 60, no. 1, pp. 310–317, Jan. 2013.

[23] Y. Yao, Y. Wang, X. Liu, F. Lin and D. G. Xu, "A novel parameter tuning method for double-sided LCL compensated WPT system with better comprehensive performance," *IEEE Trans. Power Electron.*, vol. 33, no. 10, pp. 8525–8536, Oct. 2018, doi: 10.1109/TPEL.2017.277825

[24] H. Hao, G. Covic and J. Boys, "A parallel topology for inductive power transfer power supplies," *Power Electronics, IEEE Transactions*, vol. 29, no. 3, pp. 1140–1151, Mar. 2014.

[25] F. Lu, H. Zhang, H. Hofmann and C. C. Mi, "An inductive and capacitive combined wireless power transfer system with LC-compensated topology," *IEEE Trans. on Power Electron.*, vol. 31, no. 12, pp. 8471–8482, Dec. 2016.

[26] S. Weerasinghe, U. K. Madawala and D. J. Thrimawithana, "A matrix converter-based bidirectional contactless grid interface," *IEEE Trans. Power Electron.*, vol. 32, no. 3, pp. 1755–1766, Mar. 2017.

[27] S. Weearsinghe, D. J. Thrimawithana and U. K. Madawala, "Modeling bidirectional contactless grid interfaces with a soft DC-link," *IEEE Trans. Power Electron.*, vol. 30, no. 7, pp. 3528–3541, Jul. 2015.

[28] H. H. Wu, A. Gilchrist, K. D. Sealy and D. Bronson, "A high efficiency 5 kw inductive charger for EVs using dual side control," *Industrial Informatics, IEEE Transactions on*, vol. 8, no. 3, pp. 585–595, Aug. 2012.

[29] J. L. Villa, J. Sallan, J. F. S. Osorio and A. Llombart, "High-misalignment tolerant compensation topology for ICPT systems," *IEEE Trans. Ind. Electron.*, vol. 59, no. 2, pp. 945–951, Feb. 2012.

[30] B. X. Nguyen, W. Peng and D. M. Vilathgamuwa, "An overview of power circuit topologies for inductive power transfer systems," *2016 IEEE International Conference on Sustainable Energy Technologies (ICSET)*, pp. 256–263, 2016.

[31] K. Colak, E. Asa, M. Bojarski, D. Czarkowski and O. C. Onar, "A novel phase-shift control of semibridgeless active rectifier for wireless power transfer," *IEEE Trans. Power Electron.*, vol. 30, no. 11, pp. 6288–6297, Nov. 2015.

[32] D. J. Thrimawithana and U. K. Madawala, "A novel matrix converter based bi-directional IPT power interface for V2G applications," *IEEE Int. Energy Conf. Exhibition 2010*, pp. 495–500, 2010.

[33] M. Moghaddami, A. Anzalchi and A. I. Sarwat, "Single-stage three-phase AC–AC matrix converter for inductive power transfer systems," *IEEE Trans. Ind. Electron.*, vol. 9, no. 5, pp. 997–1008, Oct. 2016.

[34] N. Xuan Bac, D.M Vilathgamuwa and U. K Madawala., "A SiC-based matrix converter topology for inductive power transfer system," *IEEE Trans. Power Electron.*, vol. 29, no. 8, pp. 4029–4038, Aug. 2014.

[35] H. L. Li, A. P. Hu and G. A. Covic, "A direct AC–AC converter for inductive power-transfer systems," *IEEE Trans. Power Electron.*, vol. 27, no. 2, pp. 661–668, Feb. 2012.

[36] N. S. González-Santini, Y. Y. H. Zeng and F. Z. Peng, "Z-source resonant converter with power factor correction for wireless power transfer applications," *IEEE Trans. Power Electron.*, vol. 31, no. 11, pp. 7691–7700, Nov. 2016.

[37] J. Liu, K. W. Chan, C. Y. Chung, N. H. L. Chan, M. Liu and W. Xu, "Single-stage wireless-power-transfer resonant converter with boost bridgeless power-factor-correction rectifier," *IEEE Trans. Ind. Electron.*, vol. 65, no. 3, pp. 2145–2155, March 2017.

[38] S. Samanta, A. K. Rathore and D. J. Thrimawithana, "Bidirectional current-fed half-bridge (C) (LC)–(LC) configuration for inductive wireless power transfer system," *IEEE Trans. Ind. Applications*, vol. 53, no. 4, pp. 4053–4062, Jul. 2017.

[39] J. Shin, S. Shin, Y. Kim, S. Ahn, S. Lee, G. Jung, S.-J. Jeon and D.-H. Cho, "Design and implementation of shaped magnetic-resonance-based wireless power transfer system for roadway-powered moving electric vehicles," *IEEE Trans. Ind. Electron.*, vol. 61, no. 3, pp. 1179–1192, Mar. 2014.

[40] M. Budhia, G. A. Covic and J. T. Boys, "Design and optimization of circular magnetic structures for lumped inductive power transfer systems," *IEEE Trans. Power Electron.*, vol. 26, no. 11, pp. 3096–3108, Nov. 2011.

[41] M. Budhia, J. T. Boys, G. A. Covic and C.-Y. Huang, "Development of a single-sided flux magnetic coupler for electric vehicle IPT charging systems," *IEEE Trans. Ind. Electron.*, vol. 60, no. 1, pp. 318–328, Jan. 2013.

[42] F. Musavi and W. Eberle, "Overview of wireless power transfer technologies for electric vehicle battery charging," *Power Electronics, IET*, vol. 7, no. 1, pp. 60–66, Jan. 2014.

[43] S. Li and C. Mi, "Wireless power transfer for electric vehicle applications," *IEEE J. Emerging Sel. Topics Power Electron.*, vol. 3, no. 1, pp. 4–17, Mar. 2015.

[44] K. E. Koh, T. C. Beh, T. Imura and Y. Hori, "Impedance matching and power division using impedance inverter for wireless power transfer via magnetic resonant coupling," *IEEE Trans. Ind. Applications*, vol. 50, no. 3, pp. 2061–2070, May 2014.

[45] P. Changbyung, L. Sungwoo, G.-H. Cho and C. T. Rim, "Innovative 5-m-off-distance inductive power transfer systems with optimally shaped dipole coils," *IEEE Trans. Power Electron.*, vol. 30, no. 2, pp. 817–827, Feb. 2015.

[46] Z. N. Low, R. A. Chinga, T. Ryan and L. Jenshan, "Design and test of a high-power high-efficiency loosely coupled planar wireless power transfer system," *IEEE Trans. Ind. Electron.*, vol. 56, no. 5, pp. 1801–1812, May 2009.

[47] J. M. Miller, O. C. Onar, C. White, S. Campbell, C. Coomer, L. Seiber, R. Sepe and A. Steyerl, "Demonstrating dynamic wireless charging of an electric vehicle: The benefit of electrochemical capacitor smoothing," *IEEE Power Electron. Magazine*, vol. 1, no. 1, pp. 12–24, Mar. 2014.

[48] F. Y. Lin, S. Kim, G. A. Covic and J. T. Boys, "Effective coupling factors for series and parallel tuned secondaries in IPT systems using bipolar primary pads," *IEEE Trans. Transport. Electrific.*, vol. 3, no. 2, Jun. 2017.

[49] Y. Li, T. Lin, R. Mai, L. Huang and Z. He, "Compact double-sided decoupled coils based WPT systems for high power applications: Analysis, design and experimental verification," *IEEE Trans. Transport. Electrific.*, vol. 33, no. 7, pp. 5565–5577, July 2018, 10.1109/TPEL.2017.2750081

[50] P. Xuewei and A. K. Rathore, "Small-signal analysis of naturally commutated current-fed dual active bridge converter and control implementation using cypress PSoC," *IEEE Trans. Vehicular Tech.*, vol. 64, no. 11, pp. 4996–5005, Nov. 2015.

[51] U. R. Prasanna and A. K. Rathore, "Small-signal modeling of active-clamped ZVS current-fed full-bridge isolated DC/DC converter and control system implementation using PSoC," *IEEE Trans. Ind. Electron.*, vol. 61, no. 3, pp. 1253–1261, Mar. 2014.

[52] A. Rathore, A. K. S. Bhat, S. Nandi and R. Oruganti, "Small signal analysis and closed loop control design of active-clamped ZVS two-inductor current-fed isolated dc-dc converter," *IET Power Electron.*, vol. 4, no. 1, pp. 51–62, 2011.

[53] S. Samanta and A. K. Rathore, "A new inductive power transfer topology using direct AC-AC converter with active source current waveshaping," *IEEE Trans. Power Electron.*, 2017.

[54] S. Samanta and A. K. Rathore, "A new current-fed CLC transmitter and LC receiver topology for inductive wireless power transfer application: Analysis, design, and experimental results," *IEEE Trans. Transport. Electrific.*, vol. 1, no. 4, pp. 357–368, Dec. 2015.

[55] S. Samanta and A. K. Rathore, "Analysis and design of current-fed half-bridge (C) (LC)–(LC) resonant topology for inductive wireless power transfer application," *IEEE Trans. Ind. Applications*, vol. 53, no. 4, pp. 3917–3926, Jul. 2017.

7 Power Grid Impacts of Electric Vehicle (EV) Integration

D. S. Kumar, A. Sharma, and
C. D. Rodríguez-Gallegos

CONTENTS

DOI: 10.1201/9781003330134-7

7.1 INTRODUCTION

Climate change has evolved as one of the most critical urban challenges in the 21st century. The change in climate has been attributed to greenhouse gas (GHG) emissions that come from transport activities, burning of fossil fuels, industrial processes, changes in land use, and deforestation [1]. As the global transport sector is one of the major polluters, with road transport contributing around 80% of the emissions, many countries have come up with policies promoting sustainable means of transportation. Electric vehicles (EVs) have gained immense traction as a zero-emission* means of transportation due to the massive subsidies provided by the governments. According to IEA Global EV Outlook 2021 [2], the global electric car stock just in the passenger section has risen to more than 10 million in 2020 in the last decade (see Figure 7.1). It is noted that EVs are also replacing internal combustion engine (ICE) vehicles in other categories such as trucks, three-wheeler, and two-wheeler vehicl es.

The increase in EV numbers is not only linked to government subsidies but can also be attributed to the reduction in battery costs and safety concerns, although there is still scope for improvement in these aspects.

Although the increase in the electric fleet will bring major advantages to the environment, it will require a major infrastructure deployment in terms of charging stations. Moreover, it will also pose a major challenge for the grid operators as the EVs would require electricity to charge the batteries. This additional requirement for electrical power would change the load demand observed by the grid [4]. The non-linear and mobile nature of the EV load demand will also affect the power quality of the distribution system. Moreover, the EVs can act not only as a load but also as a source and deliver power back to the grid using the vehicle-to-grid (V2G) feature. Thus, the charging strategies and infrastructure will play a vital role in determining the various grid impacts of EV integration as their penetration in the transport sector increases.

The aim of this chapter is to highlight the various grid impacts of EV integration as their penetration in the transport sector increases. We will present the different EV models, charger types, charging strategies, and how their usage affects the grid. The chapter will also discuss the mitigation methods to counter the different impacts and describe the challenges associated with the widespread development of the EV ecosystem.

7.2 EV MODELS, CHARGER TYPES, AND CHARGING MODES

In order to compete with ICE vehicles, various models of EVs have been released in the market with different battery sizes, traveling ranges, and charging requirements (see Table 7.1). Kindly note that the required charging power is aimed to match the

---

* Zero-emission here refers to the emissions from the vehicles. The authors understand that there are potential emissions during the manufacturing of the EV battery, and during the generation of power that is required to charge the batteries.

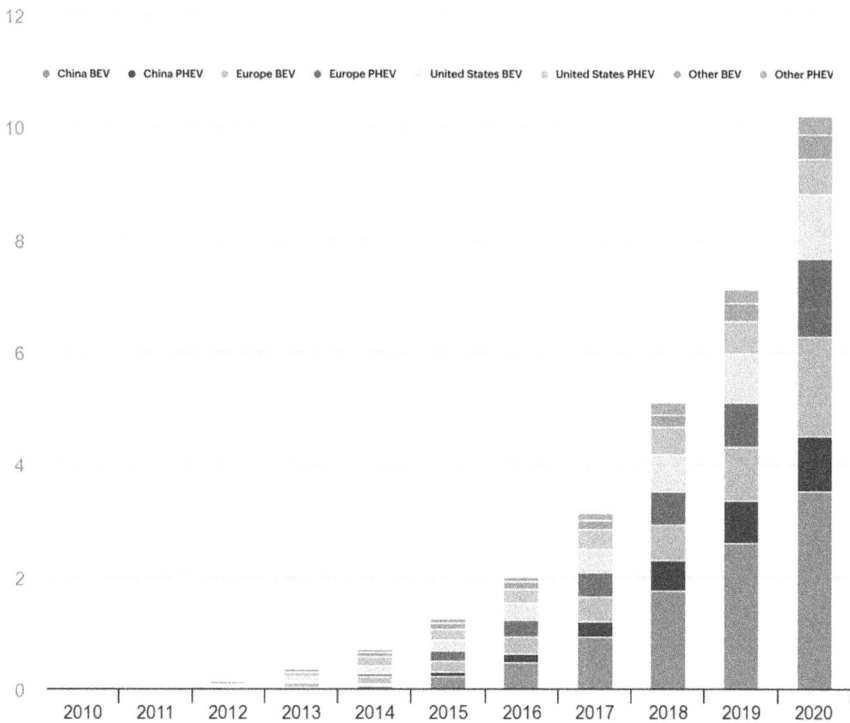

FIGURE 7.1   The global electric passenger car stock, 2010–2020 [2].

TABLE 7.1
Available Car Models in the Market Till 2020 [3]

Car Model	Battery Capacity (kWh)	Range (km)	Charging Power Required (kW)	Actual Maximum Charging Power (kW)
Mini Cooper SE	28.9	180	69	49
BMW i3	37.9	235	91	49
Hyundai Kona	64	400	154	77
Tesla Model 3	72.5	450	174	250
Porsche Taycan Turbo S	83.7	425	201	262
Tesla Model S	95	515	228	250

* to reach from 20% to 80% SOC in 15 min

refill time in a fuel station which is approximately 15–20 minutes within which the state of charge (SOC) of the vehicle should reach from 20% to 80%.

While driving range is one of the concerns, the major deterrence for car owners to make the transition to EV is due to the inadequate availability of charging

TABLE 7.2

Classification of Charger Types Based on Power Level, Location, Input Signal, etc

Power Level	Charger Type	Location	Expected Power Level (kW)	Tentative Charging Range (hours)
Level 1 chargers	On-board 1-phase	Home or Office	1.4 (12 A)	4–11
			1.9 (20 A)	11–36
Level 2 chargers	On-board 1- or 3-phase	Private or Public Outlet	4 (17 A)	1–4
			8 (32 A)	2–6
			19.2 (80 A)	2–3
DC fast chargers	Off-board	Commercial or Public Outlet	50	0.4–1
			100	0.3–0.6
			250	0.2–0.5

stations (CS) and the time taken for the EV to be fully charged. Therefore, a variety of charging options are available depending on factors such as the EV owner's needs, available electrical infrastructure, space, etc. The charging infrastructures can be categorized in different ways, based on factors such as input current type, power level, charging location, charger type, etc. (see Table 7.2). According to the various Standards available for EV charging, such as IEC 61851-1 [5], SAE J1772 [6], etc., the charging stations can operate in different modes based on the categories as shown in Figure 7.2, individually or in their combinations. Though extreme fast charging is also being developed, the research on the topic is in the nascent stage and there are no commercially available extreme fast chargers yet in the market.

The methods of charging EVs are categorized into different power levels and the power levels reflect the charger power, equipment, effect on the grid, charging time, cost, and location. The main components of an Electrical Vehicle Supply Equipment consist of Cords for EVs, attachment plugs, ports for the electrical vehicle and Charge stands. The specific configuration will vary accordingly depending on location, country, the connection to the electrical grid, the electric potential difference, frequency, and transmission standards. Level 1 and level 2 charging equipment are the primary options because most EV owners are expected to charge their vehicles overnight at home. Moreover, level 1 and level 2 chargers are convenient and cost-effective.

Level 1 AC, which uses the slowest charge method, is commonly used in a household where EV can be plugged into any convenient power outlet which requires 120/230 V. The charger uses a standard J1772 which connects to the ac port of the EV or a wall plug, Nema 515 Nema 520 to charge the EV [6].

Whereas, level 2 AC, a semi-fast charge method, requires 240/400 V and the charging infrastructure can be off-board or on-board. Level 3 AC-DC uses a three-phase commercial fast charge method which requires a voltage of 208/415 V. The

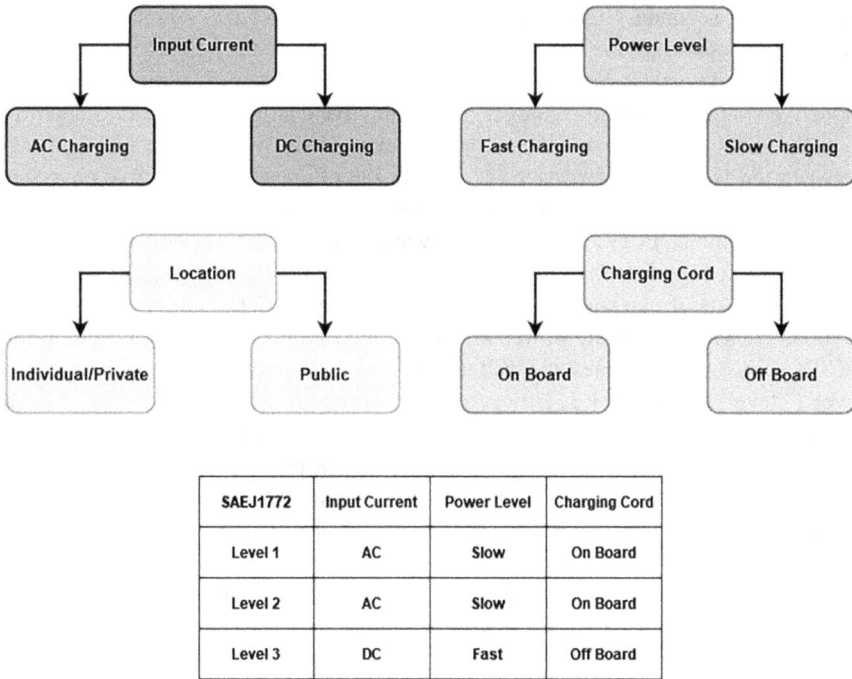

SAEJ1772	Input Current	Power Level	Charging Cord
Level 1	AC	Slow	On Board
Level 2	AC	Slow	On Board
Level 3	DC	Fast	Off Board

FIGURE 7.2   Factors defining charging infrastructure. Example: SAEJ1771 Standard.

charging takes less than an hour, however, it requires an off-board charger conversion to regulate AC to DC. Among the 3 levels, the cost of level 3 infrastructure is the most expensive as compared to the other power level chargers.

There are currently several types of electric vehicle charging infrastructure such as conductive and inductive charging. Conductive charging requires direct contact between the utility grid and the EV whereas inductive charging involves the ideology of mutual induction for the power transfer between the utility grid and the electric vehicle. Currently, conductive charging is more established than inductive charging.

Also, there are on-board and off-board EV chargers [7]. On-board chargers are integrated into the EV thus, they have a size and weight constraint resulting in power level limitation. Off-board chargers are installed on various infrastructures such as shopping centers and hospitals where they are not integrated into the EVs. Conductive chargers require a cable connection between the charge inlet and the EV connector. The cable is fed from a charging station of level 2 or level 3 or fed from an electrical outlet of level 1 or level 2. Inductive charging does not require any physical contact between the utility grid and electric vehicles; hence isolation transformers might not be required for safety purposes allowing for the reduction of the charger's size. In addition, inductive chargers, are used in level 1 and level 2 chargers both methods can either be static or dynamic. The static charger is a stationary inductive charger that consists of a primary inductor at the charging station and a secondary inductor located at the bottom of the vehicle. The power transfer happens when the primary paddle is inserted into the vehicle. The dynamic charger is a contactless roadway for charging the EV system. The idea of Roadbed inductive charging (dynamic charger) uses inductive charging to have the capability to charge EVs on the move which reduces battery size. However, the misalignment between the power-transferring coils reduces the efficiency of inductive charging.

In a unidirectional charger, the system power only transferred its power from the grid to the vehicle, also known as grid-to-vehicle (G2V). Whereas the bi-directional charger, allows the power to flow from the V2G and G2V. A typical bidirectional charger consists of two stages. Firstly, a bi-directional AC-DC converter connected to an active grid boosts the power factor of the system. Secondly, a bi-directional DC-DC converter regulates the battery current. Unidirectional chargers apply to level 1, level 2, and level 3. However, bidirectional only applies to level 2 infrastructures, this is due to level 1 being targeted at low cost and its power limits and maximizing the system flexibility. A reverse in the power flow of a level 3 fast-charge system creates conflict with the basic purpose, minimizing connection time, and delivering substantial energy.

In this work, to understand the grid impacts of EV integration, we will mainly divide the chargers into AC and DC types. It can be observed from the literature, that AC chargers are mainly used for low-power applications, take more time to charge, and are usually deployed at homes or offices. DC chargers are fast chargers with less charging time and are installed mainly on commercial sites (see Table 7.2). Slow charging would allow the grid operator to plan and regulate the load demand and with the V2G feature, EVs can be utilized even for other ancillary services. Fast or extreme fast charging, on the other end, can lead to severe grid impacts in terms of voltage fluctuations, harmonic distortions, etc., due to the high-power level and fast plug-and-play. The various impacts of EV grid integration are highlighted next.

7.3   IMPACTS OF EV INTEGRATION

Before understanding the potential impacts of the integration of EVs in the grid, it is important to understand the control architecture and market framework in which the EVs will be deployed. It is evident from Table 7.2 (considering the power levels) that an individual EV might not have a significant impact on the grid, especially for low-power chargers. However, the cumulative impact of EVs can be significant and they can even play a role in the electricity markets. As EVs are deployed on a large scale, a new entity called the "Aggregator" is emerging in the grid. An Aggregator manages and controls the charging of a group of EVs considering the requirements of the EV owners. At the same time, it utilizes the flexibility of charging and discharging the EV batteries in the electricity market based on the constraints laid out by the distribution system operator. A good example of an Aggregator can be charging infrastructures provided at public locations, such as malls or multi-story car parks. Figure 7.3 highlights the EV integration architecture from the medium voltage to the low voltage level.

From Figure 7.3, it can be observed that EVs can either be charged at individual houses as shown by the yellow box, or be charged with other EVs at an Aggregator.

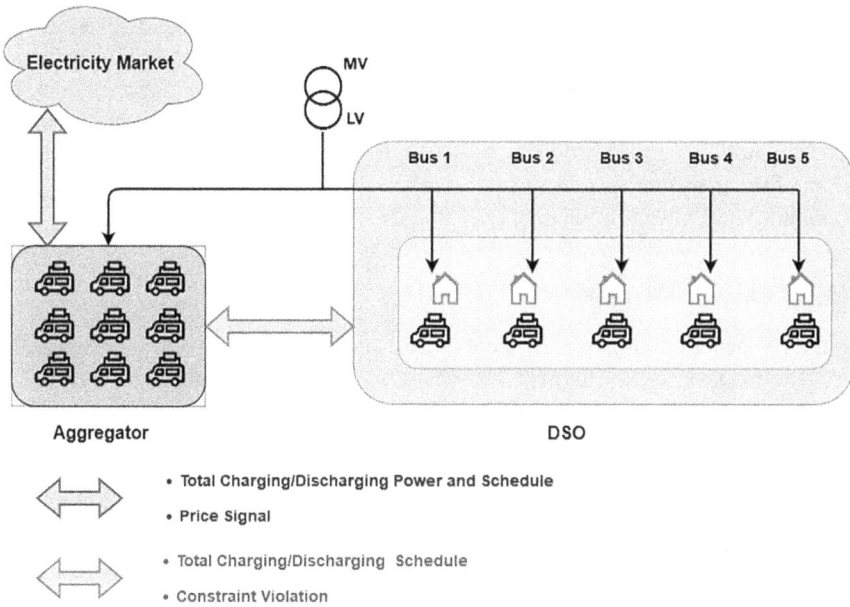

FIGURE 7.3 EV integration architecture.

The EVs charging/discharging at the individual houses would primarily have slow charging facilities. They might choose to be controlled (or not) by the DSO and would ideally not participate in the market. The uncontrolled EVs would generally be free to charge at their will and would lead to an increase in load demand, voltage drop, increase in line current, or transformer loading. The DSO needs to be prepared to deal with this situation. While the controllable EVs could allow the DSO to manage their charging or discharging to support the grid in terms of peak load management for some incentives or subsidies. It is important to highlight that in some works of literature, the controllable and uncontrolled charging aspect of EVs is referred to as smart charging and dumb charging, respectively.

The EVs getting charged at the Aggregator are mainly controllable and because of their cumulative power can participate in the electricity market and support the grid for various ancillary services. As observed from Figure 7.3 (green arrow) before an Aggregator commits in the market, it needs to check with the DSO regarding any constraint violations (such as line current, transformer loading, etc.). It is noted that for efficient control and management, there must be channels of communication sharing information among the various players (see Figure 7.3 for the potential information which could be shared between the parties).

7.4 POSITIVE IMPACTS OF EV INTEGRATION

In this chapter, the focus is on highlighting the impacts of the large-scale integration of EVs. As mentioned before, individual EVs would have minimal impact on the grid and the DSO can include them partially in the control and management of the distribution system by providing some incentives or subsidies. However, at the Aggregator, the cumulative impact of EVs can be significant and they can partic- ipate in the market to provide various ancillary services. These ancillary services are considered positive impacts of EV integration in this chapter and will be discussed next. It is noted that the participation of EVs for ancillary services depends on their availability, therefore, the Aggregator relies on forecast tools to estimate their availability and bid in the market.

7.4.1 DEMAND RESPONSE

As the EV penetration increases, the load demand of the system will increase as the EV batteries need to be charged. The charging strategy plays an important role in load management. As seen in Figure 7.4, for the dumb charging scenario it is observed that the peak load increases during the night and the difference between peak load and minimum load is around 22.5 MW, whereas, for the smart charging strategy, this difference is reduced to 5 MW. This is very advantageous for the DSO, as they do not have to depend on reserves to support the peak load. Here, the aggregated load of EVs can act as a reserve which can assist to reduce the peak load. The V2G feature of EVs plays a vital role to develop a good load manage- ment/demand response strategy.

7.4.2 REGULATION

The aggregated EVs not only can reduce the dependence on reserves, but at the same time, they can also act as reserves and support the grid in frequency and voltage regulation. EVs as controllable loads and their V2G feature can support the grid as primary, secondary, or tertiary frequency control using the droop

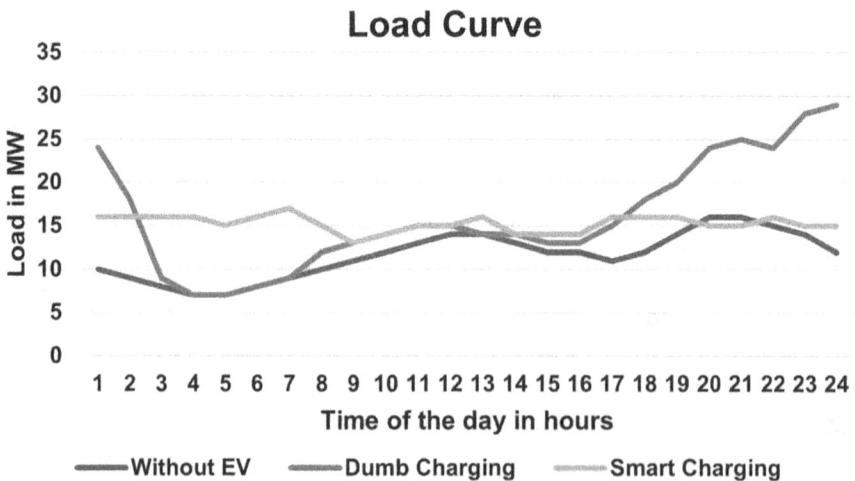

FIGURE 7.4 Variation in load demand with different charging strategy.

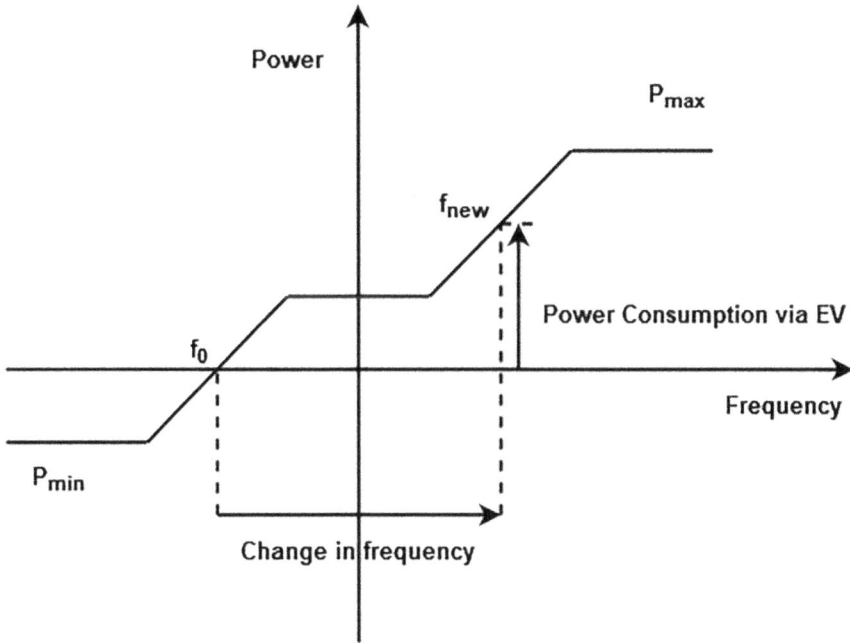

FIGURE 7.5   Droop characteristics (Active Power v/s Frequency).

characteristics by injecting or consuming active power as shown in Figure 7.5. It is noted that the use of EVs as a primary or secondary reserve would depend on if the system is grid-connected or isolated [4]. Similarly, EVs can support voltage regulation by supplying or absorbing reactive power. This mode of ancillary service assists the DSO to reduce investment in Battery Energy Storage Systems and improve the system performance even during an emergency.

It is noted that the load management and regulation strategy can also support the integration of renewable energy sources by mitigating their uncertainty or providing an energy buffer in both steady-state and dynamic operations. As the integration of EVs in the grid increases, it is the economics of the ancillary services which would allow the Aggregators and EV owners to participate in the electricity market and offset the charging cost of EVs.

7.5   NEGATIVE GRID IMPACTS WITH EV (PHEV, DC FAST CHARGING, ULTRA/EXTREME FAST CHARGING)

EV charging adds new/futuristic loads to the grid infrastructure and will be seen to impact grid reliability, resiliency, and security. With increasing EV loads in the grid, issues like feeder voltage regulation, harmonic regulation, power quality concerns, etc. are anticipated to significantly exacerbate. The EV interfaces can be three-phase or single-phase systems depending upon the type of chargers used. Signal phase systems will introduce additional issues like phase imbalance.

7.5.1 NEED FOR GRID INFRASTRUCTURE UPGRADES

With the increase in penetration of EVs, an increase in peak loading is anticipated with unconstrained charging. Changes in load profiles owing to EV penetration will impact the operation grid equipment including transformers, capacitor banks, and on-load tap changers (OLTC), bringing about reliability and lifetime concerns for the same. Though the increase in loading can be managed through coordinated charging of vehicles up to certain penetration levels, the differences in consumer charging patterns, grid constraints, etc., will drive the need for grid infrastructure upgrade.

7.5.2 VOLTAGE FLUCTUATION

Voltage fluctuation can be defined as the degree of deviation to terminal voltage i.e., the voltage at the point of common coupling (PCC) when current is injected/absorbed at any power factor or the percentage change in output voltage from no-load to full load. The deviation in feeder voltage is minimal closer to the substations and the deviation becomes significant along the feeder with the addition of feeder impedance. The voltage drops with EV penetration in a system can be understood from Figure 7.6, where the green dots suggest that the voltage is within the standard limits while the red dots suggest that the limits are violated (according to ANSI C84.1-2020 Standard [8]). It is noted that as you move away from the DSO, the

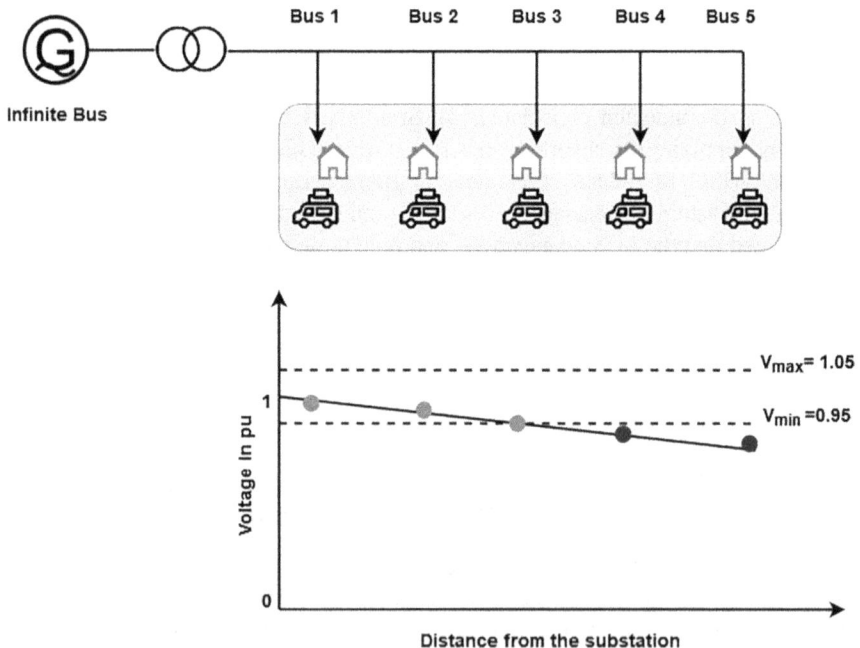

FIGURE 7.6 Voltage Drop along the feeder due to EV load demand.

voltage level would reduce compared to the voltage at the DSO. Significant grid impact with regards to EVs would be voltage regulation along with the distribution feeders. In [9], the authors have analyzed the voltage profile in a substation for a given distribution network with minimum and maximum loading of EVs and with increasing penetration. The feeder voltage profile with EV penetration is heavily dependent on the network model, distribution/number of charging infrastructures, type of EV charging infrastructures (i.e., power, 1-ph/3-ph system, etc.) and is also based on whether the EV is being used as a load or a source. Thus, the impact of voltage regulation owing to PHEV is to be aggregated while an independent xFC station can have a direct impact on the PCC.

7.5.3 GRID STABILITY

Charging infrastructures are typically interfaced to the grid via power electronic circuits made of voltage source converters (VSC) as from-end stages and dc/dc converters. These converter systems are designed to be operated under specific conditions therefore, the converter control gains are linearized for the same. Thus, the system stability is a function of the grid impedance (Z_g) and the Aggregator's (charging infrastructure) input impedance (Z_a) as shown in Figure 7.7. With impedance changes, the stability of the system can be easily compromised. Conditions such as a weak grid, parallel operation of converters, grid voltage events, etc., impact the system stability. Generally, impedance-based modeling is used to analyze the stability of the grid as in [3].

7.5.4 PHASE IMBALANCE

IEEE defines voltage imbalance as the ratio of negative or zero sequence components to the positive sequence component. Simply explained, phase imbalance exists when one or more line-line voltages in a three-phase system are mismatched. As per the definition, phase imbalance is a bigger concern with PHEVs which are connected to single-phase lines of the distribution system. Uneven distribution and

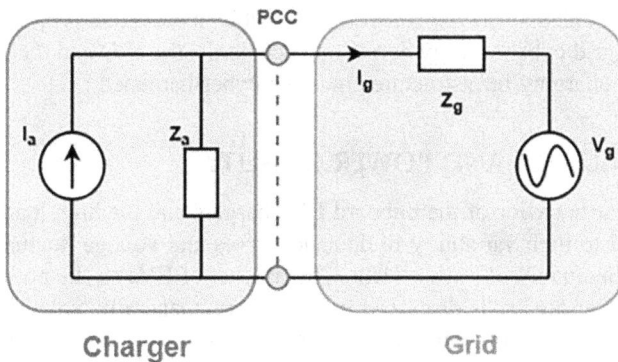

FIGURE 7.7 Impedance model of the grid with an Aggregator (lumped charger load).

capacity of PHEVs along the three-phase system feeder can lead to phase im-balances leading to different phases of the feeder with different voltage profiles. This problem is alleviated with DC fast charges and extreme fast charging (XFC) infrastructures which are directly connected to the 3-ph system with or without the 3-phase transformer.

7.5.5 PROTECTION AND FAULT MANAGEMENT

Connecting EVs to the power grid imposes numerous challenges with respect to the protection and management of fault events, be it external or internal to the PCC. For example, a grid under-voltage event can impact the EV chargers and lead to their malfunctioning. As per recommendations, the EV is designed to trip when the grid voltage reaches 0.8 p.u. in case of an under-voltage event. Cascaded trips in a section of the electrical network can suddenly reduce the load on the grid and impact the grid's stability. The same holds good for tripping under grid fault events. Additionally, existing grid infrastructures are not designed for bidirectional power flow capability from EV integration. V2G operation will require new bidirectional relays in the power grid for protection and fault identification.

7.5.6 CYBER SECURITY CONCERNS

It is evident that with the increase in EV penetration i.e., with the increase in power electronics interfaces in the grid, the cyber-attack surface leads to a weak cyber-security posture. Security breaches in the charging stations can not only endanger consumer privacy but also the stability of the grid infrastructure [10]. With EV penetration, the communication networks/infrastructures with a vast amount of resourceful data/messages including state of charge, payment information, etc., can be compromised. A detailed representation of the current and the futuristic EV charging infrastructure with the potential entry pathways for cyber-attacks can be observed in Figure 7.3 where communication channels associated between Aggregator, DSO, and electricity markets are provided. It is noted that communi-cation of individual EVs with different Aggregators would also be a feature and an IOT-enabled Vehicular Energy Network is possible in the future [10]. With the PCC governed by grid cyber security norms and standards, the EVs and their connection points to the charging infrastructures must be cyber-hardened [11].

7.6 HARMONICS AND POWER QUALITY

The non-linear behavior of the onboard EV chargers and the high load demand of EVs can lead to high variability in demand curves and voltage fluctuations which can disrupt the quality of power. Hence, the impact of EVs on the power grid with respect to power quality is discussed in detail here. Generally, voltage fluctuation. Voltage unbalances, harmonic current distortion, and harmonic voltage distortion falls under the umbrella of power quality issues. The harmonic standards, EV modeling methods, the impact of EV chargers, and standard mitigation measures are discussed in the following sub-sections.

7.6.1 Harmonics Standards

The charging infrastructures are designed to comply with harmonic standards (IEEE–519). However, violations in harmonics injected from these sources can occur in conditions involving the parallel operation of converters, weak grid situations, and grid voltage distortions. In [3], the authors have formulated the harmonic current emission (considering the circuit shown in Figure 7.8) as

$$I_g(s) = \frac{Z_a(s)I_a(s)}{Z_a(s) + Z_g(s)} - \frac{V_g(s)}{Z_a(s) + Z_g(s)} \tag{7.1}$$

Total harmonic distortion (THD) of end grid current is the parameter used to analyze the harmonic content in the current. THD can be expressed as both voltage THD (THD_v) and current THD (THD_i). THD can be defined as the ratio of the root means square of the harmonic voltage/current content expressed as a percentage of the fundamental voltage/current,

$$THD_v = \frac{\sqrt{\sum_{n=2}^{\infty} V_{n\_rms}^2}}{V_{fund\_rms}} \times 100\% \tag{7.2}$$

$$THD_I = \frac{\sqrt{\sum_{n=2}^{\infty} I_{n\_rms}^2}}{I_{fund\_rms}} \times 100\% \tag{7.3}$$

Total demand distortion (TDD) is another parameter used to analyze the grid harmonics. While in the former, the harmonics are a percentage of the fundamental current, in the latter, it is expressed as a percentage of the root mean square (RMS) maximum demand current, the latter is a better metric to analyze the grid harmonics.

$$TDD = \frac{\sqrt{\sum_{n=2}^{\infty} I_{n\_rms}^2}}{I_{max\_rms}} \times 100\% \tag{7.4}$$

The allowable THD and TDD as per IEEE-519 [12] are shown in Table 7.3. The limits in the IEEE519 vary according to the voltage at the point-of-coupling (PCC), a point where the load is connected to the utility, the larger the voltage the stricter the limits.

For bus voltages <1 kV at PCC, the voltage individual harmonics and THD are limited at 5% and 8%, respectively. It is to be noted that harmonics impacts should be analyzed at the PCC only and not the individual components as there would be an occurrence that may or may not reduce the harmonics via phase cancellations and diversity of load. Fourier analysis can be used to determine the harmonic characteristics (THD_v, THD_i, TDD).

TABLE 7.3

The TDD and THD Levels as Per IEEE 519 [3]

	Maximum Harmonic Distortion of the Individual Harmonic Order in Percent of I_L					
I_{SC}/I_L	$3 \leq h < 11$	$11 \leq h < 17$	$17 \leq h < 23$	$23 \leq h < 35$	$35 \leq h < 50$	TDD
<20	4.0	2.0	1.5	0.6	0.3	5.0
20<50	7.0	3.5	2.5	1.0	0.5	8.0
50<100	10.0	4.5	4.0	1.5	0.7	12.0
100<1000	12.0	5.5	5.0	2.0	1.0	15.0
>1000	15.0	7.0	6.0	2.5	1.4	20.0

Note:
- Limits for even harmonics are 25% of the odd harmonic limits
- DC offset in current is not allowed
- I_L = maximum demand load current and I_{SC} = maximum short circuit current at PCC

7.6.2 EV MODELING METHODS

Power quality assessment requires appropriate representations of the charging infrastructure to capture its non-linear behavior. There are two main methods in analyzing the impact of the harmonic of EV chargers, a) Mathematical models—Model the harmonic injection into the grid as a current source and compute harmonic load flow analysis, b) Circuit-based method—Model the EV charger as the actual power electronics and measure harmonic characteristics at the PCC.

When using mathematical models, the magnitude and phase variables can be determined by the combination of actual data from the EV, analytical models, and randomized using probabilistic models [13]. The individual harmonic contents could be determined by measuring data from actual chargers as in [13] and [14] where odd harmonics of level 1 and 2 industry chargers (Nissan Leaf (Gen 1 & 2) and Tesla-S) were used in the harmonic load flow analysis. Alternatively, harmonic contents can be obtained by the analytical time model by considering the difference in voltage/current at different stages of charging [15].

For the circuit-based method, the actual schematic of the different levels of chargers and EV batteries is designed to measure the actual harmonics. Due to the difficulty in exact modeling of all the power electronics components, some authors such as in [16], simplified the model by using the equivalent circuit of the nearest stage/component of the chargers that is nearest to the grid. This is because the circuit located on the AC side has the greatest harmonic content as illustrated in [16].

7.6.3 MODELING OF EVs USING ACTUAL SCHEMATICS

Modeling EVs as harmonic current injecting sources or using data from actual EV charging stations may not be accurate as EVs are considered as static loads in these

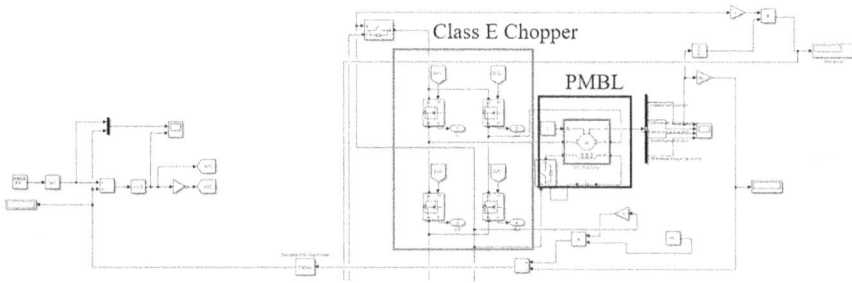

FIGURE 7.8   Simulated model of EV (Class E chopper with PMBL motor) [17].

cases. An accurate EV modeling using the actual EV schematics and charger using a Class C chopper, and permanent magnet brushless motor (PMBL) design is done in [17]. Here, the authors have connected a PMBL to two class C choppers—that operate in four quadrants (to have forward and reverse direction) and simulated the EV as shown in Figure 7.8. Moreover, they have used a Level 1 unidirectional charger shown in Figure 7.9. The level 1 unidirectional charger used here consists of an EMI filter, rectifier, and power factor correction (that can either be on or off-board), a full bridge converter, and rectifier (which is also known as unidirectional), a series resonant DC to DC converter which is an on-board circuit that connects to the car battery.

7.6.4   HARMONIC LOAD FLOW ANALYSIS

The purpose of the harmonic load flow or "Nodal Harmonic Analysis" is to obtain the harmonic voltage for computing harmonic characteristics like THD_v and compare it with the IEEE 519 limits. The harmonic load flow analysis involves computing the system admittance matrix along by modeling the different power system components. The system admittance $Y^{(h)}$ requires modeling of all transmission lines, generators, transformers, and loads that are connected to each node in a network [19]. We know that harmonic current $I^{(h)}$, is either obtained by the actual EV charger or the analytical model. Hence, once $I^{(h)}$ and $Y^{(h)}$, are known, the corresponding harmonic voltage $V^{(h)}$ can be computed using,

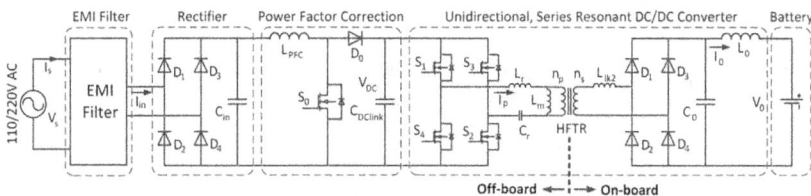

FIGURE 7.9   Circuit Model of on-board unidirectional EV charger for Level 1 system [18].

$$Y^{(h)}V^{(h)} = I^{(h)} \tag{7.5}$$

7.6.5 IMPACT OF EV CHARGERS

Harmonics impacts from EV chargers depend on the non-linear characteristics of power electronics schematics in different types of chargers, the initial state-of-charge (SOC) of the EV's battery, charging pattern, and duration. Chargers can also lead to serious harmonic and voltage fluctuations. Moreover, growth in home electronic devices can cause an impact on the power quality, as well as the increase in non-linear loads used at home such as refrigerators which require more reactive power from the grid. Hence, one of the solutions to improve the power quality and meet the requirement of reactive power is to design single-phase EV chargers that provide reactive power while charging or discharging the EV battery. Moreover, the charger can be used as an active power filter to reduce and filter the harmonic in the household network i.e., created by the household's non-linear loads or EV chargers.

If the ECS is directly connected to the grid for power supply without any energy supports, such as Renewable energy (RE) and Energy Storage Systems (ESS), it will cause harm to the grid due to the irregular charging and discharging load of an EV charger. Also, Figure 7.11 shows the connection of a DC grid connected to an EV charger and Figure 7.12 shows the connection of the AC grid to an EV charger.

The SOC of the battery is found to have a significant impact on the harmonic contents. Research conducted on a standard IEEE 34-bus test system [20] deduced that higher initial SOC led to lower THD levels. Similar results were also reported in [21] where the authors tested 20 EVs with different initial SOC (0%, 20%, 30%, 50%, and 80%) and it was concluded that individual harmonic current magnitude is reduced when they charge simultaneously with different SOC. This phenomenon is explained by SOC attribute to variations in phases and magnitudes, hence, the cancellation effects.

The location of the EV chargers also plays a critical role in the harmonic contributions. EV chargers are not recommended to be placed near transformers as transformer and chargers are non-linear devices and thus would significantly increase the harmonics contents [20]. It is also reported that transformer core saturation can distort the sinusoidal wave from primary to secondary generating harmonics on the secondary side [22]. Thus, the key takeaway is the nodes near the transformers experience more harmonics when EV chargers are connected to them compared to the ones further away from the transformer. Moreover, uncoordinated EV charging [23] and fast charging [24] also increase the harmonics caused by EVs in the power grids. Thus, the impact on power quality issues due to the EV integration can vary depending on the type of grid, SOC levels, charging rate, and type of charging [25].

7.7 APPROACHES TO MITIGATE THE IMPACTS OF EVs

With the increasing problem of harmonics level in the EV chargers and a requirement to meet IEEE 519, mitigation methods are required for power system conditioning. Some of the commonly used ways to mitigate the impact of penetration of EVs into

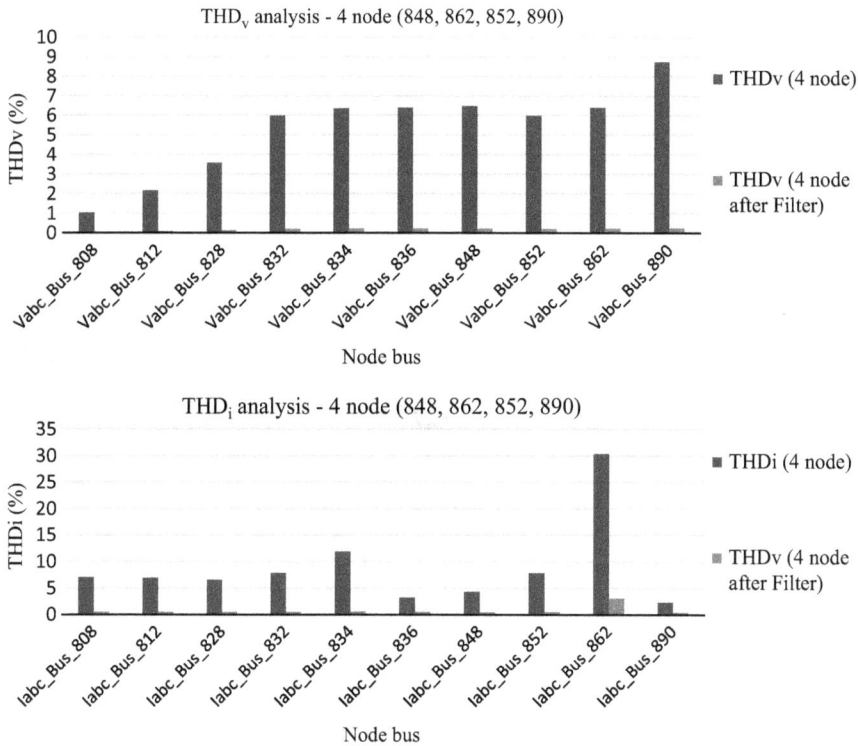

FIGURE 7.10   Voltage and current THD graphs with and without active filters [17].

the distribution system are the use energy storage system (ESS), demand response, use of external power electronic devices such as filters, and distributed–flexible alternating current transmission devices, and smart charging strategies.

7.7.1   ENERGY STORAGE SYSTEM

The ESS can be charged during the time of reduced grid power demand and discharged to the system at higher grid power demand. The main purpose is to supply both reactive and active power which will reduce the peak power demand from the grid to the EV charger.

7.7.2   DEMAND RESPONSE

This methodology enforces the end-users rather than the utility supplier to consume or charge their EV chargers at a different pattern. The purpose is to reduce the power consumption during peak demand as higher penetration of EV chargers does increase the THD values [26]. There are various alternatives for demand response such as time of use (ToU), day-ahead, and dynamic pricing. Time-of-use (ToU) is defined by having dynamic pricing during different times of the day, for example,

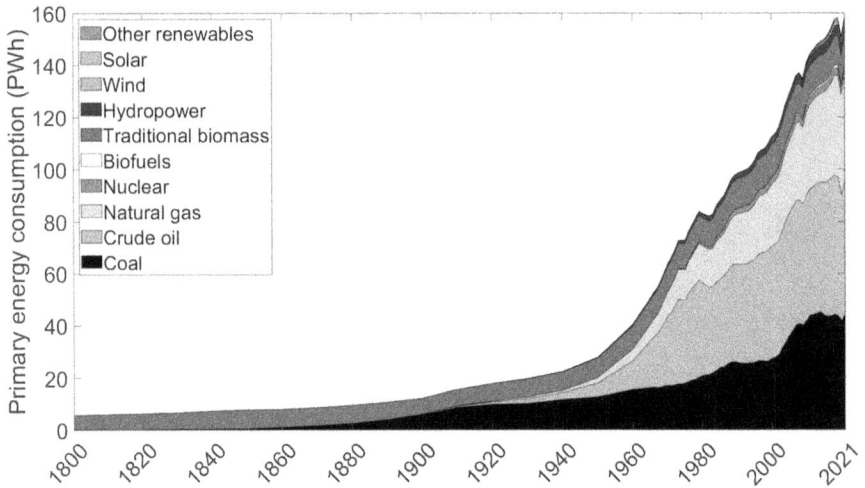

FIGURE 7.11 Shared of the primary energy consumption worldwide based on the energy source in time (adapted from [28]).

charging the vehicle during peak-period (after workhours) is relatively more expensive than during the off-peak period.

7.7.3 POWER ELECTRONIC DEVICES

There are several power electronic devices to mitigate harmonics problems such as passive filters, active filters, and Distributed Flexible AC Transmission System (DFACTS). Traditionally, harmonic mitigation is done using passive filters. However, problems such as non-ideal voltage sources could not be resolved by traditional passive filters. Hence, in recent times, devices such as shunt active filters, series active Filters, and DFACTS are implemented to stabilize and resolve power quality issues. Active filters are the simplest and most convenient choices among the discussed solutions. There are mainly two types of active filters namely shunt active filter and series active filter. Shunt active filters and series active filters resolve the harmonic problems mainly by injecting a compensated current and voltage respectively into the PCC [27]. The use of simple active filters (integrated with level 1 chargers and EVs) in mitigating the power quality issues has been demonstrated in [17] where the authors have discussed that the current/voltage injection using the shunt/series filters can effectively reduce the THD_i and THD_v. This has been demonstrated using an IEEE 34-bus distribution test system with and without active filters in Figure 7.10 where it can be seen that active filters play a significant role in reducing THD.

7.8 POTENTIAL OF PVS AND EVs

Till now, the main sources of energy are still heavily dominated by non-renewable sources such as coal, crude oil, and natural gas, as illustrated in Figure 7.11. This

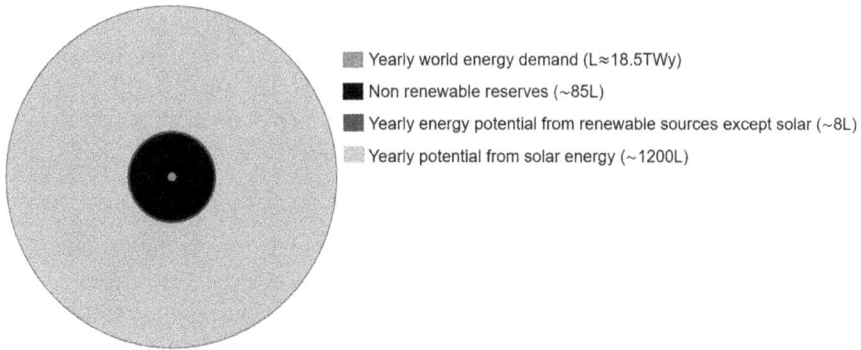

FIGURE 7.12 Energy potential based on the different energy sources (adapted from [29]).

figure also shows that renewable sources such as solar photovoltaics (PV) are still a minor player when considering the whole primary energy consumption worldwide.

Nevertheless, solar energy has the largest potential among all sources as just a small fraction of it would be required to fully satisfy all the world's energy demands. The previous can be appreciated in Figure 7.12 where the area of the inner pink circle represents the yearly world's energy demand while the black circle represents the overall known nonrenewable energy sources such as oil, coal, and natural gas (~85 times the world's yearly demand) while the green area stands for the yearly potential from all different renewable sources expect for solar energy (~8 times the world's yearly demand) and finally, the yellow area is the overall potential form solar energy for land areas only (~1200 times the world's yearly demand). The previous then highlights the strong potential of PV systems to be deployed worldwide.

Besides its energy potential, the introduction and deployment of PV systems have considerably increased in time due to many factors such as:

- Cost reductions: the cost of a solar panel was more than 20 USD/Wp in 1984 [30] while in present times it is around 0.25 USD/Wp [31].
- Concerns on the limitation on non-renewable sources: it is estimated that the known reserve left for natural gas is around 52.6 years, for oil is around 50.2 years, and for coal is around 134 years [32]. While this number might increase by finding more reserves and improving the extraction and conversion of energy processes, among others, there will be a limit nevertheless and thus, the increase of renewable sources is desired to satisfy the world's energy demand in the near future.
- Greenhouse gas emissions: it is expected that non-renewable sources have a big impact on greenhouse gas emissions which is why clean energy sources present an advantage.
- Government incentives: due to the previous points, many countries have adopted a friendly policy toward the introduction of renewable energy systems.

- Versatility: PV systems application is quite wide as it can be employed for small-scale systems in the order of a few Watts or lower (like in calculators) all the way to big systems for residential, industrial, and PV farm application where systems even beyond the GWp installation capacity have been detected (the 2.7 GW Bhadla Solar Park located in India is perhaps the largest installation to date).

Figure 7.13 the progressive adoption of PV systems in time worldwide indicating the yearly install capacity (red bars) and the accumulated installed capacity (blue bar). As it can be see from this figure, by end of 2021, around 942 GWp of PV was installed and it is expected that in the present, the 1 TWp barrier has been surpassed. This figure also shows how the installation of PV systems is increased in an exponential way and thus, considerable growth is still expected for the coming years. Furthermore, with respect to the World's electricity generation, Figure 7.13 shows the PV penetration with a solid black line where it can also be appreciated how fast the PV systems are gaining more of this share in time reaching a value of 5% by end of 2021 and based on its trend, it is also expected for PV systems to continue gaining more of the market share. This can also be noted for different countries which already have PV penetrations beyond 10% such as Australia (15.5%), Spain (14.2%), Greece (13.6%), Honduras (12.9%), Netherlands (11.8%), Chile (10.9%) and Germany (10.9%) [33].

PV systems are not the only ones that have experienced a remarkable increase in their deployment in time, but electric vehicles have followed a similar trend in the previous years as shown in Figure 7.14. This figure shows that in the year 2021, around 7 million electric vehicles were sold which is more than twice the number of sales from 2020. It is thus expected that the EV market also continue to expand in time and as a result that the electrification in the transportation sector will continue to improve.

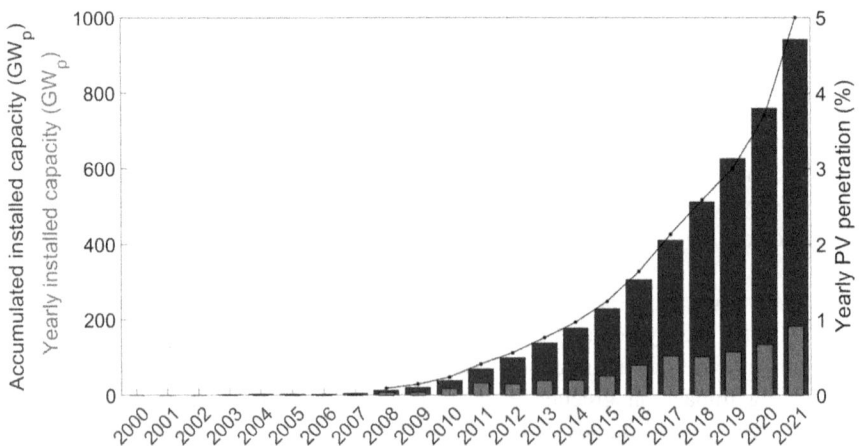

FIGURE 7.13 Yearly and accumulated PV installed capacity worldwide in black bars and gray bars, respectively. The black solid line shows the yearly PV penetration with respect to the World's electricity generation. Illustration adapted from [33–37].

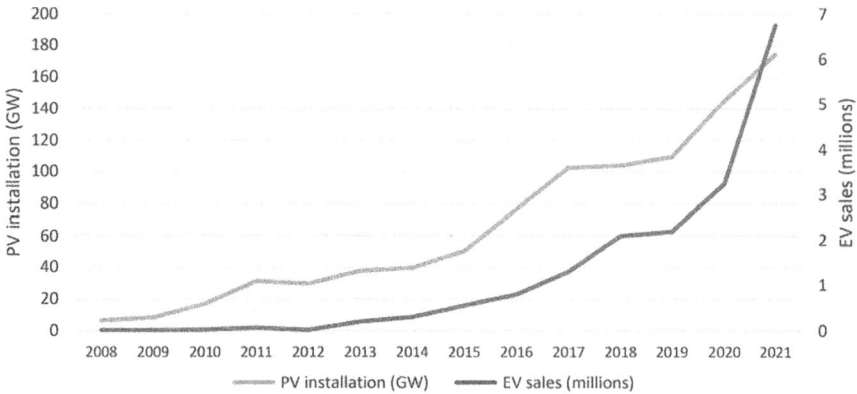

FIGURE 7.14   Yearly PV installed capacity and yearly sales of electric vehicles [33].

7.9   CHALLENGES OF HIGH PV PENETRATION

As the number of installed PV projects continues to increase, concerns about potential negative impacts on the electrical grid start to raise as, different to traditional energy sources, PV systems in principle produce power based on the weather conditions which is difficult to control as for example, big variations in the PV power are expected due to the clouds influence as this affects the amount of irradiance reaching the solar panels. Based on measured data provided by the Solar Energy Research Institute of Singapore (SERIS), Figure 7.15 provides an example of how the irradiance can experience sudden changes due to the cloud's effect. This

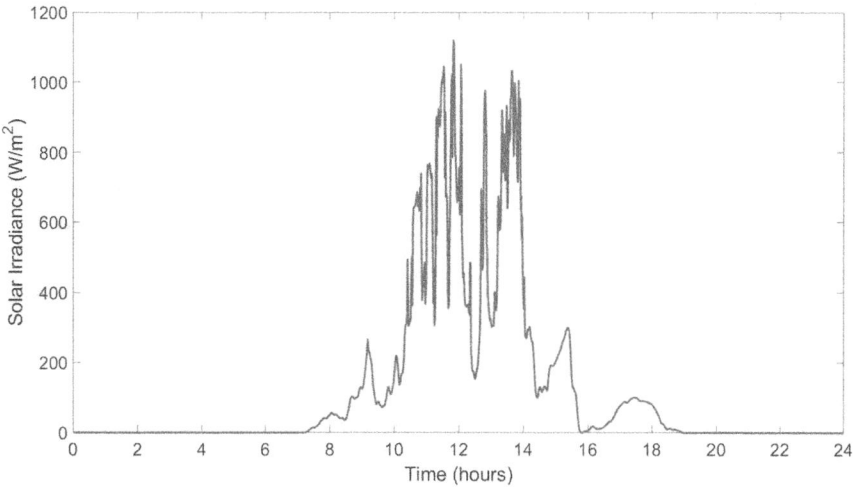

FIGURE 7.15   Irradiance variation due to clouds for a particular day in Singapore.

(Source: *SERIS*)

abrupt change in power can thus negatively affect the electrical grid quality, the electrical protections, and the load, among others.

In order to tackle the impact that a large PV penetration can cause into the grid, many solutions are proposed, such as the employment of forecasting algorithms to accurately ascertain the resource adequacy in a power system as this will also help the grid operator and generation companies to assure that enough power supply is generated to satisfy the load demand [38] and achieve an optimal reserve allocation and grid stability [39]. Another approach consists of installing extra equipment to handle different challenges such as voltage variations frequency variations and high harmonics. Examples of these equipment are capacitor banks, distributed flexible AC transmission systems, solid-state on-load tap changers, solid-state transformers, virtual synchronous generators, adaptive relays, and smart inverters, among others [40]. Among these, batteries also play a critical role as a potential solution to mitigate the negative impacts of PV on the grid as they can smooth out the power injected into the grid and thus improve the overall grid quality. As seen in Figure 7.16, the work [41] from presents an example of how, for a particular 20-node electrical grid, the voltage fluctuations are kept within the desired range (0.95–1.05 pu) only when batteries are employed.

While batteries present a great opportunity to tackle the issues induced by PV systems into the grid, one challenge is that, although the battery price keeps on reducing, it still has a considerable price which would influence the levelized cost of electricity and return of investment which has limited the adoption of these solutions in multiple projects. Thus, one potential approach could be to employ batteries

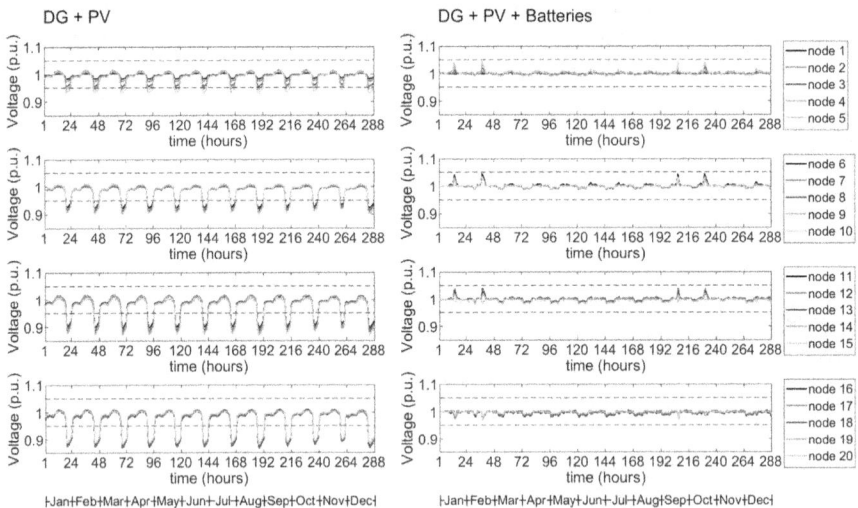

FIGURE 7.16 Grid voltage deviating from the desired range (0.95–1.05 pu) when only diesel generators and solar panels are energizing this grid (left image) and achieving desired grid voltage when batteries are added (right image).

Source [41]

whose main purpose is for other activities but that, if required, could be used to support PV systems. Such an approach can be accomplished by employing the batteries from electric vehicles. This will be discussed in the next section.

7.10   JOINT OPERATION OF PV SYSTEMS AND ELECTRIC VEHICLES

PV systems and electric vehicles can work together and support each other. On the one hand, PV systems can provide energy to charge the batteries inside the electric vehicles while on the other hand, the batteries from the electric vehicles can be used to smooth out the power fluctuations of PV into the grid and thus improving the grid quality and allowing for the further adoption of PV systems. Nevertheless, it must be considered that:

- Electric vehicles are not a static battery but a movable one which is not always connected into the grid.
- The main purpose of electric vehicles is the transportation of their passengers and thus its battery capacity should have enough level of charge to assure that the desired transportation can take place.
- There would be multiple owners of electric vehicles and thus, their approval to support the PV systems would be required.

Due to the previous challenges, while the combined operation of PV and electric vehicles has a strong potential (There is increasing deployment of these as seen in Figure 7.15), more work is still required to properly control this interaction and define the gains from the different parties.

As for example, in [42], the authors aimed to optimize the operation of electric vehicles in order to control the voltage and frequency stability in the grid by applying a volt-var control for fast vehicle-to-grid var dispatch and electric charging vehicle coordination where not only a better grid quality but also economic benefits were achieved.

The work presented in [43] then aimed to optimize the location for aggregate charging of electric vehicles in a photovoltaic power grid where 84% in reduction of the distribution losses were achieved compared to a conventional power flow strategy.

Research and applications have also been found for residential cases where the PV system and electric vehicle have the same owner and thus the aim of the optimization approach changes such as the one shown in [44] where an optimization strategy is considered for a smart house where the home appliances, rooftop PV, and electric vehicle are considered to be schedulable devices based on a controller set up. This work showed how by controlling this equipment, not only the electricity bill is reduced (economic benefits) but also the peak demand of the microgrid is reduced (technical benefits) by more than 30%.

Other applications of PV and EV can also be found such as the one where PV panels are integrated into the EV. Figure 7.17 shows examples of different models of electric cars with onboard solar panels. Here, the aim is then for the solar panels to provide as much energy as possible to the batteries in order to reduce the grid consumption if required to charge the EV. Different factors thus need to be considered

FIGURE 7.17   Different electric cars with onboard solar panels participating in the World Solar Challenge [47].

such as the PV design, technology selection, impact on the high temperature, and proper controller to transfer the PV power to the batteries with high performance and reliability, among others [45]. Different studies have aimed to estimate the amount of energy that can be provided by a PV system installed on the car itself like for example, in [46] the authors estimated the percentage of annual driving that could be provided by solar panels generated onboard under different conditions and for different locations. In their most optimistic scenario, 35% of the driving range per year was covered by solar while in their least favorable conditions, this number can go as low as 12%.

The relation and tendency to adopt solar panels and electric vehicles are clear as both technologies have the potential to support each other. This can be further seen as more than 30% owners of electric vehicles are also expected to employ solar panels [48]. Thus, further research on the common integration and operation between solar PV systems and electric vehicles is expected to increase in the years to come to design more robust and efficient systems.

REFERENCES

[1] Rajper, S. Z., and J. Albrecht. "Prospects of electric vehicles in the developing countries: A literature review." *Sustainability* 12, no. 5 (2020): 1906.

[2] IEA. *Global electric passenger car stock, 2010–2020*. Paris: IEA. https://www.iea.org/data-and-statistics/charts/global-electric-passenger-car-stock-2010-2020.

[3] Wang, L., Z. Qin, T. Slangen, P. Bauer, and T. V. Wijk. "Grid impact of electric vehicle fast charging stations: Trends, standards, issues and mitigation measures-an overview." *IEEE Open J. Power Electron.* 2 (2021): 56–74.

[4] Lopes, J. A. P., F. J. Soares, and P. M. R. Almeida. "Integration of electric vehicles in the electric power system." *Proc. IEEE* 99, no. 1 (2010): 168–183.

[5] IEC Technical Committee 69. "IEC 61851-1 Electric vehicle conductive charging system - Part 1: General requirements".

[6] Hybrid Committee. "SAE electric vehicle and plug in hybrid electric vehicle conductive charge coupler," *SAE International, Detroit, Michigan, Standard J1772* (Jan 2010).

[7] Yilmaz, M., and P. T. Krein. "Review of battery charger topologies, charging power levels, and infrastructure for plug-in electric and hybrid vehicles." *IEEE Trans. Power Electron.* 28, no. 5 (May 2013): 2151–2169.

[8] American National Standard for Electrical Power Systems and Equipment—Voltage Ratings (60 Hertz); ANSI Std. C84.1-2020; Arlington, VA, USA: National Electrical Manufacturers Association (2020).

[9] Putrus, G. A., P. Suwanapingkarl, D. Johnston, E. C. Bentley, and M. Narayana. "Impact of electric vehicles on power distribution networks." in *2009 IEEE Vehicle Power and Propulsion Conference.* IEEE (2009): 827–831.

[10] Sanghvi, A., and T. Markel. "Cybersecurity for electric vehicle fast-charging infrastructure." in *2021 IEEE Transportation Electrification Conference & Expo (ITEC).* IEEE (2021): 573–576.

[11] Graham, R. L., J. Francis, and R. J. Bogacz. "Challenges and opportunities of grid modernization and electric transportation No. DOE/EE-1473." *US Department of Energy and Allegheny Science & Technology* (2017).

[12] Recommended Practice and Requirements for Harmonic Control in Electric Power Systems, *IEEE Standard.* 519 (2014).

[13] Jiang, C., R. Torquato, D. Salles, and W. Xu. "Method to assess the power-quality impact of plug-in electric vehicles." *IEEE Trans. Power Deliv.* 29, no. 2 (2014): 958–965.

[14] Watson, J. D., and N. R. Watson. "Impact of electric vehicle chargers on harmonic levels in New Zealand," in *2017 IEEE Innovative Smart Grid Technologies - Asia (ISGT-Asia)* (2017): 1–6.

[15] Woodman, N., R. Bass, and M. Donnelly. "Modeling harmonic impacts of electric vehicle chargers on distribution network." (2018): 2774–2781.

[16] Busatto, T., S. Rönnberg, and M. Bollen. "Harmonic analysis of electric vehicle charging on the distribution system network with distributed solar generation." 18 (2020): 103–108.

[17] Kumar, D. S. et al. "Power quality assessment of electric vehicles on the distribution networks." *2021 IEEE PES Innovative Smart Grid Technologies - Asia (ISGT Asia)* (2021): 1–5.

[18] Choe, G. Y., J. S. Kim, B. K. Lee, C. Y. Won, and T. W. Lee. "A bi-directional battery charger for electric vehicles using photovoltaic PCS systems." in Proc. IEEE Veh. Power Propulsion Conf. (Sep. 2010): 1–6.

[19] Arrillaga, J., B. C. Smith, N. R. Watson, and A. R. Wood. "Direct harmonic solutions." *Power Syst. Harmonic Anal.* (17-Sep-1997): 97–131.

[20] Xu, Y., Y. Xu, Z. Chen, F. Peng, and M. Beshir. "Harmonic analysis of electric vehicle loadings on distribution system." in 2014 IEEE International Conference on Control Science and Systems Engineering (2014): 145–150.

[21] Sheng, Q., M. Chen, Q. Li, Y. Wang, and M. A. S. Hassan. "Analysis for the influence of electric vehicle chargers with different SOC on grid harmonics BT - advances in green energy systems and smart grid." *Adv. Green Energy Syst. Smart Grid. ICSEE 2018, IMIOT 2018. Commun. Comput. Inf. Sci.* 925, Singapore: Springer (2018): 284–294.

[22] Daut, I., S. Hasan, S. Taib, R. Chan, and M. Irwanto. "Harmonic content as the indicator of transformer core saturation." in 2010 4th International Power Engineering and Optimization Conference (PEOCO) (2010): 382–385.

[23] Deilami, S., A. S. Masoum, P. S. Moses et al. "Voltage profile and THD distortion of residential network with high penetration of plug-in electrical vehicles." 2010 IEEE PES Innovative Smart Grid Technologies Conf. Europe (ISGT Europe), Gothenberg, Sweden (2010): 1–6.

[24] Lucas, A., F. Bonavitacola, E. Kotsakis et al. "Grid harmonic impact of multiple electric vehicles fast charging." *Electr. Power Syst. Res.* 127 (2015): 13–21.

[25] Kim, K., C. S. Song, G. Byeon et al. "Power demand and total harmonic distortion analysis for an EV charging station concept utilizing a battery energy storage system." *J. Electr. Eng. Technol.* 8 no. 5 (2013): 1234–1242.

[26] Sivaraman, P., and C. Sharmeela. "Power quality problems associated with electric vehicle charging infrastructure," in *Power Quality in Modern Power Systems*, Elsevier Inc. (2021): 151–161.

[27] Jain, S. "Control strategies of shunt active power filter." in *Modeling and Control of Power Electronics Converter System for Power Quality Improvements*, Elsevier Inc. (2018): 31–84.

[28] Our World in Data. "Global Primary Energy." (October, 2022): [Online]. Available: https://ourworldindata.org/grapher/global-primary-energy

[29] Perez, M., and R. Perez. "Update 2015–a fundamental look at supply side energy reserves for the planet." *Natural Gas* 2, no. 9 (2015): 215.

[30] Sivaram, V., and S. Kann. "Solar power needs a more ambitious cost target." *Nature Energy* 1, no. 4 (2016): 16036.

[31] Pvinsights. "Solar PV Module Weekly Spot Price." (October, 2022): [Online]. Available: http://pvinsights.com/

[32] BP-Global. "Bp statistical review of world energy." (November, 2018): [Online]. Available: https://www.bp.com/en/global/corporate/energy-economics/statistical-review-of-world-energy.html

[33] International_Energy_Agency_(IEA). "Snapshot of global PV markets 2022." Tech. Rep. IEA PVPS T1-42:2022 (2022): [Online]. Available: https://iea-pvps.org/wp-content/uploads/2022/04/IEA_PVPS_Snapshot_2022-vF.pdf

[34] International_Energy_Agency_(IEA). "Trends 2016 in photovoltaic applications." Tech. Rep. IEA PVPS T1-30:2016 (2016): [Online]. Available: http://www.ieapvps.org/fileadmin/dam/public/report/national/Trends_2016_-_mr.pdf

[35] International_Energy_Agency_(IEA). "2018 snapshot of global photovoltaic markets." Tech. Rep. IEA PVPS T1-33:2018 (2018): [Online]. Available: http://www.iea-pvps.org/fileadmin/dam/public/report/statistics/IEAPVPS_-_A_Snapshot_of_Global_PV_-_1992-2017.pdf

[36] International_Energy_Agency_(IEA). "Snapshot of global PV markets 2020." Tech. Rep. IEA PVPS T1-37:2020 (2020):[Online]. Available: https://iea-pvps.org/wp-content/uploads/2020/04/IEA_PVPS_Snapshot_2020.pdf

[37] International_Energy_Agency_(IEA). "Snapshot of global PV markets 2021." Tech. Rep. IEA PVPS T1-39:2021 (2021): [Online]. Available: https://iea-pvps.org/wp-content/uploads/2021/04/IEA_PVPS_Snapshot_2021-V3.pdf

[38] Van der Meer, D. W., J. Widén, and J. Munkhammar, 2018. "Review on probabilistic forecasting of photovoltaic power production and electricity consumption." *Renew. Sustain. Energy Rev.* 81 (December 2016): 1484–1512. 10.1016/ j.rser.201 7.05.212

[39] Antonanzas, J., N. Osorio, R. Escobar, R. Urraca, F. J. Martinez-de Pison, and F. AntonanzasTorres. "Review of photovoltaic power forecasting." *Sol. Energy* 136 (2016): 78–111. 10.1016/j.solener.2016.06.069

[40] Kumar, D. S., O. Gandhi, C. D. Rodríguez-Gallegos, and D. Srinivasan. "Review of power system impacts at high PV penetration Part II: Potential solutions and the way forward." *Solar Energy* 210 (2020): 202–221.

[41] Rodriguez-Gallegos, C. D., O. Gandhi, D. Yang, M. S. Alvarez-Alvarado, W. Zhang, T. Reindl, and S. K. Panda. "A siting and sizing optimization approach for PV–battery–diesel hybrid systems." *IEEE Trans. Ind. Appl.* 54 no. 3 (2017): 2637–2645.

[42] Zhang, W., O. Gandhi, H. Quan, C. D. Rodríguez-Gallegos, and D. Srinivasan. "A multi-agent based integrated volt-var optimization engine for fast vehicle-to-grid reactive power dispatch and electric vehicle coordination." *Appl. Energy* 229 (2018): 96–110.

[43] Gupta, K., A. Narayanankutty, R., Sundaramoorthy, and K. A. Sankar. "Optimal location identification for aggregated charging of electric vehicles in solar photovoltaic powered microgrids with reduced distribution losses." *Energy Sources, Part A: Recovery, Utilization, and Environmental Effects* (2020): 1–16.

[44] Alilou, M., B. Tousi, and H. Shayeghi. "Home energy management in a residential smart micro grid under stochastic penetration of solar panels and electric vehicles." *Solar Energy* 212 (2020): 6–18.

[45] Wei, H., Y. Zhong, L. Fan, Q. Ai, W. Zhao, R. Jing, and Y. Zhang. "Design and validation of a battery management system for solar-assisted electric vehicles." *J. Power Sources* 513 (2021): 230531.

[46] Thiel, C., A. G. Amillo, A. Tansini, A. Tsakalidis, G. Fontaras, E. Dunlop,... & M. Yamaguchi. "Impact of climatic conditions on prospects for integrated photovoltaics in electric vehicles." *Renewable Sustainable Energy Rev.* 158 (2022): 112109.

[47] World Solar Challenge. "World Solar Challenge." (October, 2022): [Online]. Available: https://worldsolarchallenge.org/about

[48] Shahan, Z. "Electric car drivers: demands, desires & dreams (2019)." *CleanTechnica Reports* (2019): 1–14.

8 Energy Storage in EVs and Mitigating Impact of EV Charging on Power Grid

B. Sivaneasan, N. K. Kandasamy, K. T. Tan, and D. R. Thinesh

CONTENTS

One of the major sources of greenhouse gases (GHG) is the transportation sector. In highly urbanized cities, transportation can account for a significant proportion of the total emissions. Electrification of transportation is considered as the optimal solution for reducing the GHG emissions of the transportation sector. Looking beyond electric railway systems such as mass or light rapid transit, electric vehicles (EVs) are considered as an important component of future electric transportation systems and are crucial for reducing global CO_2 emissions. It is estimated by the International Energy Agency (IEA) that the electrification of transportation could reduce the transportation sector's GHG emission to net zero by 2050 [1].

Back in 2012, the California Air Resources Board's report stipulated that by 2025, 15% of all new cars and trucks sold will be powered by batteries, hydrogen fuel cells, or other technology that emits little or no air pollution [2]. In recent years, many countries have been seen to take targeted ambitious push toward EV adoption. For example, the Singapore government has targeted to have no new registration of internal combustion engine (ICE) vehicles by 2030, and no ICE vehicles on the road by 2040. Another example is India, though it has not pledged to phase out conventionally fueled vehicles by a certain year, it has an ambitious target to have 30% of all new vehicle sales be electric by 2030. EVs have the greatest potential for widespread market penetration, and it was evident that government policies are taking effect as the global EV stock increased by 43% in 2020 compared to 2019 [3]. With public transportation playing a key role in the urban transportation system, especially in urban cities like New York and Singapore, more and more EVs are also introduced into the public transportation system, serving as both feeder bus and long-haul passenger transport. Electric bus registrations increased in 2020 in China, Europe, and North America with the global electric bus stock hitting 600 000 units [3].

In the study of EV integration to the grid, the impacts of different charging methods on battery life and distribution system peak demand are the two most important factors to be considered. As batteries are charged and recharged repeatedly, it slowly loses the ability to return to its original intended capacity. Furthermore, with increased penetration of EVs, it must be noted that EV charging shifts a significant portion of transportation energy use onto buildings' and homes' electricity meters. The literature indicates that EV charging will predominantly occur at buildings and homes during off-peak hours [4]. However, the charging of EVs is unpredictable due to the various mobility patterns of EV users. Specifically, the uncertainty in the arrival of EVs which are known as flexible loads is the main cause of this problem. At large-scale penetration of EVs, these flexible loads will have a significant impact on the power system if not managed properly. Typically, it will result in problems such as an unpredictable increase in peak demand, voltage fluctuations, and increased total harmonic distortion.

8.1 ENERGY STORAGE IN ELECTRIC VEHICLES

Electric vehicles can be classified as hybrid electric vehicles (HEVs), plug-in hybrid electric vehicles (PHEVs), and battery electric vehicles (BEVs). All types of EVs primarily require some form of energy storage as they have regenerative braking systems to recover energy otherwise lost as heat due to friction during braking. HEVs are powered by both an ICE and an electric motor that draws power from a battery. Unlike other electric vehicles, however, HEVs use regenerative braking to charge their batteries. PHEV is a hybrid electric vehicle with a battery pack that can be recharged both internally and externally by plugging a charging cable into an external electric power source, as well as by its on-board ICE-powered generator. This allows the battery to store enough energy to operate the electric motor, reducing fuel use by up to 60%.

On the other hand, BEVs do not have an ICE, a fuel tank, or an exhaust pipe, and rely solely on battery-supplied energy for propulsion. While environmentally friendly and with lower operation cost, owners may experience range anxiety since they must ensure their BEV has enough energy for their journey unless they choose

TABLE 8.1
Different EV Models Based on Battery Energy Capacity

Battery Capacity Category	Energy Range (E)	Models
Low	E ≤ 16kWh	Renault Twizy, REVAi, Buddy, Mitsubishi i MiEV and Mia electric.
Medium	16kWh > E < 30kWh	Nissan Leaf, Smart ED, Renault Fluence Z.E., Ford Focus Electric, Honda Fit EV, Renault Zoe, Roewe E50, Chevrolet Spark EV, Mercedes Clase B ED, Kia Soul EV, Honda Clarity EV and Mahindra e2o.
High	30 kWh ≥ E < 45 kWh	Wheego Whip LiFe, Volkswagen e-Golf, Bolloré Bluecar, BMW ActiveE, Renault Zoe, Ford Focus Electric, BMW i3 and RAV4 EV.
Extra-High	E ≥ 45 kWh	Tesla Model S, Tesla Model 3, Tesla Model X, Audi e-tron, Renault ZOE 2, BYD e6, Hyundai Kona e, Jaguar I-Pace, Peugeot e-208, Volkswagen ID.3, Tesla Roaster and Ford Mustang Mach-E.

a model with an optional fuel-powered generator, such as the BMW i3. The driving range of these EVs depends on the energy capacity of the batteries used and the type of vehicle. The range of energy capacity of the EVs is between 7 kWh and 85 kWh [5]. Table 8.1 categorizes different commercially available EV models based on the energy capacity of their batteries. There are also many EV motorcycles and EV bicycles commercially available [6,7].

8.1.1 TYPES OF ENERGY STORAGE USED IN EVS

A variety of electrochemical batteries are available for electric transportation. Popular electrochemical batteries include Lead-Acid, nickel–metal cadmium (Ni-Cd), nickel–metal hydride (Ni-MH), lithium-ion (Li-ion), and lithium-metal (Li-metal). Due to the different types of batteries used in today's electric vehicles, it is difficult to determine which is best suited to meet all the important requirements for electric transportation use, including energy storage efficiency, structural properties, cost, safety, and service life. EVs use various types of batteries which are classified into three widely used categories based on their chemistry, namely lead-acid batteries, nickel–metal hydride batteries, and lithium-ion-based batteries. The characteristics of the three battery types are compared in Table 8.2 and the typical charging curves are shown in Figure 8.1.

Lead-acid batteries ($Pb-PbO2$) are the oldest kind of rechargeable battery invented in 1859. Although this kind of battery is very common in conventional vehicles, it has also been used in electric vehicles. It has very low specific energy and energy density ratios. The GM EV1 and the Toyota RAV4 EV, are examples of vehicles that used this kind of battery. Nickel-metal-hydride (Ni-MH) type batteries use an alloy that

TABLE 8.2

Comparison of Three Battery Types Based on Their Chemistry [8–14]

Characteristics	Lead-Acid	Nickel-Metal Hydride	Lithium-Ion
Working Temperature (°C)	−20–45	0–50	−20–60
Specific Energy	30–60 Wh/kg	60–120 Wh/kg	100–275 Wh/kg
Specific Power	75–100 W/kg	250–1000 W/kg	350–3000 W/kg
Charge/discharge efficiency	50%–92%	66%	80–90%
Self-discharge rate	3–20%/month	30%/month	15% /month
Cycle durability	500–800	500–1000	400–3000

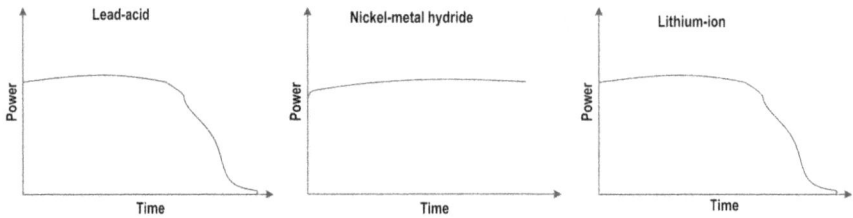

FIGURE 8.1 Typical charging curves of different battery types.

stores hydrogen for the negative electrodes instead of cadmium (Cd). Although they present a higher level of self-discharge than those of nickel-cadmium, these batteries are used by many hybrid vehicles, such as the Toyota Prius and the second version of the GM EV1. The Toyota RAV4 EV, apart from having a lead-acid version, also had another with nickel-metal-hydride. Lithium-ion batteries (Li-Ion) employ, as electrolyte, a lithium salt that provides the necessary ions for the reversible electrochemical reaction that takes place between the cathode and anode. Lithium-ion batteries have the advantages of the lightness of their components, their high loading capacity, their internal resistance, as well as their high loading and unloading cycles. In addition, they present a reduced memory effect.

When comparing different technologies of energy storage, it's crucial to consider their working temperatures, as this can limit their acceptance. In this regard, lead-acid and lithium batteries are the best at withstanding low temperatures, as they can load temperatures of up to −20 degrees Celsius, however, low temperatures severely reduce the capacity of Li-Ion batteries, causing self-discharge [15]. When comparing the specific power, lead-acid batteries offer only up to 100 W/kg, while Ni-MH with a maximum of 1000 W/kg, and Li-ion, which offers up to 3000 W/kg gives the best results. The Ni-MH and lead-acid batteries, on the other hand, perform poorly in terms of life cycles. Lithium batteries have a cycle life of up to 3000 cycles, whereas Ni-MH batteries have a cycle life of up to 1000 cycles only.

The charging curves shown in Figure 8.1 are based on the assumption that the same charging rate is used. The charging rate of a battery is defined as the portion of energy that is supplied to a battery in one hour and is usually denoted as C or C-rate. For example, if a battery with a nominal capacity of 16 Ah is charged using 2 A current then the charging rate is 1/8 C and it will approximately take 8–10 hours for charging the battery depending on its chemistry. Both nickel-metal hydride batteries and lithium-ion batteries can be charged at a comparatively higher C-rate such as 1C–3C but this is not possible with lead-acid batteries. Furthermore, both Lead-acid batteries and Lithium-ion batteries have two distinct regions namely

1. Constant current region where the charging power remains constant and
2. Constant voltage region where the charging power changes with time.

Among all the various energy storage technologies, lithium-ion battery is the most suitable technology as it has the best performance in practically all the evaluated qualities. This is mainly due to their high specific energy and high energy density. As of now, Li-ion batteries are the lightest and long-lasting batteries available for electric vehicles [16]. Furthermore, it can meet the requirements of both the electricity grids and the electric vehicles sector. There are several advantages to utilizing a Li-ion battery cell and as a result, the technology is rapidly being employed for a wide range of applications. Owing to the advantages of Li-ion technology, these batteries are finding an increasing number of applications, resulting in a lot of research and development.

The advantages of Lithium-ion battery include [17,18]:

• *Self-discharge*: The rate of self-discharge rate for Lithium-ion cells is much lower than other rechargeable cells like NiMH and Ni-Cad forms. The self-discharge rate is approximately 5% in the first four hours after getting charged, but it falls to 1% to 2% monthly.
• *High energy density*: Due to the continuous demand for batteries with much better energy density as electronic devices like mobile phones need to function longer between charges while amid consuming more power. Electric cars also require high-energy-density battery technologies.
• *Low maintenance*: Lithium-ion battery do not require maintenance to sustain their performance, unlike lead acid cells, which require topping up of battery acid periodically.
• *Load characteristics*: The load characteristics of Lithium-ion battery are constantly 3.6 volts per cell, which is great.
• *Availability of different types*: Different forms of Lithium-ion batteries can provide different current densities which is ideal for consumer mobile electronic equipment including electric vehicles and power tools.
• *No priming required*: Unlike other rechargeable cells which require priming after receiving their first charge, Lithium-ion battery do not have such a requirement.

As with all technology, lithium-ion batteries have their own drawbacks. Nonetheless, this does not rule out the potential of overcoming or at the very least minimizing these drawbacks and achieving superior performance.

The disadvantages of Lithium-ion battery include [17,18]:

- *Protection required*: Lithium-ion batteries and cells are less durable than certain other rechargeable technologies. They must be protected against being overcharged and discharged too far. In addition, the current must be kept below safe limits. As a result, one downside of lithium-ion batteries is that they require protective circuitry to guarantee that they remain within their acceptable working limits.
- *Cost*: Lithium-ion battery cost 40% more than Nickel Cadmium cells to manufacture.
- *Developing technology*: Even though lithium-ion batteries have been around for a long time, some people still regard them as a developing technology. This has the potential to be a disadvantage since technology changes. However, because new lithium-ion technologies are always being researched, this can be a benefit when better alternatives become available.
- *Ageing*: One of the most significant issues of lithium-ion batteries for electronics is that they age. This is based not only on the time or calendar but also on the amount of charge-discharge cycles that the battery has gone through. Typically, batteries can only tolerate 500 to 1000 charge-discharge cycles before their capacity degrades. This number is rising as lithium-ion technology advances, but batteries will eventually need to be replaced, which might be a problem if they are integrated with equipment. However, it is to be noted that Li-ion batteries have the best performance in terms of ageing compared to any viable technology currently available.

8.1.2   CYCLE LIFE OF ENERGY STORAGE

The first uses of energy storage technology only required a short lifespan. However, as a result of the proliferation of new applications and technology including electric transportation, the attention has shifted to ageing phenomena in energy storage devices while still taking into account manufacturer requirements to meet their applications' need for power and energy performances. So, in addition to the performance constraints, battery design specifies service life criteria such as 10–15 years or 20000–30000 discharges [19] clearly. As a result, battery ageing is widely used to extract both the major consequences of time and usage on a battery. Identifying ageing and degradation mechanisms in a battery is the most challenging and time-consuming task. Such processes are difficult because various components from the environment or from the method of usage interact to produce diverse ageing effects. This makes it difficult to comprehend battery ageing, and many researchers have sought to explore battery ageing over the years [20].

8.1.2.1   Calendar Ageing

The period of time a battery may be left idle or with low activity while still maintaining a capacity greater than 80% of its initial capacity is referred to as calendar life. Any ageing mechanisms that cause a battery cell to degrade without being subjected to a charge-discharge cycle are referred to as calendar ageing. This is a major concern as the operating hours of energy storage devices are substantially shorter than the idle intervals, especially in many lithium-ion battery applications, such as EVs. Furthermore, damage owing to calendar ageing might be noticeable in ageing experiments, especially when cycle depths and current rates are low [21]. The most prevalent process of calendar ageing is the creation of passivation layers at the electrode–electrolyte contacts. During the creation, growth, or repair of passivation layers, electrolyte decomposition, i.e., reduction at the anode and oxidation at the cathode interface, consumes cyclable lithium. Furthermore, it is thought that dissolved transition-metal ions from the cathode induce the formation of the passivation layer at the anode, known as the solid electrolyte interphase (SEI), which is eventually converted back to metals at the anode [22,23]. Both the evolution of passivation layers and transition-metal dissolution are accelerated by a high state of charge (SoC) and temperature [24,25].

8.1.2.2   Cycle Ageing

The number of full charge and discharge cycles a battery may go through before its capacity drops to 80% of its starting capacity is referred to as cycle life. When a battery is charged and discharged, cycle ageing occurs. Cycle ageing, like calendar ageing, causes the loss of lithium inventory (LLI), active material (LAM), and impedance to rise [26]. It was found that faster development of the SEI is observed while cycling at ambient temperature and an additional blocking mechanism of graphene layers on the anode SEI when cycling at 60 °C [27]. Cycling also affects the anode surface areas as the de-lithiation and lithiation in the active material of the electrodes cause volume expansion and potential electrode breakage. In addition to SEI development and active mass wear, lithium plating at low temperatures and/or high currents can cause rapid ageing [28]. In cycle ageing studies, the effects of cycle ageing are frequently layered atop those of calendar ageing. As a result, the calendar ageing effects must be analyzed and subtracted from overall ageing to achieve the pure cycle ageing effects. Pure cycle ageing refers to charge and discharge throughput in Ah or the whole equivalent cycle. The temperature, C-rate, Depth of Discharge (DoD), and SOC range of the cell all have a role in cycle ageing.

A lot of literature summarizes the ageing mechanisms of Li-ion batteries to the loss of cathode/anode active materials and the loss of lithium-ion inventory [29–31]. The intercalation/ deintercalation of lithium-ion cathode and anode active material are commonly linked to battery charging and discharging operations. As a result, the battery capacity is directly determined by the amount of active materials and available lithium ions. In addition, the battery degradation mode includes electrolyte loss (LE) and an increase in internal resistance (RI). While the charge and discharge cut-off voltages stay constant, an increase in internal resistance may cause the battery's power to fade and its capacity to shrink [32].

In [33], Kristen A. Severson's team performed analyses on the characteristics and patterns of the capacity degradation of commercial lithium-ion batteries under fast charging conditions, with varying cycle life ranging from 150 to 2300. The batteries were charged and discharged continuously till the end of its life. Since each battery is charged to the end of life, the battery characteristics can be observed from the voltages and current over its total number of cycles to study the degradation in its useful life. One particular batch of batteries was charged to 80% state of charge and left to rest for 1 minute, followed by its discharging sequence and left to rest for 1 minute after fully discharged. The battery is charged to 80% which equates to 0.88 AH. Each battery has a useful life of different cycles and each of these cycles has multiple data points for voltage and current as it charges to 80% SOC. Data such as time, voltage, current, and discharge capacity were recorded, compiled, and shared with the research community. Using the accumulated data provided, the performance of the batteries could be studied by analyzing the degradation of cycle life with respect to charging. There were 3 batches of test channels used. Batch 1 consists of 48 channels, but there were two test channels, channels 4 and 8 that did not successfully start, and hence it does not have the data in this study, whereas batch 2 consists of all 48 channels. The cells feature a 1.1 Ah nominal capacity and a 3.3 V nominal voltage. The manufacturer's suggested fast-charging technique 3.6 C constant current–constant voltage (CC-CV). The rate capability of these cells is demonstrated during the charge and discharge process.

Figure 8.2 shows the scatter plot of the last termination cycle life of all 46 channels of batteries. It can be observed that out of 46 channels, the range of the life cycle varies quite significantly. In this batch, when compared to all the other channels, channel 11 has the fastest degradation profiles with the lowest cycle number of 534, while channel 6 has the slowest degradation profile with the highest cycle number of 1227.

To further visualize the degradation of each channel based on cycle number and discharge capacity, the application of the start cycles and inter-cycle distance is used to obtain the range of the discharge capacity at the specific cycle numbers. Therefore, the start cycle is set at 50 of the cycle number, and for the inter-cycle distance, which is the distance or gap in the cycle number, is set at 10. This can be determined as:

- Dataset: Starts from 50, 60, 70, 80, 90, 100, … 1000, 1010, 1020, … 1200, 1210, 1220 (Cycle Number)
- Dataset: Ends at 1220 instead of 1227 cycle number, due to the inter-cycle distance of 10

For example, Figure 8.3 shows the results of two individual representative channels, namely 11 and 6 that are plotted for their discharge capacity versus cycle number. After the application of the start cycle and inter-cycle distance, the result of cycle number for channel 11 will be 530 instead of 534, and channel 6 will be 1220 instead of 1227 of the cycle number.

Hence, the execution from the start cycle and inter-cycle distance results will be used in the data analysis for the battery degradation on all the remaining channels of data. With the application of the start cycle and inter-cycle distance for all the channels, most of the degradation of cycle number lies on the discharge capacity that ranges from 0.8 Ah to less than 1.2 Ah.

Figure 8.4 compiles all the cycle numbers plotted against the battery discharge capacity based on the updated dataset with the start cycle and inter-cycle distance imposed. The graph indicates the degradation of the discharge capacity over the

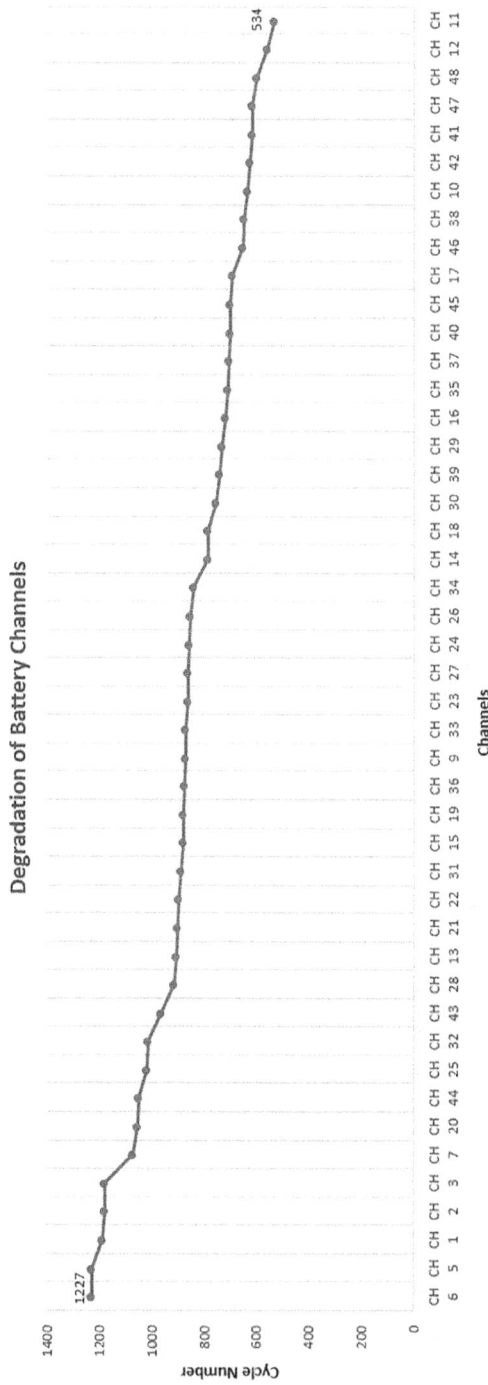

FIGURE 8.2 Scatter plot of all 46 channels.

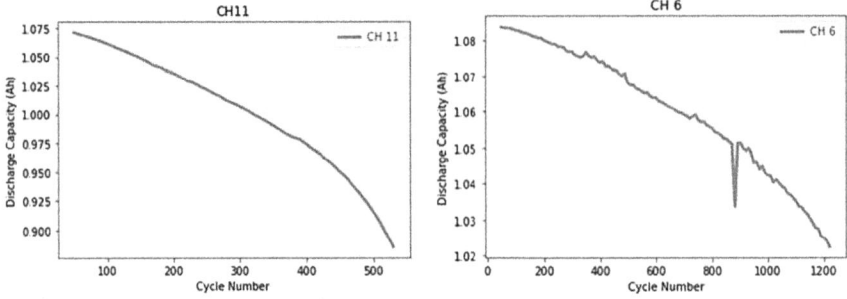

FIGURE 8.3 Channels 11 and 6 with start and inter-cycle distance.

prolonged cycle at a fast charging rate of 3.6C. All the batteries exhibit similar characteristics at either a steeper or gentler gradient. From the isolated view in Figure 8.5, showing only the distinct characteristics of the channels, it can be observed that some batteries exhibit a distinct degradation and others that are supposedly in the "well-grounded" range. Batteries of channels 11, 12, 47, and 48 exhibit rapid capacity degradation between 534 and 620 cycles with discharge capacity below 0.9 Ah. On the other end of the spectrum, it can be observed that batteries channel 2, 3, 5, and 6 has the highest number of cycles between 1150 and 1220 with discharge capacity above 0.97 Ah. The discharging capacity of these channels has a much gentler slope which indicates a higher performance of a Li-ion battery within this set of test batteries.

In order to prolong EV battery life, the most efficient charging mode is shallow charge cycling, which provides the largest cumulative capacity or cycle life [34]. The DoD has a considerable impact on the cycle life of Li-ion batteries. The number of cycles and cumulative capacity increases as DoD decreases. A battery that has

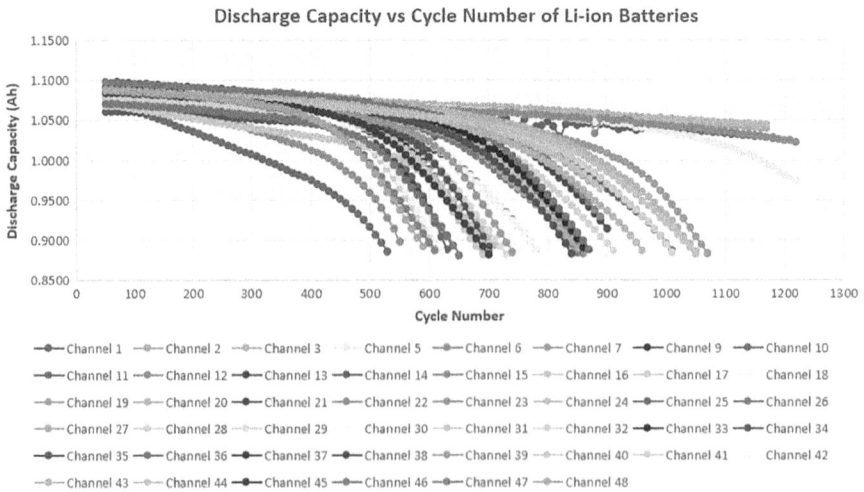

FIGURE 8.4 Overall discharge capacity.

Discharge Capacity vs Cycle Number of Li-ion Batteries

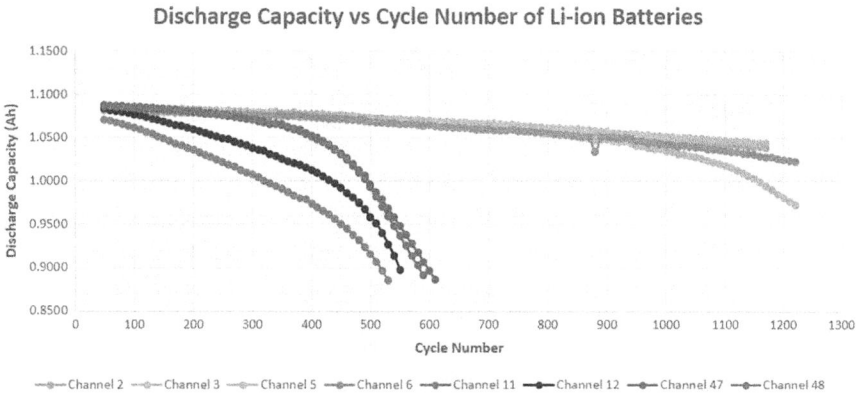

FIGURE 8.5   Isolated view of Figure 8.4 showing only the distinct characteristics.

been depleted to 20% of its full energy capacity, for example, has a far longer cycle life than one that has been discharged to with 80% DoD. Furthermore, with fast charging, lower DoD cycling is recommended with operation in a favorable DoD and SoC range, avoiding the end of charge (near to 100% SoC) region [35].

8.2   MITIGATING IMPACT OF EV CHARGING

EV integration into the grid requires proper attention to avoid unnecessary complications. The energy requirement of an EV is compared with the average residential energy consumption in major cities is noticeably higher. Various analyses on the impact of EV charging show that uncoordinated charging of EVs will result in detrimental impacts on the power system operation.

8.2.1   Impact of EV Charging on the Grid (Kuan Tak)

EVs behave like a flexible load, as such they will have a significant impact on the power system and has the potential to cause an unpredictable increase in the peak load of the power system if not managed properly. Furthermore, if the problems are not addressed properly, it will result in a decrease in power quality and an increase in system operation cost for the peak demand. Large-scale EV charging would have the following impacts on the power grid.

8.2.1.1   Unpredictable Increase in Peak Demand

The increase in energy demand due to large-scale EV charging could result in an unpredictable peak demand increase in power grids. This increase in peak demand has the potential to overload the electrical equipment in the power grid which includes transformers and feeders. If multiple EV users located within a vicinity such as multiple houses in a neighborhood or multi-storey carparks in several blocks of commercial buildings decide to charge their EVs simultaneously, the transformers and feeders might not have adequate capacity to deliver the required amount of power to charge the EVs. This issue might be more prominent with fast-charging stations which have been increasing rapidly in recent years. As shown in Figure 8.6, if the utilities do not upgrade the infrastructure of the power grid and the charging of EVs is not managed properly, it is expected that the peak demand will increase by 30% when the EV adoption reaches 50% of the vehicles on the road [36]. Additionally, the increase in peak demand will also shorten the life of transformers and cables as the life of these electrical equipment depend on the hot-spot temperature that arises from overloading.

8.2.1.2 Increased Total Harmonic Distortion

The charging of EVs involves the conversion of power using nonlinear power electronic converters which will result in the injection of harmonic currents into the power grid and increase the total harmonic distortion (THD) level of the grid voltage. The amount of harmonic currents injected from an EV charger is dependent on its circuit topology, power rating, and the impedance of the network that the charger is connected to. The distortion of grid voltage due to the injection of harmonic currents from a three-phase EV charger is demonstrated via a simulation study conducted in MATLAB/Simulink. Figure 8.7 shows the simulation model of the distribution system which is realized in Matlab/Simulink for the study. The proposed system consists of a distribution grid supplying power to a 5 kW, three-phase EV charger, and a linear load. The linear load is a three-phase RL load with a real and reactive power demand of 8 kW and 3 kVAr respectively. The system parameters are given in Table 8.3. The impedances of the distribution lines and the transformer have been obtained based on the details presented in [37]. The inverter loss resistance of the EV charger has been coarsely estimated because it is not precisely known in practice. In the simulation study, the grid only supplies real and reactive power to the linear load for $0 \leq t < 2$ s, and the EV charger is disconnected from the distribution system. Figure 8.8 shows the grid voltage for $0 \leq t < 4$ s. At $t = 2$ s, the EV charger is then connected to the distribution system. The close-up waveforms of the grid voltage for $1 \leq t < 1.1$ s (before connection of charger) and $3 \leq t < 3.1$ s (after connection of charger) during steady-state operation are shown in Figures 8.9 and 8.10, respectively. It is observed that the connection of the EV charger results in increased distortion of the grid voltage due to the injection of harmonics currents from the EV charger to the grid. The current supplied by the grid to the EV charger for $3 \leq t < 3.1$ s is shown in Figure 8.11. The THD values of the grid voltage before and after the connection of the EV charger are measured to be about 2% and 4% respectively. As the number of EV chargers increases, the increased amount of harmonics currents injected by the EV chargers would have a cascading detrimental effect on the life of transformers and cables.

8.2.2 Mitigation Measures – Coordinated Charging

If EV charging/discharging is managed properly, it is not only possible to mitigate these problems but also possible to derive many auxiliary functions such as distributed reserves for renewable energy [38–40]. Many countries have an infrastructure for parking where a large number of EVs will be available in groups for

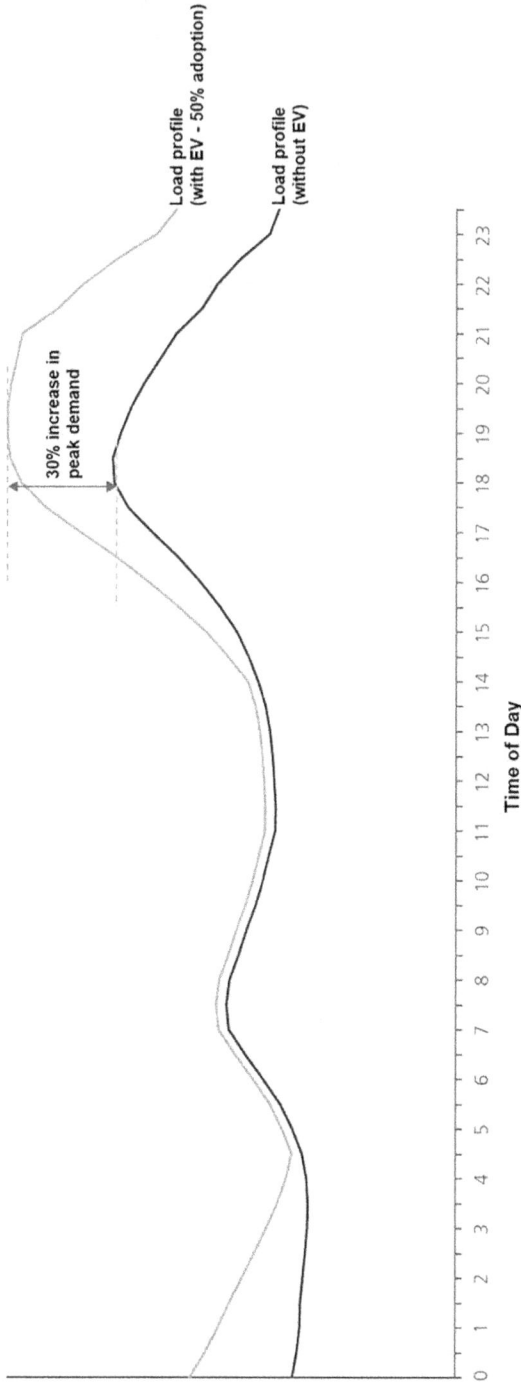

FIGURE 8.6 Estimated increase in peak demand due to EV charging (50% adoption) [34].

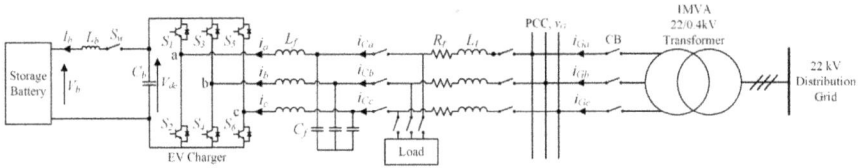

FIGURE 8.7 Overall configuration of the simulation model.

TABLE 8.3
Parameters of the Simulation Model

PARAMETER	Value
Distribution grid voltage	v_G = 230 Vrms (phase)
DC link voltage	V_{dc} = 600 Vdc
Distribution line impedance	R_ℓ = 7.5 mΩ, L_ℓ = 25.7 μH
Transformer impedance	R_T = 1.9 mΩ, L_T = 28.6 μH
LC filter	L_f = 1.2 mH, C_f = 20 μF
DG inverter loss resistance	R_f = 0.01 Ω

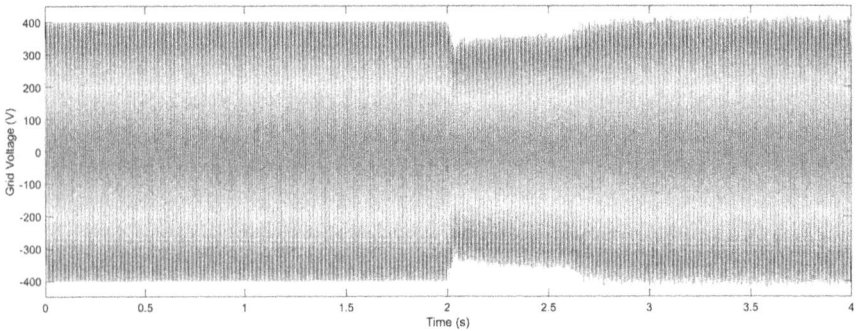

FIGURE 8.8 Waveforms of grid voltage.

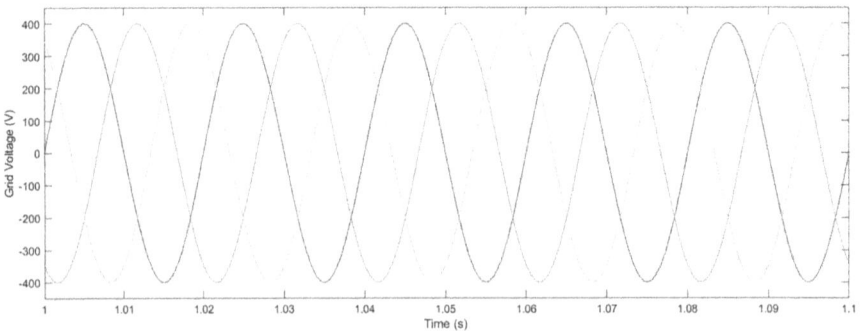

FIGURE 8.9 Closed-up waveforms of grid voltage for $1 \leq t < 1.1$ s.

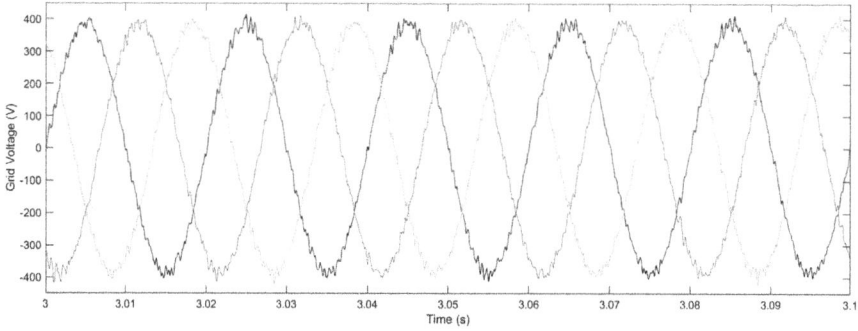

FIGURE 8.10 Closed-up waveforms of grid voltage for $3 \leq t < 3.1$ s.

charging/discharging. A well-designed charging method can therefore be utilized to achieve optimum charging/discharging with comparatively lesser infrastructure investment and improved efficiency. It also provides a better opportunity for utilizing the auxiliary services offered by the EVs and hence has the potential to decrease the power system operation cost as well as increase the power system stability.

Coordinated charging of EVs which also has other terminologies such as smart charging, regulated charging, and controlled charging is one of the most effective methods for integrating EVs into the existing power systems. In this method, the charging of EVs is carried out such that optimal performance can be obtained. There are various coordinated charging strategies available in the literature [41–46]. Coordinated EV charging can be broadly classified into two types, i.e., time-coordinated charging (TCC) and power-coordinated charging (PCC).

The difference between TCC and PCC in terms of the number of EVs being charged and the corresponding power consumed is illustrated in Table 8.4 using a typical scenario of 6 EVs connected for charging. It can be seen that in TCC, the number of EVs allowed to charge at a given time is controlled to ensure that the total EV load demand is within the total power available for EV charging. Whereas, in PCC, the power consumed by each EV is controlled to ensure that the total EV load demand is within the total power available for EV charging.

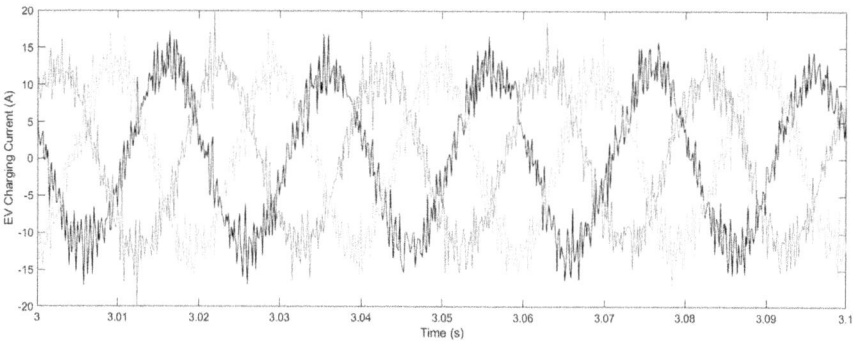

FIGURE 8.11 Closed-up waveforms of current supplied to EV charger for $3 \leq t < 3.1$ s.

TABLE 8.4

Difference Between TCC and PCC

Total Power Available for EV Charging (kW)	TCC		PCC	
	Number of EVs Charged	Average Power Consumed by Each EV (kW)	Number of EVs Charged	Average Power Consumed by Each EV (kW)
9	3	3	6	1.5
12	4	3	6	2
15	5	3	6	2.5
18	6	3	6	3

Both the TCC and PCC methods aim to control the overall EV load demand to keep within the recommended limits in order to circumvent the impacts of EV charging on the power system operation. It can be observed from Figure 8.12 that in power-controlled coordinated charging, the power supplied to each, and every vehicle is controlled in all instances such that the power limits are not violated. Whereas in the case of time controlled coordinated charging, the number of EVs charging at all instances is controlled such that the power limits are not violated.

The power limit is the constraint that is to be maintained for optimal performance. Time-controlled coordinated charging has significant advantages as the converters used for charging the EVs mostly operate near their optimal power output. Since discontinuous charging will not affect the life and performance of the batteries [47] time-controlled coordinated charging can be used without any major performance degradation. However, time-controlled coordinated charging requires an accurate load model for determining the charging demand of each EV and an effective scheduling algorithm for scheduling the EV charging with a set of constraints.

A real-time TCC is proposed in [48] for minimizing power losses and improving voltage profiles during EV charging. The authors analyze the voltage deviation due to different scenarios of uncoordinated charging for different levels of EV

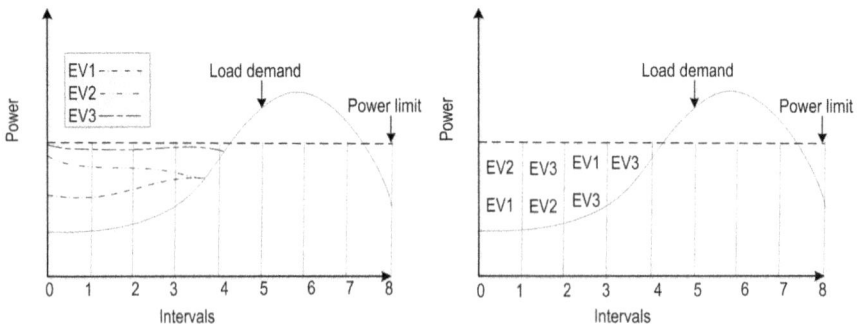

FIGURE 8.12 Typical load curves of PCC and TCC.

penetration. The authors based on their analysis proposed a real-time TCC which utilizes tariff-based priority and regulatory voltage limits as constraints. The advantage of smooth valley fill-charging with the help of a TCC is demonstrated in [42]. The authors used different proportions of Level I and Level II chargers for their analysis. The intelligent EV charging strategy proposed aims to minimize the ramping cost of the power system. Methodologies for using priority-based TCC are presented in [44] and [49] to achieve EV load leveling.

In [44] the authors introduced a centralized EV charging method that can modify the total demand of EV charging without affecting the convenience of EV users. In [49], the objective is to maximize the overall profit by increasing the number of jobs completed before their deadlines and decreasing the penalty for uncompleted jobs. It included various threshold limits for scheduling the EVs. Novel solutions of using TCC for V2G application are demonstrated in [50] and [51] where different priority criteria are employed for scheduling of EV charging. In [52] the aggregators control EV charging from 10 P.M. to 7 A.M. i.e., during off-peak period. The amount of energy that needs to be purchased in each time interval is determined by the scheduling algorithm based on the price and EV charging demands. The EVs that are to be charged during the given time interval are determined by the dispatch algorithm. In [50] the authors employed a dynamic programming algorithm to obtain the optimal charging control for each EV. The developed algorithm was applied for V2G frequency regulation services. Mathematical representation for the cost of battery charging and the revenue was also used for further investigations. In [51], the authors developed a dynamic scheduling algorithm for estimating the real-time V2G capacity that can be obtained from a group of EVs. The method has the significant advantage that the accuracy is not affected by time.

A novel PCC method for mitigating the complications in power system operation due to different levels of penetration of EV load is proposed in [41]. The authors analyzed the voltage regulation issues and power losses associated with uncoordinated charging in the residential grid. The proposed PCC minimizes power losses and maximizes the main grid load factor. Furthermore, optimal charging profiles for the EVs are computed in the proposed method using stochastic programming. In [53], a PCC method that focuses on minimizing the variance in SOC is developed for achieving fair aggregation of EVs for V2G application. The authors used state-dependent allocation and the water-filling approach both for up-regulation as well as down-regulation. The objective function of the optimization problem is to satisfy fairness criteria when the regulation service is provided by the EVs.

Research work in [54] proposes a PCC method that maximizes the total amount of energy that can be delivered to all EVs over a charging period while ensuring that network limits are not exceeded. In the above work, a linear programming-based approach is employed to determine the charging power which is optimal for each EV. The method was also tested on a section of the residential distribution network. A novel PCC algorithm to intelligently allocate electrical energy to EVs connected to the grid based on priority criteria such as energy prices, battery capacity, and charging time is proposed in [55]. Estimation of distribution algorithm (EDA) and a mathematical framework that aims at maximizing the average state-of-charge at the next time step was used by the authors. The proposed method was simulated for real-world parking deck scenarios using statistical analysis and transportation data. Another PCC method proposed in [46] uses EV charging profiles and energy prices as priority criteria for scheduling EV charging. The algorithm proposed uses the elasticity of EVs as loads for valley filling. Optimal charging profiles for all the EVs are obtained pertaining to the different conditions such as valley filling power, maximum charging rate, and deadline.

The dynamic scheduling algorithm proposed in [51] uses a data-driven scheduling approach to overcome the disadvantages of other scheduling methods while providing a robust algorithm compared to those available in the literature. The load demand of the power network consisting of m number of residential units without EVs is represented as

$$p_j^{load\ demand} = \sum_{x=1}^{m} p_{x,j}^{res\ unit} \quad \forall \quad j = 1, 2, 3, \dots 48 \tag{8.1}$$

$$p_{x,j}^{res\ unit} = rand\left(\mu_{x,j}, \sigma_{x,j}^2, pdf_{x,j}\right) \tag{8.2}$$

where $\mu_{x,j}$, $\sigma_{x,j}^2$, and $pdf_{x,j}$ are the mean, variance, and probability density function of respective units in a given interval. 'x' and 'j' represent the residential unit number and half-hour interval, respectively. The load demand data is collected from the smart meters installed in a few residential units of Nanyang Technological University (NTU) campus. Load demand data for a period of one year is used to determine the mean, variance, and probability density function of the load demand. Four-hundred and fifty units are considered for obtaining the base load without EVs and the number of units represents a typical group of Housing Development Board (HDB) blocks in Singapore. The values obtained are then used for the stochastic simulations as given by (8.2). The power limit p^{limit} for ensuring that the EV load demand does not exceed a predetermined value can be obtained using the method discussed in [56]. The method proposed in [56] uses the transformer's hotspot temperature and loss of life for determining the value for p^{limit}.

Another method to obtain p^{limit} is to find the power limit required for smooth valley filling which is essential for optimal power system operation [42] and the average power required for having a smooth valley filling is obtained by

$$p^{limit} = \frac{\sum_{j=1}^{m_{max}} (p_j^{load\ demand} + p_j^{uEV})}{m_{max}} \tag{8.3}$$

where

$$p_j^{uEV} = \sum_{i=1}^{n} p_j^{uEV,i} \tag{8.4}$$

The contracted capacity or maximum power demand limit p_{CC} can also be used as the power limit. In addition, p^{limit} can be obtained from the power system operators to account for the uncertainties in the network. The power available for EV charging is then given by

$$p_j^{avai'ble} = p_j^{limit} - p_j^{load\ demand} \qquad \forall \ j = 1, 2, 3, \ldots 48 \tag{8.5}$$

where subscript j represents each half-hour interval in the next 24-hour period. The initial and final SOC required is obtained from the distance traveled by personal vehicles in Singapore [57]. The authors of [58] used vehicle usage data to determine that the distance traveled by personal vehicles will follow a normal distribution. In this paper, the energy required/km for different types of EVs is also used for calculating the initial and final SOC. The arrival time and departure time of EVs are assumed to be normally distributed over 18:30 hours and 07:30 hours respectively with a standard deviation of 30 minutes.

The final required SOC is generally around 90–100% as the EV users will prefer to charge their EVs to the desired SOC when the cost of electricity is lowest (off-peak period). Using the final required SOC of the previous day and the distance traveled by the EVs (both obtained from stochastic samples), energy required/km the initial SOC is calculated. Thus, the uncertainties in the initial and final SOCs of the EVs are accounted into the simulations. An EV system consisting of three EVs having 7 kWh battery capacity, three EVs having 24 kWh battery capacity, and nine EVs having 16 kWh battery capacity (total 15 EVs) is considered in this research. Different battery capacities are considered to ensure that the impact of all the priority parameters on EVs with different battery capacities and average daily energy requirements can be analyzed precisely. It is important to have different types of battery capacities as it has a huge influence on all the priority criteria. The battery capacities used in the paper are the standard ones available in the EV market such as Smart ED, Nissan Leaf, Mitsubishi i-MiEV, Renault Twizy, Renault Kangoo Z.E., etc.

It is important to analyze the impact of each priority criterion as well as the impact of various combinations of different priority criteria on the chargeability of the EV to the required SOC and fairness in the allocation of charging time/energy. The analysis can serve as a guideline while selecting different parameters as priority criteria for any scheduling algorithm. In addition, the requirement for weighted priority criteria is also evaluated and optimal weights required for each case are determined. The analysis which determines the impact of different criteria with their corresponding weights is vital for tuning the performance of the scheduling algorithms. However, such studies are not yet available in the literature. As the onset of the smart grid will provide a large data which is earlier unavailable, the proposed data-driven simulations are an attractive option. The use of data-driven simulations is an effective solution for the evaluation of EV charging strategies. Furthermore, it is feasible to accurately evaluate the overall performance of both the charging strategies, namely TCC and PCC. The analysis is carried out using stochastic building load demand and EV charging demand.

Data-driven simulations offer a wide variety of advantages. Updating the model parameters to account for the changes due to ageing is one of the major advantages. Furthermore, scaling up the system to account for new additions is relatively simple. The effectiveness of data-driven simulations and controller design using data-driven modeling has been studied by many authors [59–62]. However, in this research, the data-driven simulations are used for evaluating and comparing the performance of different charging strategies of EVs with different combinations of priority criteria. The impact of all possible combinations of the three priority criteria which are widely used in literature is studied in this research.

A prospective residential power network in cities will consist of a few high-rise buildings connected together. It is assumed that a typical residential power network will serve about 450 residential units. Hence stochastic load demand data is generated. Using the stochastic load demand data of the power network and the load demand data of the EVs, the scheduling process is carried out. The estimated SOC is obtained at the end of the scheduling process. The scheduling process is carried out using individual priority criteria and various combinations of different priority criteria as listed in Table 8.4. The probability that an EV is not charged to the desired SOC is used for analyzing the impact of each of the three priority criteria and their different combinations. The analysis is carried out for both TCC and PCC charging strategies. Table 8.4 illustrates the overall probability of an EV not being charged to the desired SOC with different priority criteria. It can be observed from Table 8.4 that in the case of TCC, the probability that an EV is not charged is lowest at 0.0653 when only slack time is used as priority criterion. Whereas the probability that an EV is not charged is highest at 0.1264 when SOC and allotted time are used as priority criteria. Furthermore, it can be observed that whenever SOC is used along with any of the other two priority criterion, the probability that an EV is not charged increases, e.g., when slack time is used as the only priority criterion, the probability that an EV is not charged is 0.0653, but when it is used along with SOC the probability increases to 0.0804. It can also be observed from Table 8.5 that in the case of PCC, the probability that an EV is not charged is not significantly affected by the choice of priority criteria and the probability value is between 0.0217 and 0.0269. Furthermore, the probability that an EV is not charged is considerably less in PCC due to better utilization of the available power $p_j^{avai'ble}$. However, the overall probability that an EV is not charged just gives the overall performance of the scheduling algorithm and does not give the complete picture of the impact of different priority criteria on individual EV [63].

Hence, detailed analyses of the percentage of failure for each EV with reference to the total number of charging events are presented in Figures 8.13 and 8.14. It can be observed from Figures 8.13 and 8.14 that using SOC as one of the priority criteria (4 out of 7 cases) results in a reduction of failures with a decrease in battery capacity and energy requirement (i.e., EV1, EV2, and EV3). The result can be correlated by taking an example of two EVs having the same SOC but different battery capacity, i.e., 7 kWh and 24 kWh. Since the SOC is the same for both the EVs, an equal priority value for α_i^1 is obtained from (3.12). Hence, equal resources,

TABLE 8.5

Overall Probability of an EV Not Being Charged with TCC and PCC

Priority Criterion/Criteria	TCC	PCC
	Probability that an EV Is Not Charged	Probability that an EV Is Not Charged
SOC + Slack time +Allotted time	0.0705	0.0217
Allotted time	0.1187	0.0268
Slack time + Allotted time	0.0891	0.0220
Slack time	0.0653	0.0269
SOC	0.1148	0.0231
SOC + Allotted time	0.1264	0.0223
SOC + Slack time	0.0804	0.0245

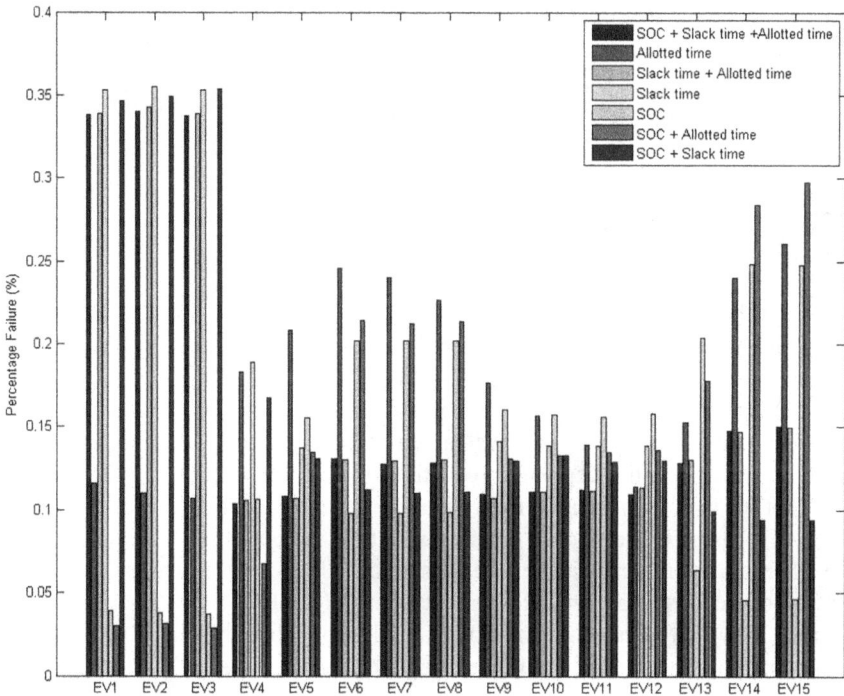

FIGURE 8.13 Percentage failure for different EVs in PCC while using different priority criteria.

i.e., either equal charging time or equal charging power or both will be allotted. With equal resources (say 3 kW power for 30 minutes), the EV with 7 kWh battery capacity will be charged to a higher SOC value compared to the EV with 24 kWh battery capacity.

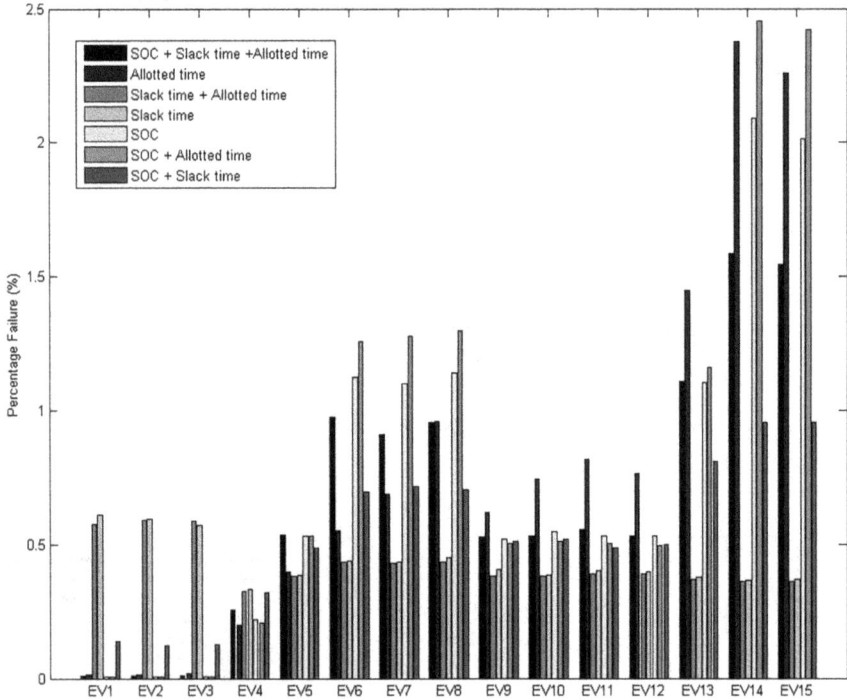

FIGURE 8.14 Percentage failure for different EVs in TCC while using different priority criteria.

Using allotted time as one of the priority criteria (4 out of 7 cases) also results in a similar phenomenon as observed from Figures 8.13 and 8.14. This is because EVs which have higher allotted time/energy will have lesser value for its priority criteria. With the same amount of allotted time/energy, an EV with a lower battery capacity will be charged to a higher SOC value compared to an EV with a higher battery capacity. However, the degree of reduction due to SOC is not the same as the allotted time which can be seen from Figures 8.12 and 8.13. Using slack time (total 4 out of 7 cases) as the priority criterion results in the opposite effect compared to the other two priority criteria, i.e., the number of failures reduces with an increase in battery capacity and energy requirement (i.e., EV13, EV14, and EV15). This is because EVs with higher battery capacity and higher energy requirements will have a low value for slack time and hence a higher priority value.

8.3 SUMMARY

In this chapter, the energy storage requirement and technologies used in EVs were reviewed with a greater focus on Li-ion batteries that are widely used in current EVs. The cycle life of Li-ion batteries and their ageing mechanisms were also discussed in detail. Analysis of the Li-ion battery performance over its lifetime was performed to understand the impact of fast charging and discharging cycles on the battery ageing profile. Beside the impact on the battery life cycle performance, uncontrolled EV charging can have a different impact on the power grid. A comprehensive review of the potential impact of EV charging on the grid was conducted. As a mitigation measure to reduce the impact of EV charging on the grid, coordinated EV charging strategies based on priority criteria were introduced for scheduling EV charging. The dynamic EV charge scheduling algorithm was proposed for both time and power-coordinated charging techniques. The impact of various combinations of different priority criteria, namely SOC, slack time, and allotted time/energy on the chargeability of EVs was studied for both types of coordinated EV charging. The concepts presented in this chapter are based on the author's publications [51,56,63].

REFERENCES

[1] IEA, *Net Zero by 2050*, Paris: IEA, 2021. https://www.iea.org/reports/net-zero-by-2050

[2] D. Clegern and S. Young, "California air resources board approves advanced clean car rules," *California Air Resources Board*, no. 05, 2012.

[3] IEA, *Global EV Outlook 2021*, Paris: IEA, 2021. https://www.iea.org/reports/global-ev-outlook-2021

[4] C. A. Palo Alto, "Environmental assessment of plug-in hybrid electric vehicles nationwide greenhouse gas emissions," *Electric Power Res. Inst.*, vol. 1, 2007.

[5] Wikipedia, List of Commercially Available Electric Vehicle. Available online: https://en.wikipedia.org/wiki/Electric_vehicle (accessed on 10 April 2022).

[6] Wikipedia, List of Commercially Available Electric Motorcycles and Scooters. Available online: https://en.wikipedia.org/wiki/Electric_motorcycles_and_scooters (accessed on 10 April 2022).

[7] Wikipedia, List of Commercially Available Electric Bicycles. Available online: https://en.wikipedia.org/wiki/Electric_bicycle (accessed on 10 April 2022).

[8] Institute of Transport Economics, Norwegian Centre for Transport Research. Available online: https://www.toi.no/ (accessed on 14 April 2022).

[9] Electric Car Use by Country. Available online: https://en.wikipedia.org/wiki/Electric_car_use_by_country (accessed on 14 April 2022).

[10] European Alternative Fuels Observatory. Available online: http://www.eafo.eu/ (accessed on 14 April 2022).

[11] Electric Car Use by Country. The Electric Vehicles world Sales Database. Available online: http://www.ev-volumes.com/ (accessed on 14 April 2022).

[12] Statista. Electric Vehicles Worldwide. Available online: https://www.statista.com/study/11578/electric-vehicles-statista-dossier/ (accessed on 14 April 2022).

[13] Hong Kong Bussiness. EV Dossier. Available online: https://hongkongbusiness.hk/transport-logistics/news/ev-sales-surge-in-2020-analyst Smart Cities 2021, 4 401 (accessed on 14 April 2022).

[14] EVAdoption.com. Analyzing Key Factors That Will Drive Mass Adoption of Electric Vehicles. 2019. Available online: https://evadoption.com/ev-market-share/ (accessed on 14 April 2022).

[15] A. Du Pasquier, I. Plitz, S. Menocal and G. Amatucci, "A comparative study of Li-ion battery, supercapacitor and nonaqueous asymmetric hybrid devices for automotive applications," *J. Power Sources*, vol. 115, pp. 171–178, 2003.

[16] J. M. Tarascon and M. Armand, "Issues and challenges facing rechargeable lithium batteries", *Nature*, vol. 414, no. 6861, pp. 359–367, Nov. 2001.

[17] Y. Miao, P. Hynan, A. von Jouanne and A. Yokochi, "Current Li-ion battery technologies in electric vehicles and opportunities for advancements," *Energies*, vol. 12, no. 6, p. 1074, Mar. 2019.

[18] Electronics Notes, "Li-Ion Battery Advantages/Disadvantages," Available online: https://www.electronics-notes.com/articles/electronic_components/battery-technology/li-ion-lithium-ion-advantages-disadvantages.php (accessed on 14 April 2022).

[19] Oak Ridge National Laboratory, "Plug-in hybrid electric vehicle value proposition study," *U.S. Department of Energy*, Jul. 2010. Available online: https://afdc.energy.gov/files/pdfs/phev_study_final_report.pdf (accessed on 20 April 2022).

[20] V. Pop, H. J. Bergveld, P. H. L. Notten and P. P. L. Regtien, "State-of-the-art of battery state-of-charge determination," *Inst. Phys. Publishing*, vol. 12, no. 16, pp. 93–110, 2005.

[21] A. J. Smith, H. M. Dahn, J. C. Burns and J. R. Dahn, "Long-term low-rate cycling of $LiCoO_2$/Graphite Li-ion cells at 55°C," *J. Electrochemical Soc.*, vol. 159, no. 6, pp. A705–A710, 2012.

[22] M. Broussely et al., "Main aging mechanisms in Li ion batteries," *J. Power Sources*, vol. 146, no. 1–2, pp. 90–96, Aug. 2005.

[23] I. Buchberger et al., "Aging analysis of Graphite/$LiNi_{1/3}Mn_{1/3}Co_{1/3}O_2$ cells using XRD, PGAA, and AC impedance," *J. Electrochemical Society*, vol. 162, no. 14, pp. A2737–A2746, 2015.

[24] M. Wohlfahrt-Mehrens, C. Vogler and J. Garche, "Aging mechanisms of lithium cathode materials," *J. Power Sources*, vol. 127, no. 1–2, pp. 58–64, Mar. 2004,

[25] J. Vetter et al., "Ageing mechanisms in lithium-ion batteries," *J. Power Sources*, vol. 147, no. 1–2, pp. 269–281, Sep. 2005.

[26] A. Barré, B. Deguilhem, S. Grolleau, M. Gérard, F. Suard and D. Riu, "A review on lithium-ion battery ageing mechanisms and estimations for automotive applications," *J. Power Sources*, vol. 241, pp. 680–689, Nov. 2013.

[27] D. Li, D. L. Danilov, L. Gao, Y. Yang and P. H. L. Notten, "Degradation mechanisms of the graphite electrode in C_6/LiFePO$_4$ batteries unraveled by a non-destructive approach," *J. Electrochemical Society*, vol. 163, no. 14, pp. A3016–A3021, 2016.

[28] S. F. Schuster et al., "Nonlinear aging characteristics of lithium-ion cells under different operational conditions," *J. Energy Storage*, vol. 1, pp. 44–53, Jun. 2015.

[29] C. R. Birkl, M. R. Roberts, E. McTurk, P. G. Bruce and D. A. Howey, "Degradation diagnostics for lithium ion cells," *J. Power Sources*, vol. 341, pp. 373–386, Feb. 2017.

[30] J. Belt, V. Utgikar and I. Bloom, "Calendar and PHEV cycle life aging of high-energy, lithium-ion cells containing blended spinel and layered-oxide cathodes," *J. Power Sources*, vol. 196, no. 23, pp. 10213–10221, Dec. 2011.

[31] E. Sarasketa-Zabala, F. Aguesse, I. Villarreal, L. M. Rodriguez-Martinez, C. M. López and P. Kubiak, "Understanding lithium inventory loss and sudden performance fade in cylindrical cells during cycling with deep-discharge steps," *J. Phys. Chem.*, vol. 119, no. 2, pp. 896–906, 2015.

[32] X. Han et al., "A review on the key issues of the lithium ion battery degradation among the whole life cycle," *eTransportation*, vol. 1, p. 100005, Aug. 2019.

[33] K. Severson, P. Attia and et al, "Data-driven prediction of battery cycle life before capacity degradation," *Nature Energy*, no. May 2019, pp. 383–391, 2019.

[34] V. Svoboda, "Batteries I Fast Charging," *Encyclopedia of Electrochemical Power Sources*, Elsevier, pp. 424–442, 2009.

[35] M. Qadrdan, N. Jenkins and J. Wu, "Chapter II-3-D - Smart grid and energy storage," *McEvoy's Handbook of Photovoltaics* (Third Edition), Academic Press, pp. 915–928, 2018.

[36] T. McGrath, N. Santha, D. Finn and N. Dunlop, "Executive insights on preparing the grid for the uptake of electric vehicles," *L.E.K. Consulting*, Nov. 2018.

[37] C. Y. Teo, *Principles and Design of Low Voltage Systems*, Singapore: Byte Power Publications, 1997.

[38] P. Chengzong, V. Aravinthan and W. Xiaoyun, "Electric vehicles as configurable distributed energy storage in the smart grid," in *Proc. Power Systems Conference (PSC)*, pp. 1–5, 2014.

[39] J. Soohyeong, H. Sekyung, H. S. Hee and K. Sezaki, "Optimal decision on contract size for V2G aggregator regarding frequency regulation," in *Proc. 12th International Conference on Optimization of Electrical and Electronic Equipment*, pp. 54–62, 2010.

[40] E. Sortomme and M. A. El-Sharkawi, "Optimal scheduling of vehicle-to-grid energy and ancillary services," *IEEE Trans. Smart Grid*, vol. 3, no. 1, pp. 351–359, Mar. 2012.

[41] K. Clement-Nyns, E. Haesen and J. Driesen, "The impact of charging plug-in hybrid electric vehicles on a residential distribution grid," *IEEE Trans. Power Systems*, vol. 25, no. 1, pp. 371–380, Feb. 2010.

[42] K. Valentine, W. G. Temple and K. M. Zhang, "Intelligent electric vehicle charging: Rethinking the valley-fill," *J. Power Sources*, vol. 196, no. 24, pp. 10717–10726, Aug. 2011.

[43] R. A. Verzijlbergh, Z. Lukszo, J. G. Slootweg and M. D. Ilic, "The impact of controlled electric vehicle charging on residential low voltage networks," in *Proc. IEEE International Conference on Networking, Sensing and Control (ICNSC)*, pp. 14–19, 2011.

[44] H. Yano, K. Kudo, T. Ikegami, H. Iguchi, K. Kataoka and K. Ogimoto, "A novel charging-time control method for numerous EVs based on a period weighted prescheduling for power supply and demand balancing," in *Proc. IEEE PES Innovative Smart Grid Technologies (ISGT)*, pp. 1–6, 2012.

[45] A. Hoke, A. Brissette, D. Maksimovic, A. Pratt and K. Smith, "Electric vehicle charge optimization including effects of lithium-ion battery degradation," in *Proc. IEEE Vehicle Power and Propulsion Conference (VPPC)*, pp. 1–8, 2011.

[46] L. Gan, U. Topcu and S. Low, "Optimal decentralized protocol for electric vehicle charging," *IEEE Trans. Power Syst.*, vol. 28, no. 2, pp. 940–951, May 2013.

[47] F. Savoye, P. Venet, M. Millet and J. Groot, "Impact of periodic current pulses on Li-ion battery performance," *IEEE Trans. Ind. Electron.*, vol. 59, no. 9, pp. 3481–3488, Sep. 2012.

[48] S. Deilami, A. S. Masoum, P. S. Moses and M. A. S. Masoum, "Real-time coordination of plug-in electric vehicle charging in smart grids to minimize power losses and improve voltage profile," *IEEE Trans. Smart Grid*, vol. 2, no. 3, pp. 456–467, Sep. 2011.

[49] C. Shiyao, T. Lang and H. Ting, "Optimal deadline scheduling with commitment," in *Proc. 49th Annual Allerton Conference on Communication, Control, and Computing*, pp. 111–118, 2011.

[50] H. Sekyung, H. Soohee and K. Sezaki, "Development of an optimal vehicle-to-grid aggregator for frequency regulation," *IEEE Trans. Smart Grid*, vol. 1, no. 1, pp. 65–72, Jun. 2010.

[51] K. N. Kumar, B. Sivaneasan, P. H. Cheah, P. L. So and D. Z. W. Wang, "V2G capacity estimation using dynamic EV scheduling," *IEEE Trans. Smart Grid*, vol. 5, no. 2, pp. 1051–1060, Mar. 2014.

[52] W. Di, D. C. Aliprantis and Y. Lei, "Load scheduling and dispatch for aggregators of plug-in electric vehicles," *IEEE Trans. Smart Grid*, vol. 3, no. 1, pp. 368–376, Mar. 2012.

[53] J. J. Escudero-Garzas, A. Garcia-Armada and G. Seco-Granados, "Fair design of plug-in electric vehicles aggregator for V2G regulation," *IEEE Trans. Veh. Technol.*, vol. 61, no. 8, pp. 3406–3419, Oct. 2012.

[54] P. Richardson, D. Flynn and A. Keane, "Optimal charging of electric vehicles in low-voltage distribution systems," *IEEE Trans. Power Syst.*, vol. 27, no. 1, pp. 268–279, Feb. 2012.

[55] S. Wencong and C. Mo-Yuen, "Performance evaluation of an EDA-based large-scale plug-in hybrid electric vehicle charging algorithm," *IEEE Trans. Smart Grid*, vol. 3, no. 1, pp. 308–315, Mar. 2012.

[56] K. N. Kumar, B. Sivaneasan, P. L. So, H. B. Gooi, N. Jadhav, R. Singh and C. Marnay, "A sustainable campus with PEVs and microgrids," in *Proc. ACEEE Summer Study on Energy Efficiency in Buildings*, USA, pp. 128–139, Jul. 2012.

[57] Land Transport Authority of Singapore, *Singapore Land Transport Statistics in Brief*. Available online: http://www.lta.gov.sg/

[58] A. Ashtari, E. Bibeau, S. Shahidinejad and T. Molinski, "PEV charging profile prediction and analysis based on vehicle usage data," *IEEE Trans. Smart Grid*, vol. 3, no. 1, pp. 341–350, Mar. 2012.

[59] J. Michopoulos, P. Tsompanopoulou, E. Houstis, C. Farhat, M. Lesoinne, J. Rice and A. Joshi, "On a data-driven environment for multiphysics applications," *Future Gener. Comput. Syst.*, vol. 21, no. 6, pp. 953–968, Jun. 2005.

[60] H. Zhongsheng and Z. Wang, "From model-based control to data-driven control: Survey, classification and perspective," *Inf. Sci.*, vol. 235, pp. 3–35, Jun. 2013.

[61] C. C. Douglas and Y. Efendiev, "A dynamic data-driven application simulation framework for contaminant transport problems," *Comput. Math. Appl.*, vol. 51, no. 11, pp. 1633–1646, Jun. 2006.

[62] H. Zhongsheng and J. Shangtai, "A novel data-driven control approach for a class of discrete-time nonlinear systems," *IEEE Trans. Control Syst. Technol.*, vol. 19, no. 6, pp. 1549–1558, Nov. 2011.

[63] K. N. Kumar, B. Sivaneasan and P. L. So, "Impact of priority criteria on electric vehicle charge scheduling," *IEEE Trans. Transp. Electrific.*, vol. 1, no. 3, pp. 200–210, Oct. 2015.

9 Electric Vehicle Development in Singapore and Technical Considerations for Charging Infrastructure

Elsa Feng Xue

CONTENTS

9.1 INTRODUCTION

With increasing widespread concerns on global warming and recurrent extreme weather events reported around the world, Singapore actively participated in the Paris Agreement for Climate Change and committed to reducing its overall carbon emission intensity from its 2005 level by 36% by the year 2030. Efforts can be made on various fronts including power generation, manufacturing industry, and transport in

DOI: 10.1201/9781003330134-9

the adoption of cleaner and more sustainable technologies which promise enhancement of energy efficiency and harnessing of green growth opportunities. In the context of Singapore, it is documented that 20% of the total carbon emission arises from land transport which is a significant amount considering the compact size of the city country. Moreover, from the perspective of the environmental impact, 75% of the air pollutants come from motorized traffic in Singapore.

Singapore, with its compact size and excellent urban infrastructure development, has positioned itself as a "Living Laboratory" that enables trialing new technologies that can bring more livability and sustainability. In line with the innovative drives and in response to the pledge of carbon emission reduction, the Singapore government has outlined a series of sustainable transport strategies. Among these strategies, one important direction is to promote a greener form of motorized land transport, such as electric vehicles (EVs). The introduction and application of EVs bring about various benefits including reducing carbon emissions and pollutants and advancing sustainable transport-related technologies in the process of getting Singapore EVs ready for the future.

9.1.1 National Strategies on Sustainable Transport

The subsection will give an introduction of a few key national strategies dedicated to sustainable transport.

The Land Transport Authority (LTA) of Singapore partnered with the Intelligent Transportation Society of Singapore to develop the first Intelligent Transport System Master Plan named "Smart Mobility 2030" for Singapore. Under the masterplan, focal areas are outlined that aim to tackle the transportation challenges for Singapore considering its increasing population and limited space. The masterplan congregated and consolidated views from agencies and industry partners. Green Mobility has been identified as one of the four focal areas under Smart Mobility 2030. It is emphasized that motorized vehicles that are powered by electricity or alternative fuels which rely on fully or partially powered by electric motors are inherently more energy efficient.

Furthermore, the Land Transport Master Plan (LTMP) 2013 presented aspirations and key strategies for a more sustainable future transport network. In LTMP 2013, one of the key strategies still lies with the promotion and wider adoption of public transport as much as possible as it is considered one of the most energy-efficient modes of commuting around the city as compared to privately owned cars. To achieve this, more comprehensive coverage of the city by public transport network and easier access to public transport will further encourage the take-up of public transport. For privately owned vehicles, Fuel Economy Labelling Scheme (FELS) has been mandatorily enforced. It is a label required to be displayed for all vehicles on sale. This label provides information on fuel efficiency and how it is compared to the average efficiency level. Potential buyers can then make informed choices in choosing the vehicle model. The label also potentially nudges buyers toward more environmentally responsible choices. Building upon the FELS scheme, a Carbon Emission-based Vehicle Scheme (CEVS) was introduced in the year of 2013 which particularly incentivizes fuel-efficient vehicles that entail lesser carbon emission and environmental impacts. While the initial CEVS solely focused on carbon dioxide emission, an amendment to the scheme was made in 2018 to form an updated version called Vehicular Emission Scheme (VES). The VES includes four more pollutants in addition to carbon dioxide which guarantees a fair coverage of pollutants from vehicles and therefore a guaranteed preference for truly environmentally friendly vehicles.

Sustainable Singapore Blueprint (SSB) 2015 laid out ambitious targets and associated plans to reach these targets. The overall goal is to enhance the livability and sustainability of the city. EVs have been highlighted in SSB 2015 as an alternative mode of green transport and part of the strategies in overcoming resource constraints for a sustainable economy growth. One initiative called "Sharing Electric Vehicle at our Doorstep" highlighted the idea of car sharing using EVs which aims to make EVs more widely accessible to the general public without the need of EV ownership. The later part of the chapter will detail a piloting project on EV car sharing.

More recently, EV Early Adoption Incentive was mandated from 1 January 2021 to 31 December 2023 to further encourage the adoption of cleaner and greener vehicles [1]. A rebate of 45% off the Additional Registration Fee will be given to owners who register full electric cars during this period. The incentive will help to lower the upfront cost by an average of 11% which helps to close the gap between EVs and Internal Combustion Engine cars. Along with the incentive, EV road tax structure is also revised in terms of its calculation to account for tiered power ratings in EVs. Relevant schemes are also adjusted to better reflect the efficiency of EVs which weren't addressed in the past [2].

9.1.2   EV Phase 1 and the Transition to EV Phase 2

A multi-agency and multi-ministry EV taskforce which is co-chaired by the Energy Market Authority (EMA) and LTA was established in 2009 to spearhead the EV test-bedding Phase 1. Members across agencies and ministries provide their support in the roll-out of the initiative. The EV Phase 1 test-bedding started in June 2011 and concluded in 2013. The Phase 1 test-bedding provided a "Living Laboratory" platform to enable EV manufacturers, charging infrastructure providers, and related solution providers in testing their products and ideas in Singapore. It also acted as a fusion platform for players to work together in achieving a common goal which is to assess the practical feasibility of using EVs under the urbanized traffic conditions and tropical weather conditions of Singapore. The test results provided valuable insights into the performance of EVs and related charging solutions. A total of 89 EVs participated in the EV Phase 1 testing bedding along with 68 units of normal-speed charging stations and 3 fast DC charging stations.

The EV Phase 1 pilot trial conducted a post-survey on the market perception of EVs. The findings from the survey pointed out that the high price of EVs remains as the biggest hindrance to the wider adoption of EVs in Singapore. Other concerns include range anxiety, battery performance deterioration, long charging waiting time due to the low charging power, and a lack of easily accessible charging infrastructure. To gauge the economic viability of EVs, a cost-benefit analysis was conducted. The key conclusion is that EVs are proven to be technically feasible owing to their dri-vability and feasible charging patterns, however, they are yet to be economically viable for mass adoption due to the high upfront cost. It is suggested in the report that EVs are about three times more costly than their internal combustible engine coun-terparts even taking into account all the monetary incentives for their social and environmental benefits available at the time [3]. The report also highlighted that new business models such as fleet operations and car-sharing programmes have the potential to address concerns with the economic viability of EVs.

In July 2014, the LTA collaborated with the Energy Research Institute of Nanyang Technological University (ERI@N) to develop an Electromobility Roadmap [4] which outlines the future development trends and serves as a guideline for technological prioritization in the electrification journey leading up to the year of 2050. The roadmap gives a landscape overview of the trends of EV global up-take, development in battery technologies, and charging technologies/standards. Future trends of local EV development were then projected in the context of Singapore. A carbon emission modeling tool was developed as part of the roadmap. Based on the outputs from this model, three scenarios of low, medium, and high fleet electrification were laid out for the near-, mid-, and long-term targets of electrification leading to 2050. Factors including the growth of vehicle population, pricing trends of EVs, development of battery technologies, increase in charging infrastructure development as well as the sound policies that can encourage support EV uptake are taken into account in formulating the recommendations of key policies and research focuses.

Based on the findings and recommendations from the EV Phase 1 test-bedding and the electromobility roadmap, EV Phase 2 with a particular focus on fleet electrification was commissioned in 2014.

The rest of the chapter is dedicated to detailing the development and findings in the EV Phase 2 test-bedding and more recent development beyond EV Phase 2. The overall content includes a few piloting projects that focus on fleet electrification including a nationwide EV car-sharing programme, and electric bus and electric taxi trials. To support EV proliferation, a national charging standard was introduced in EV Phase 2 and the rationale behind the adoption of the standards is analyzed.

9.2   EV PHASE 2 TEST BEDDING AND BEYOND

Built upon the findings from the EV Phase 1 test beddings and directions outlined in the roadmap, the EV Phase 2 with a focus on fleet electrification was commissioned. Fleet electrification was predicted to ensure greater economic viability by reaping bigger economic returns due to the overall higher mileage and more extensive use of EVs. It also at the same time guarantees a bigger potential for carbon emission reduction.

9.2.1   EV Car-Sharing Programme

A 10-year EV car-sharing programme kick-started as the flagship project of EV Phase 2. A Request for Information (RFI) was announced at the end of 2014 which aimed to garner information from the industry on a plan to introduce 1000 EVs and a charging infrastructure network with 2000 charging points. The RFI was imbued with the "car-lite" spirit in the electrification journey and well-positioned public use of EVs as a priority in fleet electrification. The car-sharing programme mainly targets first-mile and last-mile connectivity, as a greener mode of transport to complement the public transport network. In the meantime, it allows for the private use of environmentally friendly cars without the need to own one. It also has the potential to reduce the car population in the long run when the public experience the convenience of using cars on demand. The car-sharing programme once rolled out, will help to gain first-hand insights into the feasibility of large-scale EV adoption in Singapore and the market uptake of the one-way car-sharing mode that offers greater convenience and flexibility. A total of 13 submissions were received by the agencies in 2015 when the RFI closed. The proposal put up by the Bolloré Group was the top proposal received and assessed by an inter-agency-and-ministry consortium.

Following the RFI and after rounds of discussions and negotiations, the first large-scale EV car-sharing programme in Singapore was awarded to the Bolloré Group. A local subsidiary of the group called BlueSG Pte. Ltd. was founded to operate the EV car-sharing programme, which thereafter was named BlueSG programme. BlueSG has up to four years to complete the introduction of 1000 EVs and 2000 charging points across the island. In the 2000 charging points, 80% will be dedicated to spreading across residential areas around Singapore and 20% will be located in public car parks, commercial development, and industry business parks which are not well covered by public transport just yet. Twenty percent of the charging points will form a part of a national public charging network that can be used by cars other than those in the EV car-sharing programme. Given the scale of the charging infrastructure to be built in the programme, the 2000 charging points will form the foundation of a nationwide charging network that future-proofs the EV uptake in Singapore. The BlueSG programme was officially launched in 2017 and as of today, it has over 80,000 users and has reached its one-millionth rental [5].

9.2.2 ELECTRIC BUS TRIALS

Before the introduction of electric buses, there were more than 300 public bus service routes in Singapore operated by four local bus operators which in total own a fleet size of around 6000 diesel buses. As one of the main forms of public transport, buses total enormous mileage on a daily basis contributing around 60% of the PM2.5 emissions in Singapore at the same time [6]. In the electromobility roadmap, fleet electrification of bus services promises to be one of the best ways to bring down carbon emissions by up to 56% per vehicle and generate near-to-no pollutants [4].

The first electric bus trial saw the actual operation of a single unit of a BYD K9 e-bus on Singapore roads. The e-bus being trialed is powered by BYD's iron-phosphate batteries. It was a joint effort between BYD, S Dreams, and Go-Ahead Singapore. The trial lasted for 6 months with the main objective to test the technical feasibility of electric buses under the tropical climate and also the frequent start-

and-stop urban transport. The K9 electric bus was tested on a collection of routes of various natures available in actual bus service routes overseen by Go-Ahead Singapore. Data and information were collected including energy efficiency, charging time required, charging infrastructure requirements, and feedback from the commuters.

Following the promising trial results and guidance from the electromobility roadmap, the LTA laid out plans to scale up the adoption of electric buses. The LTA committed to having a 100% cleaner energy bus fleet by 2040 and procuring cleaner energy buses thereafter. A contract of 60 units of electric buses was awarded in 2019 which consisted of buses with a mixture of different types of charging technologies including depot overnight charging and pantograph opportunistic charging [7]. The results of this larger-scale electric bus operation will give insights into the operational and technical challenges of both the buses and the charging infrastructure which provides guidance for the future full electrification of public buses.

9.2.3 Electric Taxi Trials

Taxies are another form of public transport that provides more private services to commuters. A subsidiary of BYD in Singapore named HDT was awarded a special license to run a fleet of 100 units of fully electric cars of Model E6 from BYD. The operation was the first commercial electric taxi trial to last for 8 years. There are a total number of 100 units of Model E6 progressively introduced into operation since the end of 2016. The e-taxi trial pilots a new employer-employee business model in which the company HDT hired the drivers full-time and pays them a fixed base salary, which is different from the current practice of taxi companies leasing taxis to drivers and the drivers' income solely relies on the amount of income from covering trips. The end of the trial will provide insights into the feasibility of such a new business model and whether it can benefit the operation of electric taxis in the long run. In 2018, the company obtained a 10-year license to further introduce more electric taxis to a minimum of 800 units in 4 years' time. Sufficient charging stations that can well support the e-taxies are to be put into place across the island.

At a broader level, the taxi industry has recently committed to electrifying at least half of the taxis by 2030 [8]. As a special grant from the agencies to allow taxi operators more time in strategizing their electrification plan, the statuary lifespan of electric taxis is extended from 8 years to 10 years. Private hire cars are also following above-mentioned footsteps to achieve electrification.

9.3 TECHNICAL CONSIDERATIONS BEHIND ADOPTION OF A NATIONAL STANDARD FOR PUBLIC CHARGING INFRASTRUCTURE

9.3.1 Background

There are two dominant charging couplers in the IEC 62196: Type 1 and Type 2.

Type 1 which is the IEC62196/SAE J1772 refers to a type of single-phase charging coupler detailing the plug specifications of the SAE J1772. SAE J1772 originated from North America and was designed for electrical connectors used for EVs. Type 1 is compatible with single-phase electrical power supplies generally available in North America and Japan. The EV Phase 1 mentioned in the introduction used Type 1 chargers which was the then prevailing standard accepted by the EV manufacturers.

Type 2 which is IEC 62196-2 can be used in either single or three-phase power supplies. Type 2 standard details the technical specifications of the VDE-AR-E 2623-2-2 plug. It was originally proposed in 2009 by Mennekes Elektrotechnik GmbH & Co. KG which is a leading manufacturer of industrial plugs and connectors with headquarters in Kirchhundem/Sauerland region and Neudorf/Erzgebirge. The system was later standardized by the German Association of the Automotive Industry (VDA) as VDE-AR-E 2623-2-2 plug. It was then recommended as the optimal charging infrastructure solution in Europe in 2011 [9,10]. Prior to this recommendation, Type 1 was recognized as the default charging coupler due to the faster development of the EV industry in North America and Japan where the Type 1 coupler originated.

The Phase 2 piloting projects including the BlueSG programme along with their large number of charging stations to be adopted, entailed an opportunity for charging standardization in Singapore which will ensure inter-operability and future-proof Singapore for a wider EV uptake. A multi-agency study was conducted to compare the two main existing charging standards and decide which standard to be adopted as the national charging standard. In the process of the evaluation, views were sought from local EV stakeholders which showed that the industry players expressed no obvious preferences for Type 1 or Type 2 at that point of time but they urged for the agencies to mandate a uniform charging standard in order to better regulate the EV market and therefore support potential benefit-reaping.

The evaluation of the agencies concluded that the IEC Type 2 will be adopted as the national charging standard and the considerations behind this decision will be discussed in detail subsequently. On 30 June 2016, the decision was announced to adopt the Type 2 AC and Combo 2 charging systems according to the IEC 61851 series and 62196 series charging standard as the national public charging standard for all new public EV charging infrastructure. All 2000 charging points to be rolled out in the BlueSG programme will comply with this new standard.

9.3.2 Technical Considerations Behind Adoption of IEC Type 2

This section details the technical considerations behind the adoption of the IEC Type 2 standard. Comparison between Type 1 and Type 2 charging plugs and power ratings are shown. The advantages of Type 2 over Type 1 are analyzed from perspectives of the charging time and other aspects.

9.3.2.1 Connector

Configurations of Type 1 and 2 connectors used for alternative current (AC) system and Combined Charging System (CCS) used for both AC and direct current (DC) system, are compared in Table 9.1. As shown in the picture, a distinctive difference between Type 1 and Type 2 plugs is the number of power pins. Type 1 has two

TABLE 9.1

Connector configuration of Type 1 and Type 2

	Type 1/USA, JAPAN	Type 2/ Europe
Alternative current (AC)	IEC 62196 / SAE J1772	IEC 62196 -2
Combined AC/DC charging system(CCS)	CCS Combo 1	CCS Combo 2

power pins which are the two bigger-sized pins on the top row of the plug. Type 2 comes with three power pins in which the first pin is at the left in the central row and the other two pins at the very bottom of the plug. Type 1 is only compliant with single-phase power supplies while Type 2 has three power pins that can work in both single and three-phase power supplies. For locations where a three-phase supply is yet to be introduced, Type 2 can future-proof the adoption of three-phase supplied charging stations and ensure a convenient upgrade to a higher power level. CCS Type 1 and Type 2 are extensions to AC Type 1 and Type 2 plugs, respectively, by adding two more power pins at the bottom that works for DC systems. CCS provides greater flexibility that works with both DC and AC. If the AC part of the plug is used, the power is sourced from the domestic low-voltage distribution network through the integrated AC/DC converter in the vehicle to charge the batteries. If the DC part of the plug is used, the power is sourced from the AC/DC converter circuit which is normally located in the charging station itself.

9.3.2.2 Charging Time/Speed

Originating from the design thinking, Type 1 charging standard is only compliant with single-phase or split-phase power supplies, which are widely available in North America and Japan. The Type 1 J1772 defines two levels of AC charging mode shown in Table 9.2 based on the current levels offered.

Type 2 connector can work for both single-phase and three-phase power supplies and they are adopted in a broader range of regions including Singapore. Type 2 plugs enable much higher power levels due to the additional power pins and inherently higher voltage levels of the three-phase supplies. The different power levels are shown in Table 9.3.

TABLE 9.2

Two charging levels for IEC Type 1 [11,12]

Charging Level	Voltage Level	Power Level
AC single phase	120V AC, single-phase	1.4 kW @ 12 amp
		1.9 kW @ 16 amp
AC single phase	240V AC, split-phase	7.68 kW @ 32 amp
		19.2 kW @ 80 amp

TABLE 9.3

Charging levels for IEC Type 2 [13]

Charging Level	Voltage Level	Power Level
AC three phase	240/415V AC	12 kW @16 amp
AC three phase	240/415V AC	23 kW @ 32 amp
AC three phase	240/415V AC	45 kW @ 63 amp

We use an EV with a moderate battery size of 25 kWh to carry out a comparative assessment of the charging time needed using Type 1 and Type 2 respectively, at various charging current levels. The results are shown in Table 9.4, which gives a more straightforward comparison of the charging time needed. It is shown that Type 2 can significantly shorten the charging time needed. At a comparable charging current level, Type 2 is able to shorten the charging duration by about 70%. Even at a low current level, Type 2 is able to ensure reasonable charging time needed due to its higher voltage compared to that using single-phase power supplies. Compared to Type 1, Type 2 is able to shorten the charging time by 56% at a lower current level of 63 A than Type 1 at a higher current level of 80 A.

9.3.2.3 Cost Effectiveness

One of the main considerations in the roll-out of charging infrastructure has to do with cost implications. In addition to the cost of assets, installation cost constitutes one of the main cost components. Conduit and wiring systems that connect the charging stations, control hardware, and power supplies take up a large portion of the installation cost. Type 1 chargers incur less upfront asset cost compared with Type 2 due to their lesser complicated hardware and lower power level therefore cheaper associated wiring cost. Based on the information gathered in 2015 from industrial charging service providers, a cost analysis for Type 1 and Type 2 chargers was conducted to compare the cost differential between the two.

Shown in the diagram of Figure 9.1 is a typical layout of a charging system with all associated components. The cost analysis took into account the cost of assets, and the cost of the installation which enables the power connection from the

TABLE 9.4

Comparison of charging time between IEC Type 1 and Type 2

Type of Plug	Current Level	Charging Time
IEC Type 1	12 amp	18 hrs
	16 amp	13 hrs
	32 amp	3 hrs 20 mins
	80 amp	1 hrs 20 mins
IEC Type 2	16 amp	2 hrs
	32 amp	1 hr
	63 amp	35 mins

Note: Charging time is calculated as the time needed to fully charge an EV with a usable battery size of 25 kWh

FIGURE 9.1 Typical layout of a charging system.

substation to the Electrical Vehicle Supply Equipment (EVSE as shown) which are essentially a cluster of chargers.

Reasonable assumptions are made below in the analysis without losing the fairness in the comparison:

- The distance between the substation and the power intake point is 100 meters as also shown in Figure 9.1.
- Type 1 charging station takes power from one of the three phases while Type 2 charging station from three phases.
- The cost analysis does not include the cost of insurance endorsement or maintenance.
- One unit of EVSE comprises of 5 chargers.

The cost breakdown of setting up EVSEs using Type 1 and Type 2 chargers is shown in Table 9.5:

It is noted that Type 2 chargers have a longer lifespan of 7 years compared with 5 years of Type 1 chargers. Taking into account the difference in the lifespan, the cost per unit per annum of the two types of chargers is calculated as shown in the last row of Table 9.5. It is noted that the cost difference of a mere 7% is far from significant to become a major inhibiting factor in choosing Type 2 chargers.

EVSE using Type 2 chargers may incur a slightly higher upfront cost but the cost difference should not be a main concern considering the multiple benefits Type 2 chargers can offer in the longer term.

TABLE 9.5
Breakdown of cost of EVSE using Type 1 and Type 2 chargers

S/No.	Classification	Type 1 Cost in SGD	Type 2 Cost in SGD
1	Cabling to Meter Box	$ 4500	$ 13500
2	Main supply meter box	$ 900	$ 1620
3	Cabling to panel control	$ 9000	$ 22500
4	Panel Control DB	$ 5000	$ 7500
5	Sub-DB boxes	$3450	$ 4800
6	EVSE hardware	$ 50740	$ 63425
7	Mounting brackets	$ 12500	$ 12500
8	Services	$ 3500	$ 7500
9	LEW	$ 2000	$ 4000
	Installation Subtotal	$ 91590	$ 137345
	Cost per unit per annum	$ 3,663	$ 3,924

9.3.2.4 Charging Efficiency

It is known that the majority of Singapore's electricity is generated from natural gas. Undoubtedly, once the charging network is scaled up and the number of charging events and associated energy consumption significantly rise, the charging efficiency of charging systems plays an important factor in reducing the power losses during charging events and influences energy sustainability.

A study was conducted to compare the efficiency and sensitivity to the ambient temperature of charging systems when using a lower and high charging power level [14]. It was concluded that the overall efficiency of using a high power level is higher than that when using a lower power level. An efficiency increment of 13% was found in some of the case studies while using a higher power level. It is therefore safe to say that using a Type 2 charging system that uses a high voltage level has the potential to achieve higher charging efficiency and incur fewer power losses which in a way further reduces carbon emission considering the source of the electricity in Singapore.

On the level of sensitivity toward the ambient temperature, it was found that charging systems that come with higher charging power shows higher efficiency at all temperature levels [14]. This is mainly attributed to lesser temperature rise during the shorter periods of charging time when using a higher power, which also helps to preserve battery lifetime. The resilience toward temperature levels when using higher charging power is critical for Singapore considering the hot and humid weather. Type 2 chargers can ensure overall better performance in such an environment. Greater benefits arising from ambient temperature resilience are even more salient when using Type 2 chargers at public charging locations where charging events occur more frequently and weather conditions are more uncontrollable. This further justifies the mandate to use Type 2 for all public charging facilities.

9.3.2.5   Ability to Support EV Proliferation and V2G Future Proof

It is no doubt to say that fast turn-over rates are crucial for the profitability of fleet electrification. Surveys show that taxi companies are conservative about using EVs due to concerns about long charge time and hence low turn-over rates. This is believed to be an inhibiting factor for sensible business models. With this hindrance in mind, it is difficult for taxi companies who invest in a high upfront cost in EV fleets to expect a reasonably short pay-back period. In general, operations of most types of fleet vehicles demand long operating hours and intensive coverage of mileage, in view of which, a sufficiently fast charging speed is indispensable for a sustainable and profitable business model upon switching to fleet electrification. In the case of car-sharing programmes, faster charging speed enables higher rates of car utilization and lesser customer frustration in waiting time. In the case of commercial fleets, faster turn-over rates help to maximize economic returns, minimize disruptions to service during the transitional period, and support long-term strategic and sustainable business plans.

With the EV population taking up in a major way in the near-and-medium term future, there is plenty of space in exploring Vehicle-to-Grid (V2G) applications where idle EVs with their onboard energy storage can participate in electrical grid services. Functions including demand response, peak load shavings, energy arbitrage, and so on can be performed and experimented with in V2G applications. With a unified V2G platform, EV owners have the potential to get economic returns from providing V2G services. Research has shown that faster charging speed, which is enabled by Type 2 charging standard, allows for V2G functions to execute in a timely manner [13]. This is critical in nearly all V2G applications considering the nature of electrical grid services is time sensitive. In the case of a charging network that has renewable energy integrated, faster charging speed enables timely and opportunistic utilization of intermittent renewable energy.

Compared with Type 1 slower charging, Type 2 which comes with a faster charging speed can offer a series of benefits in a wide range of applications.

9.4   CONCLUSION

This book chapter gives an overview of the electromobility journey of Singapore starting from the year of 2011 [15]. The chapter covers the piloting trials, analysis, and roadmaps in EV Phases 1 and 2, as well as the charging infrastructure standard. More specifically, the technical considerations behind the national public charging standard adopted during EV Phase 2 with a focus on fleet electrification are presented. It is also worth mentioning that in March 2022, the national EV charging standard has been further updated and expanded to better support the EV proliferation [16]. The update includes the introduction of two new modes of low-power charging which is lower than 2.3 kW; standards for pantograph charging; increase of the power limit of up to 400 kW for the CHAdeMO charging system and up to 500 kW for the CCS and battery swapping for motorcycles. The updates to the charging standard include various types of charging modes which effectively support land-based transport electrification in multiple ways.

REFERENCES

[1] https://www.lta.gov.sg/content/ltagov/en/newsroom/2020/2/news-releases/ Supporting_cleaner_and_greener_vehicles.html.

[2] https://www.lta.gov.sg/content/ltagov/en/newsroom/2021/3/news-release/ Encouraging_the_adoption_of_electric_cars.html.

[3] EV Phase 1. 2014. *Key findings from EV Phase 1 Test-bed.* Singapore. https://www. lta.gov.sg/data/apps/news/press/2014/20141208_EVPh2-TestBed.pdf.

[4] E-MOBILITY. 2016. "E-MOBILITY Technology Roadmap." https://www.nccs. gov.sg/sites/nccs/files/Roadmap_E-M_1.pdf.

[5] https://www.straitstimes.com/singapore/transport/electric-car-sharing-service- reaches-its-one-millionth-rental.

[6] MEWR. 2015. ...Factsheet on More Measures to Improve Singapore's Air Quality Factsheet, Singapore MEWR

[7] https://www.lta.gov.sg/content/ltagov/en/newsroom/2020/10/news-releases/update- on-deployment-of-electric-buses.html.

[8] https://www.lta.gov.sg/content/ltagov/en/newsroom/2022/3/news-releases/ reducing-peak-land-transport-emissions-by-80-.html.

[9] ACEA. 2011. *ACEA Position and Recommendations for the Standardization of the Charging of Electrically Chargeable Vehicles.* Position paper, Brussels: ACEA.

[10] Standard for New Public Charging Infrastructure for EVs https://www.lta.gov.sg/ content/dam/ltagov/news/press/2016/20160630_Public_charging_standard_ Annex.pdf.

[11] SAE. 2011. "SAE Charging Configurations and Ratings Terminology." http://www. sae.org/smartgrid/chargingspeeds.pdf.

[12] EVConnectors. n.d. "EVConnectors." *Electric Vehicle Charging Products & Connectors.* https://evconnectors.com/EV_brochure.pdf.

[13] Stuart Speidel, Fakhra Jabeen, Doina Olaru, David Harries, Thomas Bräunl. Sep. 2012. "Analysis of western Australian electric vehicle." Australian Transport Research Forum ATRF.

[14] Justine Sears, Evan Forward, Eric Mallia, David Roberts, Karen Giltman. March 20, 2013. "Assessment of level 1 and level 2 electric vehicle charging efficiency." . Transportation Research Record: Journal of the Transportation Research Board, 2454. 92-96. 10.3141/2454-12.

[15] Feng Xue, Evan Gwee. 2017. Electric vehicle development in Singapore and technical considerations for charging infrastructure, *Energy Procedia*, Volume 143, Pages 3–14.

[16] https://www.lta.gov.sg/content/ltagov/en/newsroom/2022/3/news-releases/ introduction-of-updated-national-electric-vehicle-charging-stand.html.

Index

For Product Safety Concerns and Information please contact our EU
representative GPSR@taylorandfrancis.com
Taylor & Francis Verlag GmbH, Kaufingerstraße 24, 80331 München, Germany

www.ingramcontent.com/pod-product-compliance
Lightning Source LLC
Chambersburg PA
CBHW060346220326
41598CB00023B/2823